Cutaneous Atlas of Ex Vivo Confocal Microscopy

Manu Jain • Anthony Rossi •
Kishwer Nehal • Mercedes Sendín-Martín
Editors

Cutaneous Atlas of Ex Vivo Confocal Microscopy

Springer

Editors
Manu Jain
Dermatology Service, Department of Medicine
Memorial Sloan Kettering Cancer Center
(MSKCC)
New York, NY, USA

Anthony Rossi
Dermatology Service, Department of Medicine
Memorial Sloan Kettering Cancer Center
(MSKCC)
New York, NY, USA

Kishwer Nehal
Dermatology Service, Department of Medicine
Memorial Sloan Kettering Cancer Center
(MSKCC)
New York, NY, USA

Mercedes Sendín-Martín
Dermatology Department
Hospital Universitario Virgen del Rocío
Sevilla, Spain

ISBN 978-3-030-89315-6 ISBN 978-3-030-89316-3 (eBook)
https://doi.org/10.1007/978-3-030-89316-3

This Springer imprint is published by the registered company Springer Nature Switzerland AG
The registered company address is: Gewerbestrasse 11, 6330 Cham, Switzerland

Foreword

It is my very great pleasure to introduce a new reference text promoting confocal microscopy, this time exploring its use not with traditional methods in vivo, but with ex vivo in surgical situations and directly comparing fluorescent confocal microscopy images and digitally colored purple and pink images (DHE images) with their conventionally stained H&E stained histopathology. It has been almost a decade since our Springer book, Reflectance Confocal Microscopy for Skin Diseases, was published in 2012, and, in my estimation, it is high time for another text on such an exciting technology. This Cutaneous Atlas of Ex Vivo Confocal Microscopy (EVCM) follows the tradition of clinico-pathologic-confocal correlation by displaying the near-histopathological detail of confocal microscopic images to great effect, featuring micro-morphology of neoplasms and inflammatory diseases.

With this book, Manu Jain, Anthony Rossi, Kishwer Nehal, and Mercedes Sendín Martín bring to your reference shelf a clear guide to not only the basics of confocal microscopy and a hands-on guide for its use, but also a detailed look at normal skin, benign, and malignant lesions. They particularly highlight the use of ex vivo confocal microscopy, a technique that complements the long-established use of cryosectioning for rapid histopathological analysis during surgery. We are confident that ex vivo confocal microscopy will become as routine a part of dermatological surgery as dermoscopy is a part of the dermatological examination.

One of the confocal microscopy's strengths is the digital records it generates automatically for each examination. As well as improving immediate and ongoing patient care, this atlas provides high-quality material for continuing professional development. It also opens the door to the development of artificial intelligence clinical assistants and expansion into telehealth, two fast-growing areas of health services research and delivery.

We look forward to adding this book to our collection and using it to further improve our clinical practice.

H. Peter Soyer, MD, FACD, FAHMS
Professor and Chair in Dermatology
Dermatology Research Centre
The University of Queensland Diamantina Institute
The University of Queensland
Brisbane, Australia
e-mail: p.soyer@uq.edu.au

Preface

Ex Vivo Confocal Microscopy (EVCM) is a novel imaging technique, which is being integrated in surgical workflow for rapid evaluation of freshly excised tissues. As a relatively new technique, it is essential to train new adopters in this field, which motivated us to create the first comprehensive atlas on cutaneous lesions using EVCM. We assembled experts in the field who contributed excellent images and descriptions of neoplastic and non-neoplastic lesions.

We begin by describing the fundamentals of EVCM and provide a detailed hands-on guide for image acquisition. In subsequent chapters, we describe EVCM features of lesions in various imaging modes: fluorescent, reflectance, and purple and pink digitally colored images. We compare these images with their corresponding conventional H&E images. Lastly, we discuss emerging techniques that might further improve current EVCM imaging standards.

We hope that this atlas will serve as an essential guide for dermatologists and surgeons in integrating EVCM in their clinical practice.

Hope you will enjoy it! We loved writing it.

New York, USA — Manu Jain
New York, USA — Anthony Rossi
New York, USA — Kishwer Nehal
Sevilla, Spain — Mercedes Sendín-Martín

Acknowledgments

This atlas is dedicated to our students and fellows whose enthusiasm to learn novel techniques led to "the conception of this atlas". We wish to thank them, our fellow colleagues, and authors whose dedication to the field of confocal microscopy inspired this endeavor and whose collective expertise is detailed in this text.

We are also deeply grateful to our skin cancer and skin imaging teams at MSKCC whose commitment to clinical and research excellence raises the bar for all of us. We are truly indebted to Dr. Allan Halpern and Dr. Milind Rajadhyaksha, Drs. Salvador Gonzalez and Joseph Malvehy whose leadership and support has been instrumental to this innovation and success of this technology. Of course, we are incredibly appreciative of our families and their infinite love and support and sacrifices.

I would like to acknowledge the unconditional support of my son, Tanay Singhal, and my husband, Amit Singhal, who lovingly cooks gourmet cuisine for me every night. Most importantly, I would like to dedicate this atlas to my parents Dr. Vijay Kumar Jain and Dr. Snehlata Jain, who encouraged my childhood passions and supported me through challenging days of medical school.

Manu Jain

I would like to thank my mentor, Dr. Manu Jain, for her excellent guidance and support during the process of elaborating this atlas. I also wish to thank my family; without whose cooperation I would not have been able to finish this project. They motivate me to keep always learning and becoming a better doctor for my patients.

Mercedes Sendín-Martín

Contents

Contributors

Raquel Albero-González Department of Pathology, Hospital Clínic de Barcelona, Barcelona, España

Paola Azzoni Department of Surgical, Medical, Dental and Morphological Sciences with Interest in Transplant, Oncological and Regenerative Medicine, University of Modena and Reggio Emilia, Modena, Italy

Isin Sinem Bagci Department of Dermatology, Stanford School of Medicine, Redwood City, CA, USA

Alex Bang Dermatology Service, Department of Medicine, Memorial Sloan Kettering Cancer Center (MSKCC), New York, NY, USA

Laura Bertoni Department of Surgical, Medical, Dental and Morphological Sciences with Interest in Transplant, Oncological and Regenerative Medicine, University of Modena and Reggio Emilia, Modena, Italy

Luca Reggiani Bonetti Department of Medical and Surgical Sciences for Children and Adults, University of Modena and Reggio Emilia-AOU Policlinico of Modena, Modena, Italy

Stefania Caramaschi Department of Medical and Surgical Sciences for Children and Adults, University of Modena and Reggio Emilia-AOU Policlinico of Modena, Modena, Italy

John Carucci Ronald O. Pearlman Department of Dermatology, New York University, New York, NY, USA

Chih-Shan Jason Chen Dermatology Service, Department of Medicine, Memorial Sloan Kettering Cancer Center (MSKCC), New York, NY, USA

Marc Combaia Department of Dermatology, Hospital Clinic de Barcelona, Universitat de Barcelona, Barcelona, Spain

Nicole Doudican Ronald O. Pearlman Department of Dermatology, New York University, New York, NY, USA

Banu Farabi Department of Internal Medicine, Saint Peter's University Hospital, New Brunswick, NJ, USA

Daniel Gareau Laboratory for Investigative Dermatology, The Rockefeller University, New York, NY, USA

Ucalene Harris Dermatology Service, Department of Medicine, Memorial Sloan Kettering Cancer Center (MSKCC), New York, NY, USA

Daniela Hartmann Department of Dermatology and Allergy, University Hospital, LMU Munich, Munich, Germany

Manu Jain Dermatology Service, Department of Dermatology, Memorial Sloan Kettering Cancer Center (MSKCC), New York, NY, USA

Kivanc Kose Dermatology Service, Department of Medicine, Memorial Sloan Kettering Cancer Center (MSKCC), New York, NY, USA

Erica Lee Dermatology Service, Department of Medicine, Memorial Sloan Kettering Cancer Center (MSKCC), New York, NY, USA

Samantha Lish The Rockefeller University, New York, NY, USA

Josep Malvehy Dermatology Department, Hospital Clínic de Barcelona, Barcelona, España

Matthew Moronta Dermatology Service, Department of Medicine, Memorial Sloan Kettering Cancer Center (MSKCC), New York, NY, USA

Kishwer Nehal Dermatology Service, Department of Medicine, Memorial Sloan Kettering Cancer Center (MSKCC), New York, NY, USA

Giovanni Pellacani Dermatology Clinic, Department of Clinical Internal, Anesthesiological and Cardiovascular Sciences, Sapienza University of Rome, Rome, Italy

Susana Puig Dermatology Department, Melanoma Unit, Hospital Clínic de Barcelona, IDIBAPS, Universitat de Barcelona, Barcelona, Spain

Melissa Pulitzer Department of Pathology, Memorial Sloan Kettering Cancer Center (MSKCC), New York, NY, USA

Javiera Pérez-Anker Dermatology Department, Hospital Clínic de Barcelona, Barcelona, España

Milind Rajadhyaksha Dermatology Service, Department of Medicine, Memorial Sloan Kettering Cancer Center (MSKCC), New York, NY 10021, USA

Babar K. Rao Department of Dermatology, Rutgers Center, Somerset, NJ, USA;
Weil Cornell Medicine, New York, NY, USA

Luca Reggiani Bonetti Department of Medical and Surgical Sciences for Children and Adults, University of Modena and Reggio Emilia-AOU Policlinico of Modena, Modena, Italy

Camilla Reggiani Department of Surgical, Medical, Dental and Morphological Sciences with Interest in Transplant, Oncological and Regenerative Medicine, University of Modena and Reggio Emilia, Modena, Italy

Anthony Rossi Dermatology Service, Department of Medicine, Memorial Sloan Kettering Cancer Center (MSKCC), New York, NY, USA

Mercedes Sendín-Martín Dermatology Department, Hospital Universitario Virgen del Rocío, Sevilla, Spain

Part I
Basics of Ex Vivo Confocal Microscopy

Roles and Applications of Ex Vivo Confocal Microscopy

1

Mercedes Sendín-Martín and Manu Jain

1.1 Introduction

Confocal microscopy (CM) was introduced in dermatology in the last decade [1].

It provides a new rapid imaging approach for evaluation of neoplastic and non-neoplastic lesions, enabling bedside pathology [2]. Ex vivoconfocal microscopy (EVCM) acquires images in two modes: reflectance confocal microscopy (RCM) and fluorescence confocal microscopy (FCM). RCM (grayscale) and FCM (grayscale or green scale) images can be visualized separately [3] or in a combined (RCM + FCM) digitally pseudo- color purple and pink images also called digital H&E (DHE) mode as it resembles conventional hematoxylin–eosin (H&E)-stained images [4]. The DHE image is created by combining signal from FCM (nuclear signal) channel which is digitally converted to purple color (similar to hematoxylin stain) and signal from RCM (cytoplasmic and collagen signal) which is digitally converted to pink color (similar to eosin stain). For information about the principles of ECM device, please refer to Chap. 2.

One of the major advantages of imaging with the EVCM device is that it can image fresh tissues without the need for freezing and sectioning. The lack of tissue processing significantly shortens tissue evaluation time from 20 to 45 min (per excision) [5] with conventional frozen section to less than 5 min [6]. Detailed hands-on guide for tissue preparation and EVCM imaging is provided in Chap. 3. Furthermore, this device can be operated with minimal training and doesn't require an expensive laboratory set-up which could aid in reducing the cost of the procedure.

M. Sendín-Martín
Dermatology Department, Hospital Universitario Virgen del Rocío, Sevilla, Spain

M. Jain (✉)
Dermatology Service, Department of Medicine, Memorial Sloan Kettering Cancer Center, 530 E. 74th Street, New York, NY 10021, USA
e-mail: jainm@mskcc.org

In this chapter, we will provide a brief overview of applications of EVCM device in dermatology and also in non-dermatological field.

1.2 Applications of Ex Vivo Confocal Microscopy in Dermatology (Table 1.1)

1. *EVCM during peri-operative margin assessment for keratinocytic neoplasm during Mohs surgery*: The use of EVCM has been mostly described during Mohs micrographic surgery (MMS) for rapid evaluation of tumor margins for basal cell carcinoma (BCC) and squamous cell carcinoma (SCC) [3, 7]. EVCM has demonstrated an overall high sensitivity (~96%) and specificity (~89%) for detection of BCC [8, 9] and ~95% sensitivity and ~ 96% specificity for SCC detection [10], which is comparable with the standard frozen section analysis. EVCM images of BCC and SCC are shown in detail in Chaps. 7 and 8, respectively.
2. *EVCM for the evaluation of melanocytic lesions:* EVCM has been used to evaluate melanoma and nevi, and their main microscopic features and histopathological correlations have been reported [11]. Also, fluorescent-labeled antibodies such as S100 and Melanin A have been shown to be useful to differentiate melanoma from other tumors such as BCC, much more rapidly than traditional immunohistochemistry and conventional formalin-based tissue fixation [12]. Additionally, EVCM has been used to measure melanoma thickness with a high concordance with conventional histopathology [13]. EVCM images of melanocytic lesions are shown in detail in Chaps. 9, 10 and 11.
3. *Role of EVCM for evaluation for other neoplastic skin lesions:* Features of dermatofibrosarcoma protuberans (DFSP) have been described using EVCM with an

M. Jain et al. (eds.), *Cutaneous Atlas of Ex Vivo Confocal Microscopy*,
https://doi.org/10.1007/978-3-030-89316-3_1

Table 1.1 Application of ex vivoconfocal microscopy in dermatology

Neoplastic lesions	Non-Neoplastic lesions		Others
	Infectious diseases	Inflammatory diseases	
• Basal cell carcinoma [3, 4, 8, 34–39]	• Mucormycosis [22]	• Pemphigoid [18]	• Skin fillers [40]
• Squamous cell carcinoma [4, 7, 41]	• Aspergillosis [42]	• Vasculitis [19]	
• Melanocytic lesions [6, 11–13, 43]	• Dermatophytosis [44, 45]	• Psoriasis [17]	
• Dermatofibrosarcoma protuberans [14]	• Herpes virus [20]	• Eczema [17]	
• Eccrine syringomatous carcinoma [15]	• Molluscum contagiosum [21]	• Lichen planus [17, 46]	
• Extramammary Paget's disease [16]		• Discoid lupus erythematosus [17, 47]	
• Nail tumors [23, 24]			

excellent correlation with histopathology [14]. This suggests the potential role of EVCM to assess surgical margins of DFSP during Mohs surgery. However, studies need to be conducted to evaluate sensitivity and specificity of EVCM to detect DFSP during Mohs surgery. EVCM features of other less frequent neoplastic skin lesions, such as eccrine syringomatous carcinoma [15] or extramammary Paget's disease [16], have been described. However, the role of EVCM has not been explored for rapid, bedside evaluation of benign lesions, that mimics skin cancers, in clinical setting. In Chaps. 5 and 6, we describe features of common benign lesions including seborrheic keratosis, solar lentigo, epidermal inclusion cyst, lipoma, neurofibroma, and dermatofibroma for the first time.

4. *Role of EVCM for the evaluation of inflammatory skin lesions:* EVCM has a great potential for rapid bedside evaluation of cutaneous inflammatory lesions. Key morphological features of several inflammatory diseases such as psoriasis, lichen planus, eczema, and discoid lupus have been described [17]. Also, the use of immunofluorescence antibodies has been reported to be helpful for the diagnosis of pemphigoid [18] and vasculitis [19] with EVCM device. EVCM images of cutaneous inflammatory lesions are shown in detail in Chaps. 12, 13, 14 and 15.
5. *Role of EVCM for the evaluation of infectious diseases:* EVCM can detect specific morphological features of some infectious skin lesions such as cytopathic effect associated with herpes virus [20] and pox virus that causes molluscum contagiosum [21], and hyperreflective elongated and ramificated fungi structures of mucormycosis [22]. We have covered examples of some common infectious diseases including verruca vulgaris and molluscum contagiosum in Chap. 5.
6. *Role of EVCM for the evaluation of nail tumors:* Lastly, EVCM has shown to be a useful tool for intraoperative diagnosis of malignant nail lesions such as invasive SCC and Bowen's disease. The diagnosis of these lesions relies on identification of significant nuclear and cytological atypia, which is similar to histopathology [23]. EVCM has also demonstrated its utility for rapid intraoperative diagnosis of melanonychia striata [24] and for the diagnosis of benign epithelial tumors such as onychomatricoma and onychopapilloma [23]. However, larger scale studies showing diagnostic accuracy for nail tumors with EVCM device are lacking.

1.3 Applications of Ex Vivo Confocal Microscopy in Non-Dermatologic Field (Table 1.2)

EVCM provides a rapid approach for imaging tissues compared with conventional histopathology. Thus, EVCM has a great potential to be incorporated into surgical pathology for intraoperative assessment of tumor margins as an alternative to frozen section and for the evaluation of small biopsies such as endoscopic biopsies or core needle biopsies. Although the use of EVCM is not yet standardized for these applications in clinics, many recent studies have reported this technology to be useful for diagnosis of non-dermatological cancers such as prostatic cancer [25–28] and breast cancer [29–33], among others. As this section is out of the scope for this atlas, we have summarized publications in this field in Table 1.2.

Table 1.2 Application of ex vivoconfocal microscopy for non-dermatological cancers

	Authors	Year
Urological Cancers		
Prostatic cancer	• Puliatti et al. [26]	2019
	• Rocco et al. [28]	2020
	• Bertoni et al. [27]	2020
	• Panarello et al. [25]	2020
Bladder cancer	• Sonn et al. [48]	2009
	• Wiesner et al. [49]	2011
	• Bonnal et al. [50]	2012
Renal cancer	• Su et al. [51]	2016
	• Krishnamurthy et al. [52]	2019
	• Phung et al. [53]	2020
Gynecological Cancers		
Breast cancer	• Patel et al. [33]	2012
	• Dobbs et al. [31]	2013
	• Ragazzi et al. [54]	2013
	• Dobbs et al. [29]	2015
	• Abeytunge et al. [30]	2017
	• Krishnamurthy et al. [52]	2019
	• Elfgen et al. [32]	2019
Uterine cervix dysplasia	• Collier et al. [55]	2002
	• Carlson et al. [56]	2005
Gastrointestinal cancers		
Esophagus cancer	• Inoue et al. [57]	2000
	• Yoshida et al. [58]	2007
	• Gorospe et al. [59]	2012
Stomach cancer	• Inoue et al. [57]	2000
	• Yoshida et al. [58]	2007
Colon cancer	• Inoue et al. [57]	2000
	• Yoshida et al. [58]	2007
	• Ragazzi et al. [54]	2013
Liver cancer	• Krishnamurthy et al. [52]	2019
Oral Mucosa Dysplasia	• El Hallani et al. [60]	2013
Respiratory Tract Cancer		
Lung cancer	• Sorokina et al. [61]	2014
	• Krishnamurthy et al. [52]	2019
	• Takemura et al. [62]	2019
Larynx cancer	• Just et al. [63]	2006
Central Nervous System Cancer		
Brain Cancer	• Wirth et al. [64]	2012
	• Snuderl et al. [65]	2013
	• Martirosyan et al. [66]	2016
	• Belykh et al. [67]	2020
	• Belykh et al. [68]	2020
	• Acerbi et al. [69]	2020

(continued)

Table 1.2 (continued)

	Authors	Year
Endocrine Cancer		
Thyroid	• Ragazzi et al. [54]	2013
Others		
Lymph Node Methastasis	• Ragazzi et al. [54]	2013

References

1. Malvehy J, Pérez-Anker J, Toll A, Pigem R, Garcia A, Alos LL, et al. Ex vivo confocal microscopy: revolution in fast pathology in dermatology. Br J Dermatol. 2020;183:1011–25. https://doi.org/10.1111/bjd.19017.
2. Jain M, Rajadhyaksha M, Nehal K. Implementation of fluorescence confocal mosaicking microscopy by "early adopter" Mohs surgeons and dermatologists: recent progress. J Biomed Opt. 2017;22:24002. https://doi.org/10.1117/1.JBO.22.2.024002.
3. Bennàssar A, Carrera C, Puig S, Vilalta A, Malvehy J. Fast evaluation of 69 basal cell carcinomas with ex vivo fluorescence confocal microscopy. JAMA Dermatol. 2013;149:839. https://doi.org/10.1001/jamadermatol.2013.459.
4. Mu EW, Lewin JM, Stevenson ML, Meehan SA, Carucci JA, Gareau DS. Use of digitally stained multimodal confocal mosaic images to screen for nonmelanoma skin cancer. JAMA Dermatol. 2016;152:1335–41. https://doi.org/10.1001/jamadermatol.2016.2997.
5. Ragazzi M, Longo C, Piana S. Ex Vivo (fluorescence) confocal microscopy in surgical pathology: State of the art. Adv Anat Pathol. 2016;23:159–69. https://doi.org/10.1097/PAP.0000000000000114.
6. Bennàssar A, Vilalta A, Carrera C, Puig S, Malvehy J, Bennássar A, et al. Rapid diagnosis of two facial papules using Ex vivo fluorescence confocal microscopy: toward a rapid bedside pathology. Dermatologic Surg. 2012;38:1548–51. https://doi.org/10.1111/j.1524-4725.2012.02467.x.
7. Longo C, Ragazzi M, Gardini S, Piana S, Moscarella E, Lallas A, et al. Ex vivo fluorescence confocal microscopy in conjunction with Mohs micrographic surgery for cutaneous squamous cell carcinoma. J Am Acad Dermatol. 2015;73:321–2. https://doi.org/10.1016/j.jaad.2015.04.027.
8. Karen JK, Gareau DS, Dusza SW, Tudisco M, Rajadhyaksha M, Nehal KS. Detection of basal cell carcinomas in Mohs excisions with fluorescence confocal mosaicing microscopy. Br J Dermatol. 2009;160:1242–50. https://doi.org/10.1111/j.1365-2133.2009.09141.x.
9. Gareau DS, Karen JK, Dusza SW, Tudisco M, Nehal KS, Rajadhyaksha M. Sensitivity and specificity for detecting basal cell carcinomas in Mohs excisions with confocal fluorescence mosaicing microscopy. J Biomed Opt. 2009;14:034012. https://doi.org/10.1117/1.3130331.
10. Horn M, Gerger A, Koller S, Weger W, Langsenlehner U, Krippl P, et al. The use of confocal laser-scanning microscopy in microsurgery for invasive squamous cell carcinoma. Br J Dermatol. 2007;156:81–4. https://doi.org/10.1111/j.1365-2133.2006.07574.x.
11. Hartmann D, Ruini C, Mathemeier L, Bachmann MR, Dietrich A, Ruzicka T, et al. Identification of ex-vivo confocal laser scanning microscopic features of melanocytic lesions and their histological correlates. J Biophotonics. 2017;10:128–42. https://doi.org/10.1002/jbio.201500335.
12. Hartmann D, Krammer S, Vural S, Bachmann MR, Ruini C, Sárdy M, et al. Immunofluorescence and confocal microscopy for ex-vivo diagnosis of melanocytic and non-melanocytic skin tumors: a pilot study. J Biophotonics. 2018;11:e201700211. https://doi.org/10.1002/jbio.201700211.
13. Hartmann D, Krammer S, Ruini C, Ruzicka T, von Braunmühl T. Correlation of histological and ex-vivo confocal tumor thickness in malignant melanoma. Lasers Med Sci. 2016;31:921–7. https://doi.org/10.1007/s10103-016-1936-5.
14. Lamberti A, Cinotti E, Habougit C, Labeille B, Rubegni P, Perrot JL. Ex vivo confocal microscopy for dermatofibrosarcoma protuberans. Ski Res Technol. 2019;25:589–91. https://doi.org/10.1111/srt.12690.
15. Longo C, Ragazzi M, Gardini S, Moscarella E, Argenziano G. Ex Vivo fluorescence confocal microscopy of eccrine syringomatous carcinoma: a report of 2 cases. JAMA Dermatol. 2015;151:1034–6. https://doi.org/10.1001/JAMADERMATOL.2015.1008.
16. Debarbieux S, Dalle S, Depaepe L, Jeanniot PY, Poulalhon N, Thomas L. Extramammary paget's disease of the scalp: examination by in vivo and ex vivo reflectance confocal microscopy. Skin Res Technol. 2014;20:124–6. https://doi.org/10.1111/SRT.12087.
17. Bertoni L, Azzoni P, Reggiani C, Pisciotta A, Carnevale G, Chester J, et al. Ex vivo fluorescence confocal microscopy for intraoperative, real-time diagnosis of cutaneous inflammatory diseases: a preliminary study. Exp Dermatol. 2018;27:1152–9. https://doi.org/10.1111/exd.13754.
18. Bağcı IS, Aoki R, Krammer S, Ruzicka T, Sárdy M, French LE, et al. Ex vivo confocal laser scanning microscopy for bullous pemphigoid diagnostics: new era in direct immunofluorescence? J Eur Acad Dermatology Venereol. 2019;33:2123–30. https://doi.org/10.1111/jdv.15767.
19. Bağcı IS, Aoki R, Krammer S, Ruzicka T, Sárdy M, Hartmann D. Ex vivo confocal laser scanning microscopy: An innovative method for direct immunofluorescence of cutaneous vasculitis. J Biophotonics. 2019;12. https://doi.org/10.1002/jbio.201800425.
20. Cinotti E, Perrot JL, Labeille B, Campolmi N, Thuret G, Naigeon N, et al. First identification of the herpes simplex virus by skin-dedicated ex vivo fluorescence confocal microscopy during herpetic skin infections. Clin Exp Dermatol. 2015;40:421–5. https://doi.org/10.1111/ced.12546.
21. Cinotti E, Labeille B, Douchet C, Chovet M, Habougit C, Cambazard F, et al. Apport de la microscopie confocale in et ex vivo et de la tomographie en cohérence optique dans le diagnostic du molluscum contagiosum. Ann Dermatol Venereol. 2016;143:564–6. https://doi.org/10.1016/j.annder.2016.02.030.
22. Leclercq A, Cinotti E, Labeille B, Perrot JL, Cambazard F. Ex vivo confocal microscopy: a new diagnostic technique for mucormycosis. Ski Res Technol. 2016;22:203–7. https://doi.org/10.1111/srt.12251.
23. Debarbieux S, Gaspar R, Depaepe L, Dalle S, Balme B, Thomas L. Intraoperative diagnosis of nonpigmented nail tumours with ex vivo fluorescence confocal microscopy: 10 cases. Br J Dermatol. 2015;172:1037–44. https://doi.org/10.1111/bjd.13384.

24. Debarbieux S, Hospod V, Depaepe L, Balme B, Poulalhon N, Thomas L. Perioperative confocal microscopy of the nail matrix in the management of in situ or minimally invasive subungual melanomas. Br J Dermatol. 2012;167:828–36. https://doi.org/10.1111/j.1365-2133.2012.11013.x.
25. Panarello D, Compérat E, Seyde O, Colau A, Terrone C, Guillonneau B. Atlas of ex vivo prostate tissue and cancer images using confocal laser endomicroscopy: a project for intraoperative positive surgical margin detection during radical prostatectomy. Eur Urol Focus. 2020;6:941–58. https://doi.org/10.1016/j.euf.2019.01.004.
26. Puliatti S, Bertoni L, Pirola GM, Azzoni P, Bevilacqua L, Eissa A, et al. Ex vivo fluorescence confocal microscopy: the first application for real-time pathological examination of prostatic tissue. BJU Int. 2019;124:469–76. https://doi.org/10.1111/bju.14754.
27. Bertoni L, Puliatti S, Reggiani Bonetti L, Maiorana A, Eissa A, Azzoni P, et al. Ex vivo fluorescence confocal microscopy: prostatic and periprostatic tissues atlas and evaluation of the learning curve. Virchows Arch. 2020;476:511–20. https://doi.org/10.1007/s00428-019-02738-y.
28. Rocco B, Sighinolfi MC, Bertoni L, Spandri V, Puliatti S, Eissa A, et al. Real-time assessment of surgical margins during radical prostatectomy: a novel approach that uses fluorescence confocal microscopy for the evaluation of peri-prostatic soft tissue. BJU Int. 2020;125:487–9. https://doi.org/10.1111/bju.15000.
29. Dobbs J, Krishnamurthy S, Kyrish M, Benveniste AP, Yang W, Richards-Kortum R. Confocal fluorescence microscopy for rapid evaluation of invasive tumor cellularity of inflammatory breast carcinoma core needle biopsies. Breast Cancer Res Treat. 2015;149:303–10. https://doi.org/10.1007/s10549-014-3182-5.
30. Abeytunge S, Larson B, Peterson G, Morrow M, Rajadhyaksha M, Murray MP. Evaluation of breast tissue with confocal strip-mosaicking microscopy: a test approach emulating pathology-like examination. J Biomed Opt. 2017;22:034002. https://doi.org/10.1117/1.jbo.22.3.034002.
31. Dobbs JL, Ding H, Benveniste AP, Kuerer HM, Krishnamurthy S, Yang W, et al. Feasibility of confocal fluorescence microscopy for real-time evaluation of neoplasia in fresh human breast tissue. J Biomed Opt. 2013;18:106016. https://doi.org/10.1117/1.jbo.18.10.106016.
32. Elfgen C, Papassotiropoulos B, Varga Z, Moskovszky L, Nap M, Güth U, et al. Comparative analysis of confocal microscopy on fresh breast core needle biopsies and conventional histology. Diagn Pathol. 2019;14. https://doi.org/10.1186/s13000-019-0835-z.
33. Patel R, Khan A, Wirth D, Kamionek M, Kandil D, Quinlan R, et al. Multimodal optical imaging for detecting breast cancer. J Biomed Opt. 2012;17:066008. https://doi.org/10.1117/1.jbo.17.6.066008.
34. Bennàssar A, Vilata A, Puig S, Malvehy J. Ex vivo fluorescence confocal microscopy for fast evaluation of tumour margins during Mohs surgery. Br J Dermatol. 2014;170:360–5. https://doi.org/10.1111/bjd.12671.
35. Longo C, Rajadhyaksha M, Ragazzi M, Nehal K, Gardini S, Moscarella E, et al. Evaluating ex vivo fluorescence confocal microscopy images of basal cell carcinomas in Mohs excised tissue. Br J Dermatol. 2014;171:561–70. https://doi.org/10.1111/bjd.13070.
36. Peters N, Schubert M, Metzler G, Geppert JP, Moehrle M. Diagnostic accuracy of a new ex vivo confocal laser scanning microscope compared to H&E-stained paraffin slides for micrographic surgery of basal cell carcinoma. J Eur Acad Dermatol Venereol. 2019;33:298–304. https://doi.org/10.1111/jdv.15243.
37. Longo C, Pampena R, Bombonato C, Gardini S, Piana S, Mirra M, et al. Diagnostic accuracy of ex vivo fluorescence confocal microscopy in Mohs surgery of basal cell carcinomas: a prospective study on 753 margins. Br J Dermatol. 2019;180:1473–80. https://doi.org/10.1111/bjd.17507.
38. Hartmann D, Krammer S, Bachmann MR, Mathemeier L, Ruzicka T, von Braunmühl T. Simple 3-criteria-based ex vivo confocal diagnosis of basal cell carcinoma. J Biophotonics. 2018;11. https://doi.org/10.1002/jbio.201800062.
39. Gareau DS, Li Y, Huang B, Eastman Z, Nehal KS, Rajadhyaksha M. Confocal mosaicing microscopy in Mohs skin excisions: feasibility of rapid surgical pathology. J Biomed Opt. 2008;13:054001. https://doi.org/10.1117/1.2981828.
40. Cinotti E, Perrot JL, Labeille B, Boukenter A, Ouerdane Y, Cosmo P, et al. Identification of a soft tissue filler by ex vivo confocal microscopy and Raman spectroscopy in a case of adverse reaction to the filler. Ski Res Technol. 2015;21:114–8. https://doi.org/10.1111/srt.12166.
41. Hartmann D, Krammer S, Bachmann MR, Mathemeier L, Ruzicka T, Bagci IS, et al. Ex vivo confocal microscopy features of cutaneous squamous cell carcinoma. J Biophotonics. 2018;11:e201700318. https://doi.org/10.1002/jbio.201700318.
42. Forest F, Cinotti E, Habougit C, Ginguéné C, Perrot JL, Labeille B, et al. Rapid characterization of human brain aspergillosis by confocal microscopy on a thick squash preparation. Cytopathology. 2016;27:221–2. https://doi.org/10.1111/cyt.12258.
43. Cinotti E, Haouas M, Grivet D, Perrot JL. In vivo and ex vivo confocal microscopy for the management of a melanoma of the eyelid margin. Dermatologic Surg. 2015;41:1437–40. https://doi.org/10.1097/DSS.0000000000000517.
44. Cinotti E, Perrot JL, Labeille B, Raberin H, Flori P, Cambazard F. Hair dermatophytosis diagnosed by reflectance confocal microscopy: Six cases. J Eur Acad Dermatol Venereol. 2015;29:2257–9. https://doi.org/10.1111/jdv.12557.
45. Krammer S, Krammer C, Vladimirova G, Salzer S, Ruini C, Sattler E, et al. Ex vivo confocal laser scanning microscopy: a potential new diagnostic imaging tool in onychomycosis comparable with gold standard techniques. Front Med. 2020;7. https://doi.org/10.3389/fmed.2020.586648.
46. Bağcı IS, Aoki R, Krammer S, Vladimirova G, Ruzicka T, Sárdy M, et al. Immunofluorescence and histopathological assessment using ex vivo confocal laser scanning microscopy in lichen planus. J Biophotonics. 2020;13. https://doi.org/10.1002/jbio.202000328.
47. Bağcı IS, Aoki R, Vladimirova G, Ergün E, Ruzicka T, Sárdy M, et al. New-generation diagnostics in inflammatory skin diseases: Immunofluorescence and histopathological assessment using ex vivo confocal laser scanning microscopy in cutaneous lupus erythematosus. Exp Dermatol. 2020;30. https://doi.org/10.1111/exd.14265.
48. Sonn GA, Mach KE, Jensen K, Hsiung PL, Jones SN, Contag CH, et al. Fibered confocal microscopy of bladder tumors: An ex vivo study. J Endourol. 2009;23:197–201. https://doi.org/10.1089/end.2008.0524.
49. Wiesner C, Jäger W, Salzer A, Biesterfeld S, Kiesslich R, Hampel C, et al. Confocal laser endomicroscopy for the diagnosis of urothelial bladder neoplasia: a technology of the future? BJU Int. 2011;107:399–403. https://doi.org/10.1111/j.1464-410X.2010.09540.x.
50. Bonnal JL, Rock A, Gagnat A, Papadopoulos S, Filoche B, Mauroy B. Confocal laser endomicroscopy of bladder tumors associated with photodynamic diagnosis: an ex vivo pilot study. Urology. 2012;80:1162.e1-1162.e5. https://doi.org/10.1016/j.urology.2012.06.035.
51. Su LM, Kuo J, Allan RW, Liao JC, Ritari KL, Tomeny PE, et al. Fiber-optic confocal laser endomicroscopy of small renal masses:

toward real-time optical diagnostic biopsy. J Urol. 2016;195:486–92. https://doi.org/10.1016/j.juro.2015.07.115.
52. Krishnamurthy S, Ban K, Shaw K, Mills G, Sheth R, Tam A, et al. Confocal fluorescence microscopy platform suitable for rapid evaluation of small fragments of tissue in surgical pathology practice. Arch Pathol Lab Med. 2019;143:305–13. https://doi.org/10.5858/arpa.2018-0352-OA.
53. Phung MC, Rouse AR, Pangilinan J, Bell RC, Bracamonte ER, Mashi S, et al. Investigation of confocal microscopy for differentiation of renal cell carcinoma versus benign tissue. Can an optical biopsy be performed? Asian J Urol. 2020;7:363–8. https://doi.org/10.1016/j.ajur.2019.12.008.
54. Ragazzi M, Piana S, Longo C, Castagnetti F, Foroni M, Ferrari G, Gardini G, Pellacani G et al. Fluorescence confocal microscopy for pathologists. Mod Pathol. 2014;27:460–71. https://doi.org/10.1038/MODPATHOL.2013.158.
55. Collier T, Lacy A, Richards-Kortum R, Malpica A, Follen M. Near real-time confocal microscopy of amelanotic tissue: Detection of dysplasia in ex vivo cervical tissue. Acad Radiol. 2002;9:504–12. https://doi.org/10.1016/S1076-6332(03)80326-4.
56. Carlson K, Pavlova I, Collier T, Descour M, Follen M, Richards-Kortum R. Confocal microscopy: Imaging cervical precancerous lesions. Gynecol Oncol. 2005;9. https://doi.org/10.1016/j.ygyno.2005.07.049.
57. Inoue H, Igari T, Nishikage T, Ami K, Yoshida T, Iwai T. A novel method of virtual histopathology using laser-scanning confocal microscopy in-vitro with untreated fresh specimens from the gastrointestinal mucosa. Endoscopy. 2000;32:439–43. https://doi.org/10.1055/s-2000-654.
58. Yoshida S, Tanaka S, Hirata M, Mouri R, Kaneko I, Oka S, et al. Optical biopsy of GI lesions by reflectance-type laser-scanning confocal microscopy. Gastrointest Endosc. 2007;66:144–9. https://doi.org/10.1016/j.gie.2006.10.054.
59. Gorospe EC, Leggett CL, Sun G, Anderson MA, Gupta M, Penfield JD, et al. Diagnostic performance of two confocal endomicroscopy systems in detecting Barrett's dysplasia: a pilot study using a novel bioprobe in ex vivo tissue. Gastrointest Endosc. 2012;76:933–8. https://doi.org/10.1016/j.gie.2012.07.005.
60. El Hallani S, Poh CF, Macaulay CE, Follen M, Guillaud M, Lane P. Ex vivo confocal imaging with contrast agents for the detection of oral potentially malignant lesions. Oral Oncol. 2013;49:582–90. https://doi.org/10.1016/j.oraloncology.2013.01.009.
61. Sorokina A, Danilevskaya O, Averyanov A, Zabozlaev F, Sazonov D, Yarmus L, et al. Comparative study of ex vivo probe-based confocal laser endomicroscopy and light microscopy in lung cancer diagnostics. Respirology. 2014;19:907–13. https://doi.org/10.1111/resp.12326.
62. Takemura M, Kurimoto N, Hoshikawa M, Maeno T, Hisada T, Kurabayashi M, et al. Probe-based confocal laser endomicroscopy for rapid on-site evaluation of transbronchial biopsy specimens. Thorac Cancer. 2019;10:1441–7. https://doi.org/10.1111/1759-7714.13089.
63. Just T, Stave J, Boltze C, Wree A, Kramp B, Guthoff RF, et al. Laser scanning microscopy of the human larynx mucosa: a preliminary, ex vivo study. Laryngoscope. 2006;116:1136–41. https://doi.org/10.1097/01.mlg.0000217529.53079.59.
64. Wirth D, Snuderl M, Sheth S, Kwon C-S, Frosch MP, Curry W, et al. Identifying brain neoplasms using dye-enhanced multimodal confocal imaging. J Biomed Opt. 2012;17:026012. https://doi.org/10.1117/1.jbo.17.2.026012.
65. Snuderl M, Wirth D, Sheth SA, Bourne SK, Kwon CS, Ancukiewicz M, et al. Dye-enhanced multimodal confocal imaging as a novel approach to intraoperative diagnosis of brain tumors. Brain Pathol. 2013;23:73–81. https://doi.org/10.1111/j.1750-3639.2012.00626.x.
66. Martirosyan NL, Eschbacher JM, Yashar M, Turner JD, Belykh E, Spetzler RF, et al. Prospective evaluation of the utility of intraoperative confocal laser endomicroscopy in patients with brain neoplasms using fluorescein sodium: Experience with 74 cases. Neurosurg Focus 2016;40. https://doi.org/10.3171/2016.1.FOCUS15559.
67. Belykh E, Zhao X, Ngo B, Farhadi DS, Byvaltsev VA, Eschbacher JM, et al. Intraoperative confocal laser endomicroscopy ex vivo examination of tissue microstructure during fluorescence-guided brain tumor surgery. Front Oncol 2020;10. https://doi.org/10.3389/fonc.2020.599250.
68. Belykh E, Ngo B, Farhadi DS, Zhao X, Mooney MA, White WL, et al. Confocal laser endomicroscopy assessment of pituitary tumor microstructure: a feasibility study. J Clin Med. 2020;9:3146. https://doi.org/10.3390/jcm9103146.
69. Acerbi F, Pollo B, De Laurentis C, Restelli F, Falco J, Vetrano IG, et al. Ex vivo fluorescein-assisted confocal laser endomicroscopy (CONVIVO® System) in patients with glioblastoma: results from a prospective study. Front Oncol. 2020;10. https://doi.org/10.3389/fonc.2020.606574.

Principles of Ex Vivo Confocal Microscopy

2

John Carucci, Samantha Lish, Marc Combaia, Nicole Doudican, and Daniel Gareau

Ex vivo confocal microscopy (EVCM) is a type of advanced imaging that stands to revolutionize dermatology by incorporating digital diagnostics into pathological practice [1, 2]. The goal of this chapter is to familiarize the reader with underlying optical physics principles for laser-based EVCM imaging. The specimen will be treated as an optical element in the imaging system that can reflect light (reflectance mode) or be stained to emit light through fluorescence (fluorescence mode). Though the optics of imaging in these two modes represents the bulk of this chapter, concluding remarks will draw the reader's attention to the combination of these single-mode images into a digital color-coded "multimodal" image also called digital H&E (DHE) that can be used to replicate the appearance of standard pathology.

2.1 Historical Perspective of Reflectance Mode Confocal Microscopy

Underpinning the implementation EVCM technology for pathological use is a long history of key advances. The confocal microscope was invented by Marvin Minsky in 1957 and subsequently underwent a series of refinements that increased the speed of digital imaging implementing updated light sources and scanning mechanisms. Confocal microscopes were adapted to image human skin in vivo, using a white light source and a spinning disk of pinholes [3–6]. A laser light source and spinning polygon mirror [7–9] were added. A strength of digital confocal microscopy is the speed at which pixels are acquired. Whereas in vivo scanning needs the speed to capture images quickly (relative to breathing or heartbeat motion artifacts), ex vivo imaging needs the speed to provide pathology images of specimens that are wide (e.g., 1 cm) with high resolution (e.g., 1 μm). Pictures with this many pixels are slow to acquire unless the imaging systems are very fast and acquire tens or hundreds of thousands of pixels per second. Though there have been many confocal embodiments and ongoing device engineering aims to reduce device complexity while further increasing imaging speed [10], the core principles of confocal microscopy remain unchanged, since Minsky's original invention and are applicable to both reflectance and fluorescence modes.

Confocal microscopy is different from common microscopic techniques in that it achieves sectioning via optical rather than physical means. The common wide-field microscope illuminates and images a large volume of tissue, largely washing out cellular detail. Thus, thin sections must be prepared before imaging, as in histology, to enable cellular resolution. By comparison, the confocal microscope illuminates a small voxel, or three-dimensional equivalent of a pixel, from which reflected light is collected to produce a pixel. Scanning the voxel in two dimensions creates an illuminated plane that produces an image pixel-by-pixel.

2.2 Ex Vivo Confocal Microscopy in Practice

As with any imaging system, there are only two main quality factors without which an image cannot be acquired: contrast and resolution. If both of these factors are not optimized, image quality is compromised. Resolution is determined by the geometry of the optical lenses (geometrical optics) within the microscope, whereas contrast is defined by the sample. In context, high resolution and good contrast are required to confidently read pathology. In addition to the laser, lenses, and mirrors in EVCM, which image at industry standards, the only other variable component to consider in the optical imaging system is the specimen. The specimen is

J. Carucci · N. Doudican
Ronald O. Pearlman Department of Dermatology, New York University, 550 First Avenue, New York, NY 10016, USA

S. Lish · D. Gareau (✉)
Laboratory for Investigative Dermatology, The Rockefeller University, 1230 York Avenue, New York, NY 10065, USA
e-mail: dgareau@mail.rockefeller.edu

M. Combaia
Department of Dermatology, Hospital Clinic de Barcelona, Universitat de Barcelona, Barcelona, Spain

M. Jain et al. (eds.), *Cutaneous Atlas of Ex Vivo Confocal Microscopy*,
https://doi.org/10.1007/978-3-030-89316-3_2

an optical component because in three-dimensional (3D) imaging, light travels through the 3D specimen, forming the image. During that travel, the light-tissue interaction is complicated compared to the light path through the microscope. Those complications lead to the clinical, pathological, and medical limitations of EVCM.

Figure 2.1 shows a whole tissue specimen mounted for confocal imaging with optical sectioning. Confocal optical sectioning images thin optical sections within whole tissue samples such as skin, so physical sectioning isn't needed. A confocal microscope achieves point illumination of a voxel (volume element) in the sample focal plane by focusing a laser through an objective lens and point detection of a pixel (picture element) and by placing a pinhole in a conjugate focal plane. The pinhole collects light emanating only from the focus and blocks light from elsewhere. Point illumination of the voxel within the sample and the pinhole lie in optically **con**jugate **focal** planes, leading to the name **confocal**. This arrangement is sensitive only to the illuminated voxel in-focus and insensitive to out-of-focus light from all other locations. Scanning the voxel in the focal plane of the objective lens enables optical sectioning: non-invasive imaging of a thin (<5 μm) section within a thick sample. This optical sectioning capability of confocal microscopy is an attractive alternative or adjunct to physical sectioning that is routinely performed for histology because it can operate on a whole specimen without the need for the traditional freezing and sectioning associated with standard histopathology. Figure 2.2 shows a schematic of how optical sectioning works and illustrates the paths light rays take to achieve selective measurement within a small volume ($\sim 10^{-15}$ L) around the focal point within the sample.

Under the confocal configuration, illumination laser light is reflected into the optical path off of a beam splitter. An objective lens is used to focus the light onto a point in the sample focal plane. Light emanating back (up) from that point is collected by the objective lens and returned to the detector through a detector lens focused on a pinhole. The confocal configuration is not sensitive to light from out of focus regions such as below the focus or within the focal plane at a laterally displaced position. Scanning of the in-focus light detection point within the sample is achieved by altering (with a Scan Mirror) the angle of incidence of the laser light upon the back of the objective, which in turn displaces the focal point laterally within the specimen. The two scanned positions illustrated in Fig. 2.2 represent two pixels at different points within the image. Since the light path back from the focal point within the specimen to the pinhole and detector reflects off the same Scan Mirror as in the forward path, the elements on the opposite side of the scan mirror do not move even though the scanned point of light moves within the sample. Thus, the returning light is "de-scanned" by traveling back along the exact same path as the forward laser light.

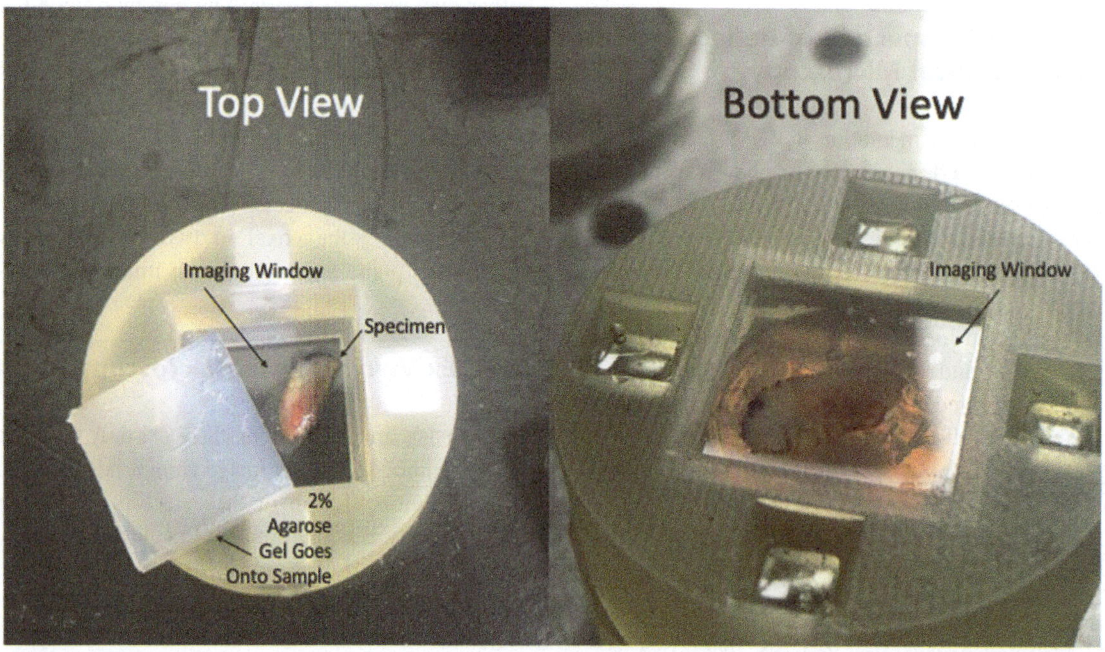

Fig. 2.1 Specimen mounted for confocal imaging. The Top View shows a circular specimen compression chamber with the specimen face-down on the Imaging Window. Agarose Gel is prepared at 2% concentration to press onto the sample. The Bottom view shows the specimen through the imaging window in the stabilized, hydrated state after the agarose gel has been compressed onto the specimen

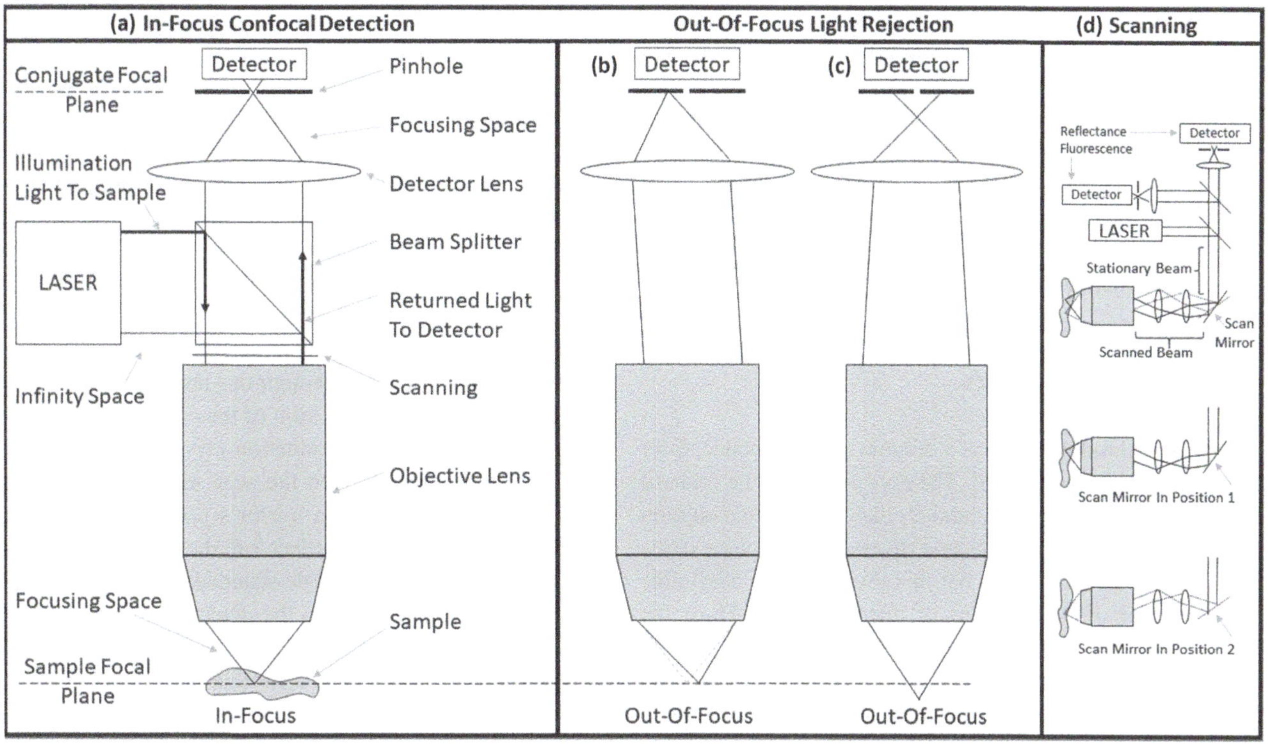

Fig. 2.2 Confocal schematic. The schematic **a** shows the key components of a confocal microscope, such as the laser and the lenses. Conditions of confocally are shown to be met (**a**) as well as not met for conditions outside of the focal spot (**b** & **c**). Motors which mechanize the optical scanning of the confocal prob point are shown (**d**) in two positions which represent two different pixels formed in the confocal image and correlate to two different voxels within the sample

2.3 The Reflectance Contrast Mechanism

The mechanism of bright contrast in reflectance confocal microscopy is backscattering. Backscattering is primarily governed by the structures' refractive index (n) compared to surrounding medium. In gray-scale confocal images, structures that appear bright (white) have components with high refractive index compared with their surroundings and are similar in size to the wavelength of light. Examples of highly reflective components in skin include melanin (n = 1.72) [11], hydrated collagen (n = 1.43) [12], and keratin as in the stratum corneum (n = 1.51) [13]. These components appear brighter than the surrounding epidermis (n = 1.34) and dermis (n = 1.41) [13]. A secondary mechanism that determines reflectivity is size. Tissue components that are similar in size to the wavelength of light appear bright in the image. Predictions of reflectivity based on size and refractive index can be made using Mie theory [14].

The mean free path of photon transport through tissue, such as skin, is defined as the average (mean) distance that any photon travels without being scattered or absorbed. Confocal microscopes work by removing scattered light, so there is an imaging depth limit at which all the light that is geometrically headed to the subsurface focus becomes depleted. At this limit, all the light traveling between the surface, and the subsurface imaging focal plane of the microscope, shrinks into the noise floor (a feature present in all digital systems) and the signal and image are lost. It is interesting to note that the difference between optical coherence tomography and confocal microscopy is the added presence of an interferometer, which effectively lowers the noise floor, permitting deeper imaging in a given tissue. Each biological tissue has its own optical properties, including the scattering coefficient, the absorption coefficient, and the scattering anisotropy. Considerable research has been conducted using skin because of its inherent optical complexity compared to other tissue types. Collagen, keratin, and especially melanin, among others, contribute to scattering, whereas blood, melanin, and to a lesser extent molecule such as bilirubin and beta-carotene contribute to absorption. In general, any biological tissue that has heterogeneous volumetric distributions of both light scattering and absorption is called an optically turbid tissue. Because the mean free path for visible photons is approximately 0.1 mm in skin, the penetration depth of ex vivo microscopy, being a specialized confocal microscope, is 2–3 mean free paths, which is approximately 0.2–0.3 mm. A surface layer of 300 μm represents a volume that contains far more evaluable cells than are visible for pathologic inspection. Pathological bandwidth is the rate at which pathological images can be acquired, whose increase will make more image data available and result in robust pathological analysis. By either using a computer algorithm to select which cells within a volume to show the pathologist making the diagnosis or by directly deriving computational models of disease through analysis of those large cell populations, ECVM will play an important role in future pathological bandwidth.

2.4 Geometric Optics

Lasers pass through space as a column of light; hence, laser light is said to be collimated. The rays are said to be focused at infinity because they are parallel, like a cylindrical column in space. In reality, laser beams diverge to various extents depending on their quality, but this divergence is negligible (on the order of milli-radians or micro-radians). Thus, the beams effectively behave as a perfect column. That column turns into a cone when laser shines through a lens and the cone comes to a point in the lenses' focal plane. Imagine a tiny perfect mirror disk centered at the point and contained in the focal plane. Since the cone focuses to a point on the mirror, the reflected light is guaranteed (by Snells Law) to be the exact same cone. The only difference is that the light propagates in the opposite direction, but light can travel right through itself, so the two reflected light paths exist in parallel. When the reflected light returns to the lens after having been focused, the "illumination path" is reversed and it gets re-collimated through the lens. The forward and reverse rays are equivalent in spatial position but opposite in the direction of light propagation.

The objective lenses used in microscopy contain sets of about 20 small lenses, each approximately the size of a pea, in a metal casing, along the same axis that is conventionally called the Z-axis. But think of this group as a single lens for simplicity and give it the name objective lens. On the side of the objective lens opposite from the sample, there exists a so- called "infinity space," because both the illumination laser and reflected light from the focus in the sample are overlapping sets of parallel rays. Illumination rays travel toward the specimen and detection rays travel from the specimen to the detector. The set of parallel rays that comprise the detection light path emanating from the sample focal point to the pinhole are also parallel in infinity space and the illumination light-ray-set is, itself, parallel to the detection light-ray-set. However, the key fact regarding the detection light rays' parallel nature is not a virtue of its being a laser (as for illumination light), but rather a consequence of all rays emanating from the same point in the sample focal plane.

A 45° beam splitter is placed in the path of both cylindrical ray sets as shown in Fig. 2.2 such that it reflects the illumination laser light into the sample and transmits the detection light to the detector.

Scanning is achieved using an electromechanically moving mirror to vary the angle at which the laser enters the backside of the objective lens, leading to the focal spot (in the sample focal plane) moving laterally in the image plane. Thus, the "point illumination" laser light is not aimed directly at the objective lens with normal incidence to its back surface but reflected off the scan mirror at an angle from the objective lens. Although this angle is varied by the scan mirror to effect the position of the confocal point in the sample focal plane, the illumination rays and detection rays have the same path between the scan mirror and the focal point regardless of the scan mirror's position. As a result, when the detection rays reflect off the scan mirror, they become stationary and get "de-scanned" as they propagate from the scan mirror back to the detector. The scan mirror thus moves the illumination and detection rays in unison in the space between the focal point and itself, and all illumination and detection light is stationary in all other space.

When the stationary detection beam backpropagates from the scan mirror toward the detector, the beam splitter transmits it so that the detection light emanating from the focal point can be isolated from the illumination light and detected.

"Point detection" occurs when that isolated, stationary, collimated, light (in infinity space) is aimed onto a second lens, the detector lens, which focuses it through a pinhole. Compared to the objective lens which focuses light into the specimen to a focal point, the detector lens focuses the reflected light column onto the detection focal plane, which contains the confocal pinhole aperture. The aperture is a thin sheet of metal (like tin foil) with a pinhole in it that has a diameter of about the width of a hair shaft (approximately 50 μm). The exact diameter must be matched to the focal spot of laser light in the sample focal plane through the ratio of the focal lengths of the detector lens and focal lens. This ratio is the magnification of the optical imaging system and is irrelevant to clinical use, provided it can be assured (from the microscope manufacturer) that the optical system has proper confocal gating and is not misaligned.

The detector sits on the other side of the pinhole sheet metal, such that the only light that gets measured to produce the electronic readout signal is the light passing through the pinhole.

Microscopes have a magnification between the focal plane in the sample and the **con**jugate **focal** (**confocal**) plane containing the pinhole, point-detector. The magnification of the system equals the ratio of the focal length of the detector lens to the focal length of the objective lens. An object in the sample focal plane casts an image on the conjugate focal plane and the ratio of object sizes in these two planes is the magnification of the optics.

To appreciate the sensitivity and spatial selectivity of confocal laser illumination and detection, consider a circular, 1 μm-diameter mirror in the sample focal plane at the laser focal point. In the absence of optical scattering, all the laser light would reflect off this first mirror back through the microscope and through the detection pinhole. More light will be detected reflecting from that mirror than from a second, similar mirror elsewhere in the sample focal plane because (1) no laser light (or only a slight amount of scattered light) is focused on that second mirror and (2) the conjugate image spot of the second mirror in the detection plane is on the pinhole sheet metal so it gets blocked from striking the detector behind the sheet metal (see Fig. 2.2b). Another consideration could be made for a different second mirror that was displaced, not within the lateral image plane as before, but along the z-axis above or below the sample focal plane (see Fig. 2.2c).

Considering the reflectance from our mirror compared to the reflectance from a mirror that is positioned in the focal plane adjacent to our point illumination-reflecting mirror, the light of the second mirror is conjugated to a position on the metal in the conjugate focal plane that is not the pinhole. This is the mechanism by which confocal microscopes filter out scattered light. The story of the second mirror can be generalized to any illumination photons that did not make it directly to the focus or propagate directly back out of the tissue without being scattered or absorbed. Light photons that do propagate directly to the focus and then directly back out of the tissue without being scattered or absorbed (and hence propagate through the pinhole to the detector) are called "ballistic photons." The dwindling number of ballistic photons in deeper tissue (e.g., > 300 μm) defines the imaging depth penetration limit, because statistically a photon can only penetrate 2–3 mean free paths before, on average, being scattered or absorbed and therefore, lost.

2.5 Confocal Image Formation

The ballistic photons that survive propagating through the gauntlet described above without being scattered or absorbed can build an optical section image when the point illumination and detection are applied sequentially to a set of points in the sample focal plane. This X–Y scan (Z is the laser axis) is called a raster scan and can be repeated at different Z depths in the tissue to make 3D images.

The optical reflectance signal from a point within a biological specimen is different than the reflectance from a mirror surface. This difference is based on a limit imposed by the laws of optical physics which prevent focus to a true point. The illumination pattern of a mirror in the focal plane to a small disk, called the Airy disc, does not converge to a simple point in the focal plane. Instead, the Airy disk is about as wide as the wavelength of light (>10X smaller than typical cells) in the focal plane. The Airy disc size determines the lateral resolution of the image, so EVCM achieves sub-cellular resolution because the Airy disc is smaller than cells.

Consider the light intensity not just on the Airy disc in the focal plane but in a three-dimensional volume bounded by an American football-shaped surface, whose circular cross-section is the Airy disk. The volumetric element (voxel) bounded by this surface has more light density than in the tissue volume surrounding the voxel.

The long Z distance within the volume (on the football axis) compared to the shorter X–Y distance (Airy disc diameter), means the light probes a finer "point" in the X–Y and as a result, the Z resolution is about 4 times worse than the X and Y directions. The implication of poorer Z resolution is that when looking at the 3D image from the side, the cells will be harder to distinguish. As a side note, more advanced fluorescence microscopes, such as 2-photon microscopes, can overcome this resolution limit and achieve super resolution down to nm, but these devices are in early research phases and aren't typically found in the medical care setting.

Tissue reflectance contrast is the result of many photons being reflected in all different directions (i.e., scattered) within the voxel of illumination. Only the fraction of scattered photons that retro-reflect back into the cone of the lens' focus get detected and because the tissue optical element is no longer a simple mirror, some light gets scattered in forward directions and absorbed deep in the tissue. The amount of ballistic light detected from skin reflectance is only a small percentage of what would reflect from a mirror in our simple model, but this amount of light is enough of a contributor to make images that have 3 important properties: (1) sub-cellular resolution which is important for evaluating cellular pathologic morphology, (2) endogenous contrast to dispel the need for additional contrast agents in reflectance mode confocal microscopy, and (3) available from thin sections within thick specimens that don't require freezing, fixing or physically sectioning tissue.

Multimodal confocal microscopy, including fluorescence mode to image fluorescently-labeled cells [15], enables better pathological evaluation than the reflectance mode alone because strong contrast of the cellular morphology is key in evaluation. Multiple modes can be combined into color-coded images that mimic the appearance of standard hematoxylin and eosin-based pathology, which is an attractive adjunct to surgery. Though "multi- " can be more than two [16–19], consider the multimodal combination of reflectance mode (described above) with fluorescence mode

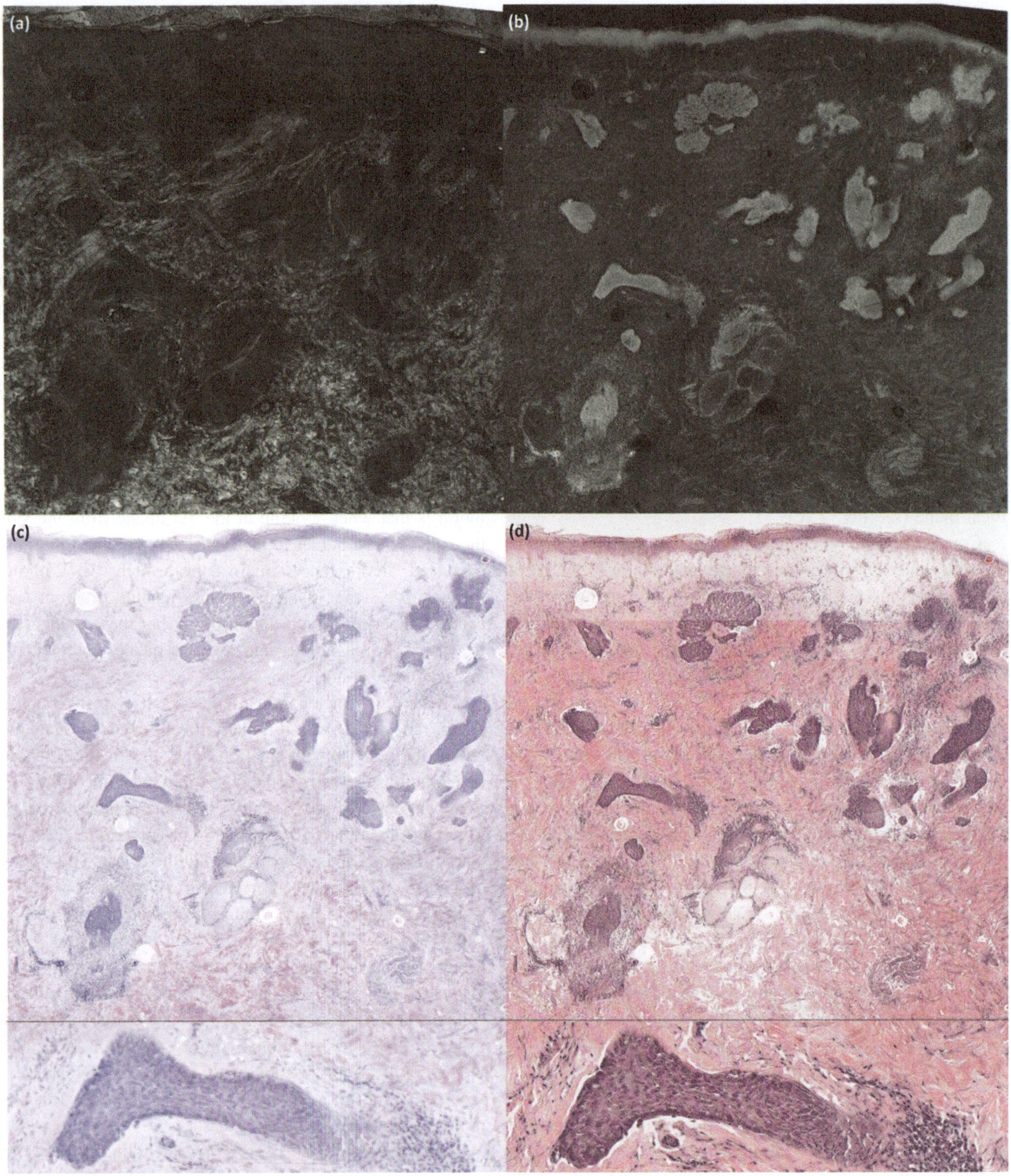

Fig. 2.3 Ex vivo confocal microscopy of a skin specimen. The reflectance mode image (**a**) shows bright contrast where there are microscopic variations in refractive index in the tissue caused by components such as collagen, keratin, and melanin. The fluorescence mode image (**b**) shows bright contrast where there is the presence of the fluorescent stain acridine orange which labels nuclei. The individual mode images in gray scale (**a** and **b**) are combined in color code (**c**) which is enhanced by an artificial intelligence filter to resemble histology more closely (**d**)

in a 2-mode configuration. Fluorescence occurs when a photon is absorbed, diminished in energy, and then readmitted at a longer wavelength. For practical purposes, scattering events and fluorescence events are all governed by the same geometrical property in that the only difference is the wavelength of the detected light. Whereas reflectance is ubiquitous nonspecific endogenous contrast in the image whose spatial pattern contains information about the source of its origin, fluorescence can be a targeted label such as intercalation of molecular acridine orange between the base pairs of DNA in cell nuclei, so its mere presence identifies the specific source of its origin (a cell nucleus).

In reflectance mode, keratinocyte nuclei are dark due to the diminutive size of chromatin filaments. Since they are smaller than the wavelength of light, they do not scatter light efficiently. In fluorescence mode, keratinocyte nuclei are bright due to acridine orange labeling. Thus, fluorescence mode provides nuclear contrast and reflectance mode provides counter contrast. In each mode, the pixel brightness is a number, so the image information content of any particular pixel (or voxel in 3D) consists of two numbers that correspond to the fluorescence and reflectance brightness at that pixel. Those two numbers can be transformed into three numbers such that the pixel can have a defined red/green and blue level and become a color pixel. The transformation algorithm [20] can be designed so that the color pixels reproduce the contrast that would be obtained from standard frozen or fixed section histopathology using hematoxylin and eosin staining. Figure 2.3 shows confocal images of a skin tissue specimen.

The reflectance mode image (Fig. 2.3a) shows bright contrast in the tissue where there are microscopic variations in refractive index caused by components such as collagen, keratin, and melanin. The fluorescence mode image (Fig. 2.3b) shows bright contrast in the presence of the fluorescent stain acridine orange which labels nuclei. The individual mode images in gray scale (Fig. 2.3a, b) are combined by an algorithm [20] that implements color-coding to match the appearance of hematoxylin and eosin (Fig. 2.3c), which is enhanced by an artificial intelligence filter [21] to better resemble histology (Fig. 2.3d).

References

1. Krishnamurthy S, Brown JQ, Iftimia N, Levenson RM, Rajadhyaksha M. Ex vivo microscopy: a promising next-generation digital microscopy tool for surgical pathology practice. Arch Pathol Lab Med. 2019;143(9):1058–68. Epub 2019/07/12. https://doi.org/10.5858/arpa.2019-0058-RA. PubMed PMID: 31295016; PubMed Central PMCID: PMCPMC7365575.
2. Malvehy J, Perez-Anker J, Toll A, Pigem R, Garcia A, Alos LL, et al. Ex vivo confocal microscopy: revolution in fast pathology in dermatology. Br J Dermatol. 2020;183(6):1011–25. Epub 2020/03/07. https://doi.org/10.1111/bjd.19017. PubMed PMID: 32134506.
3. Corcuff P, Gonnord G, Pierard GE, Leveque JL. In vivo confocal microscopy of human skin: a new design for cosmetology and dermatology. Scanning. 1996;18(5):351–5. Epub 1996/08/01. https://doi.org/10.1002/sca.1996.4950180505. PubMed PMID: 8765611.
4. Bertrand C., Corcuff P. In vivo spatio-temporal visualization of the human skin by real-time confocal microscopy. Scanning. 1994;16(3):150–4. Epub 1994/05/01. https://doi.org/10.1002/sca.4950160301. PubMed PMID: 8038914.
5. Corcuff P, Leveque JL. In vivo vision of the human skin with the tandem scanning microscope. Dermatology. 1993;186(1):50–4. Epub 1993/01/01. https://doi.org/10.1159/000247302. PubMed PMID: 8435517.
6. Rajadhyaksha M, Grossman M, Esterowitz D, Webb RH, Anderson RR. In vivo confocal scanning laser microscopy of human skin: melanin provides strong contrast. J Invest Dermatol. 1995;104(6):946–52. Epub 1995/06/01. https://doi.org/10.1111/1523-1747.ep12606215. PubMed PMID: 7769264.
7. New KC, Petroll WM, Boyde A, Martin L, Corcuff P, Leveque JL, et al. In vivo imaging of human teeth and skin using real-time confocal microscopy. Scanning. 1991;13:369–72. https://doi.org/10.1002/sca.4950130507.
8. Rajadhyaksha M, Anderson RR, Webb RH. Video-rate confocal scanning laser microscope for imaging human tissues in vivo. Appl Opt. 1999;38(10):2105–15. Epub 2008/03/06. https://doi.org/10.1364/ao.38.002105. PubMed PMID: 18319771.
9. Rajadhyaksha M, Gonzalez S, Zavislan JM, Anderson RR., Webb RH. In vivo confocal scanning laser microscopy of human skin II: advances in instrumentation and comparison with histology. J Invest Dermatol. 1999;113(3):293–303. Epub 1999/09/01. https://doi.org/10.1046/j.1523-1747.1999.00690.x. PubMed PMID: 10469324.
10. Gareau DS, Krueger JG, Hawkes JE, Lish SR, Dietz MP, Mulberger AG, Mu EW, Stevenson ML, Lewin JM, Meehan SA, Carucci J. A. Line scanning, stage scanning confocal microscope (LSSSCM). Biomed Opt Express. 2017;8(8):3807–15. Epub 2017/09/01. https://doi.org/10.1364/BOE.8.003807. PubMed PMID: 28856051; PubMed Central PMCID: PMCPMC5560842.
11. Vitkin IA, Woolsey J, Wilson BC, Anderson RR. Optical and thermal characterization of natural (Sepia officinalis) melanin. Photochem Photobiol. 1994;59(4):455–62. Epub 1994/04/01. https://doi.org/10.1111/j.1751-1097.1994.tb05064.x. PubMed PMID: 8022888.
12. Wang XJ, Milner TE, Chang MC, Nelson JS. Group refractive index measurement of dry and hydrated type I collagen films using optical low-coherence reflectometry. J Biomed Opt. 1996;1(2):212–6. Epub 1996/04/01. https://doi.org/10.1117/12.227699. PubMed PMID: 23014687.
13. Tearney GJ, Brezinski ME, Southern JF, Bouma BE, Hee MR, Fujimoto JG. Determination of the refractive index of highly scattering human tissue by optical coherence tomography. Opt Lett. 1995;20(21):2258. Epub 1995/11/01. https://doi.org/10.1364/ol.20.002258. PubMed PMID: 19862316.
14. Rajadhyaksha M, Gonzalez S, Zavislan JM. Detectability of contrast agents for confocal reflectance imaging of skin and microcirculation. J Biomed Opt. 2004;9(2):323–31. Epub 2004/04/07. https://doi.org/10.1117/1.1646175. PubMed PMID: 15065898.
15. Gareau DS, Li Y, Huang B, Eastman Z, Nehal KS, Rajadhyaksha M. Confocal mosaicing microscopy in Mohs skin excisions: feasibility of rapid surgical pathology. J Biomed Opt. 2008;13(5):054001. Epub 2008/11/22. https://doi.org/10.1117/1.2981828.

PubMed PMID: 19021381; PubMed Central PMCID: PMCPMC2662759.

16. Gareau D, Bar A, Snaveley N, Lee K, Chen N. Swanson N, Simpson E, Jacques S. Tri-modal confocal mosaics detect residual invasive squamous cell carcinoma in Mohs surgical excisions. J Biomed Opt. 2012;17(6):066018. Epub 2012/06/28. https://doi.org/10.1117/1.JBO.17.6.066018. PubMed PMID: 22734774; PubMed Central PMCID: PMCPMC3381035.
17. Karen JK, Gareau DS, Dusza SW, Tudisco M, Rajadhyaksha M, Nehal KS. Detection of basal cell carcinomas in Mohs excisions with fluorescence confocal mosaicing microscopy. Br J Dermatol. 2009;160(6):1242–50. Epub 2009/05/07. https://doi.org/10.1111/j.1365-2133.2009.09141.x. PubMed PMID: 19416248; PubMed Central PMCID: PMCPMC2693082
18. Gareau DS, Karen JK, Dusza SW, Tudisco M, Nehal KS, Rajadhyaksha M. Sensitivity and specificity for detecting basal cell carcinomas in Mohs excisions with confocal fluorescence mosaicing microscopy. J Biomed Opt. 2009;14(3):034012. Epub 2009/07/02. https://doi.org/10.1117/1.3130331. PubMed PMID: 19566305; PubMed Central PMCID: PMCPMC2705883
19. Gareau DS, Jeon H, Nehal KS, Rajadhyaksha M. Rapid screening of cancer margins in tissue with multimodal confocal microscopy. J Surg Res. 2012;178(2):533–8. Epub 2012/06/23. https://doi.org/10.1016/j.jss.2012.05.059. PubMed PMID: 22721570; PubMed Central PMCID: PMCPMC3458153.
20. Gareau DS. Feasibility of digitally stained multimodal confocal mosaics to simulate histopathology. J Biomed Opt. 2009;14(3):034050. Epub 2009/07/02. https://doi.org/10.1117/1.3149853. PubMed PMID: 19566342; PubMed Central PMCID: PMCPMC2929174.
21. Combalia M, Garcia S, Malvehy J, Puig S, Mülberger AG, Browning J, Garcet S, Krueger JG, Lish SR, Lax R, Ren J, Stevenson M, Doudican N, Carucci JA., Jain M, White K, Rakos J, Gareau DS. Deep learning automated pathology in ex vivo microscopy. Biomed Optics Express. 2021;12(6):3103–16. https://doi.org/10.1364/BOE.422168.

Hands-On Guide for Ex Vivo Confocal Imaging

3

Ucalene Harris, Matthew Moronta, Javiera Pérez-Anker, and Manu Jain

3.1 Introduction

Ex vivo confocal microscopy (EVCM) is a novel and rapid imaging technique that has revolutionized the fields of dermatology and surgical pathology [1, 2]. New generations of EVCM microscopes combine two different simultaneous diode lasers (reflectance and fluorescence), to create a fusion pseudo-colored image of the scanned tissue called Fusion confocal microscopy (fuCM) with digital H&E (DHE), as it mimics conventionally stained hematoxylin and eosin (H&E) tissue sections. Thus, EVCM allows for a complete image evaluation at architectural and cellular level, akin to histopathology. However, unlike histopathology or frozen section analysis, EVCM can image fresh tissues without the need for tissue fixation or processing; thus, it's being used as an alternative to frozen section for rapid evaluation of tissues.

In this chapter, we will detail tissue handling and imaging techniques using the new generation EVCM, Vivascope 2500 (Caliber ID, Rochester, NY, USA) and the most commonly used fluorescent dye, acridine orange [1, 2]. The technique detailed here is based on the methods described by the experts in the literature [1, 3, 5] and on our personal experience of imaging and handling tissues in dermatology research laboratory at Memorial Sloan Kettering Center.

U. Harris · M. Moronta
Department of Dermatology, Memorial Sloan Kettering Cancer Center (MSKCC), New York, NY 10021, USA

J. Pérez-Anker
Department of Dermatology, Hospital Clinic de Barcelona, Barcelona, Spain

M. Jain (✉)
Dermatology Service, Department of Medicine, Memorial Sloan Kettering Cancer Center (MSKCC), 530 E. 74th Street, New York, NY 10021, USA
e-mail: jainm@mskcc.org

3.2 Material required for imaging (Fig. 3.1)

Before you begin imaging, assemble the following reagents and materials (Fig. 3.1):

1. Biopsy foam pads
2. Standard glass slides
3. 0.9% Sodium Chloride (saline)
4. 0.6 M Acridine Orange dye
5. Conical tubes/Petri-dish
6. Forceps

3.3 Tissue Type, Source, and Preparation (Fig. 3.2)

- Tissue type: EVCM images can be acquired from freshly excised tissues or from frozen-thawed fresh tissue. If the tissue is received in a fresh state (unprocessed or not-frozen) just proceed to acridine orange staining (Section C) and EVCM imaging (Section D, E). However, often times fresh tissue is frozen using an 'Optimal Cutting Temperature' (OCT) compound for the purpose of preserving the tissue during transportation. In this scenario, the frozen tissue block needs to be first thawed before staining. See details on frozen tissue preparation below
  - Step 1: Place OCT compound embedded tissue in a standard Petri-dish with saline (enough to submerge the entire tissue).
  - Step 2: After approximately 20 min, check to see if the surrounding OCT compound has completely melted. If not, gently remove residual OCT compound with forceps.

M. Jain et al. (eds.), *Cutaneous Atlas of Ex Vivo Confocal Microscopy*,
https://doi.org/10.1007/978-3-030-89316-3_3

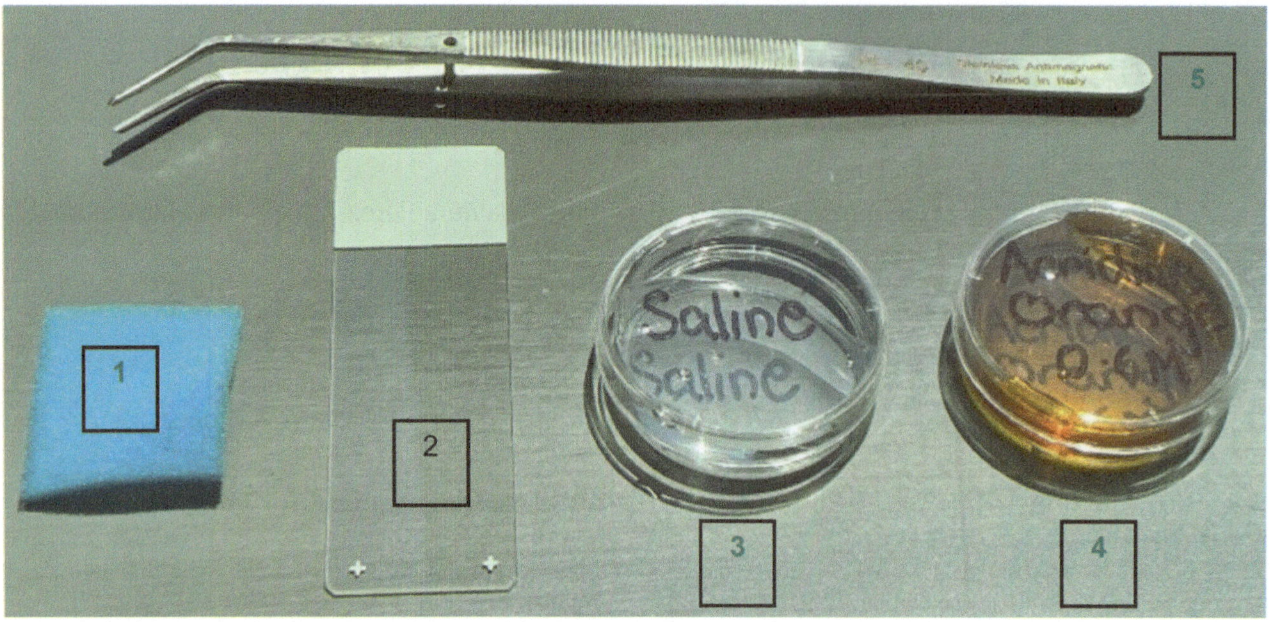

Fig. 3.1. Materials and reagents required for staining and imaging the tissue with an ex Vivo confocal microscopy device: (1) Biopsy foam pads for tissue flattening; (2) Standard glass slides for tissue mounting; (3) 0.9% Sodium Chloride, and (4) 0.6 M Acridine Orange in Petri dishes (3 and 4) for tissue staining; and (5) Forceps for tissue manipulation

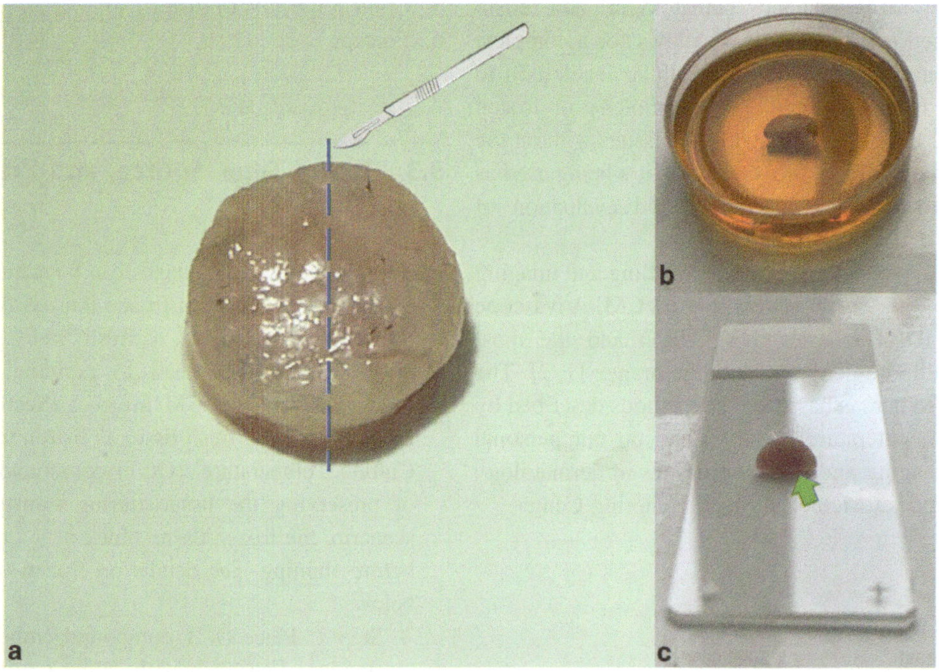

Fig. 3.2. Fresh biopsy or excision sample tissue measuring >3 mm preparation for imaging with an EVCM device. **a** Vertically bisect (blue dashed line) the fresh tissue from the epidermal side. **b** Bisected tissue is submerged in acridine orange dye for imaging. **c** Stained bisected tissue is mounted on a glass slide with the epidermis (green arrow) placed face down for imaging

- Tissue source: Tissue for imaging can be from a Mohs surgical margin or from an excision or a biopsy sample. If the tissue is received from a Mohs surgical margin no further grossing of the specimen is required. For the tissues from an excision or a biopsy additional grossing of the specimen maybe required as detailed below (Fig. 3.2):
  - Step 1: Measure the size of the specimen.
  - Step 2: Bisect the specimen perpendicular to the skin's surface across the shortest axis. Specimens < 3 mm should not be bisected.
  - Step 3: Remove excess blood from the surface of the tissue by rinsing in saline to prevent interference during imaging [5].

3.4 Tissue Staining (Fig. 3.3)

Step 1: Gently submerge the tissue into 0.6 μM Acridine Orange (AO) for 30–60 s, depending on the tissue size: 60 s for > 2 cm, 45 s for 1–2 cm, and 30 s for < 1 cm

Step 2: Remove tissue from AO and place and place into a Petri dish with saline for 15–30 s, depending on the tissue size: 30 s for >2 cm, 20 s for 1–2 cm, and 15 s for <1 cm [1].

Different staining protocols are described in Chap. 16.

3.5 Tissue Mounting and Flattening (Fig. 3.4)

Step 1: Place tissue on a glass slide with epidermis facing downwards [3, 4]. Ensure that the epidermis is as flat as possible. To obtain a flat imaging surface following steps maybe used:

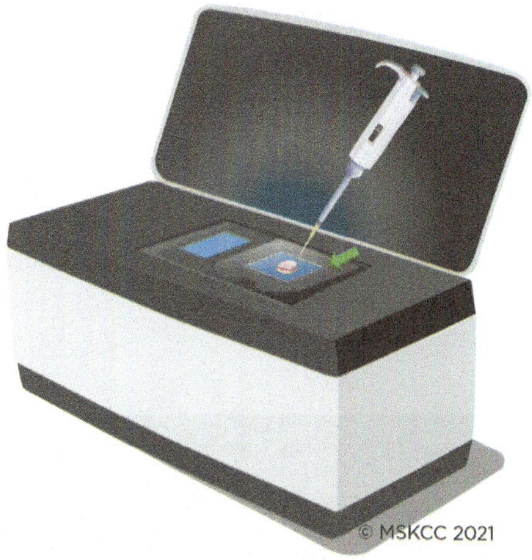

Fig. 3.4 A cartoon of the EVCM device showing tissue mounting procedure. Ex vivo tissue placed on a glass slide (green arrow) and inserted in the grooves of the EVCM device. The blue area reflection seen in the glass slide is from the blue laser light coming from the objective lens (not visible in this cartoon) underneath. Use a pipette (white arrow) to put a few drops of saline on the tissue surface

- Roll-out epidermis using gentle manipulation with forceps—Apply pressure on the tissue with a biopsy foam pad
- Applying additional foam pads to a sample which is uneven in thickness to level the entireSpecimen [3]

Step 2: Mount the glass slide on the device by inserting it into the groves above the light source.

Step 3: Add a drop of saline or acridine orange to the tissue then place a biopsy foam pad over the tissue to flatten the tissue.

Step 4: Apply gentle pressure to the tissue with the magnetic lid to manually flatten.

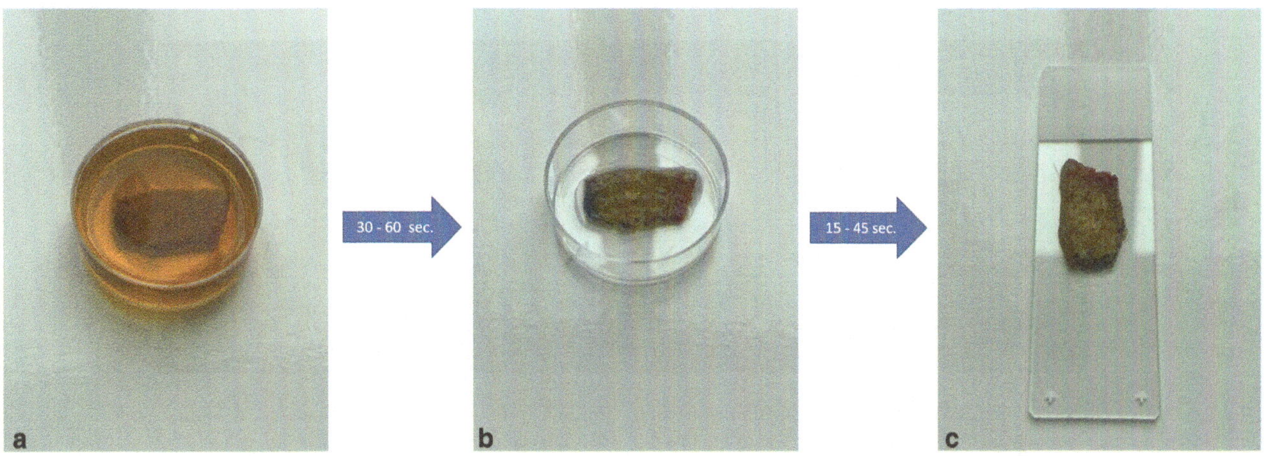

Fig. 3.3. Steps for tissue staining for EVCM imaging: **a** Step 1: Tissue submerged in 0.6 M Acridine Orange for 30–60 s. **b** Step 2: Tissue is then submerged tissue in saline for 15–45 s to wash-off excess dye. **c** Place the stained tissue on a standard glass slide in preparation for tissue mounting

3.6 Image Acquisition (Figs. 3.5, 3.6, 3.7, 3.8, 3.9)

Step 1: Center the tissue by moving the stage using joystick on the device.

Step 2: Once the tissue is centered, take a macroscopic picture of the tissue by pressing "Take Picture" on the imaging program (Fig. 3.5).

Step 3: Apply gel to the objective lens and press "OK." The 488 nm (fluorescent blue light) and 785 nm (reflectance infrared light) channels will come on automatically (Fig. 3.6). Pseudo-color or DHE mode should be automatically activated by the program.

Step 4: Find the tissue by coarsely adjusting the Z-depth with the sliding scale bar (Fig. 3.6).

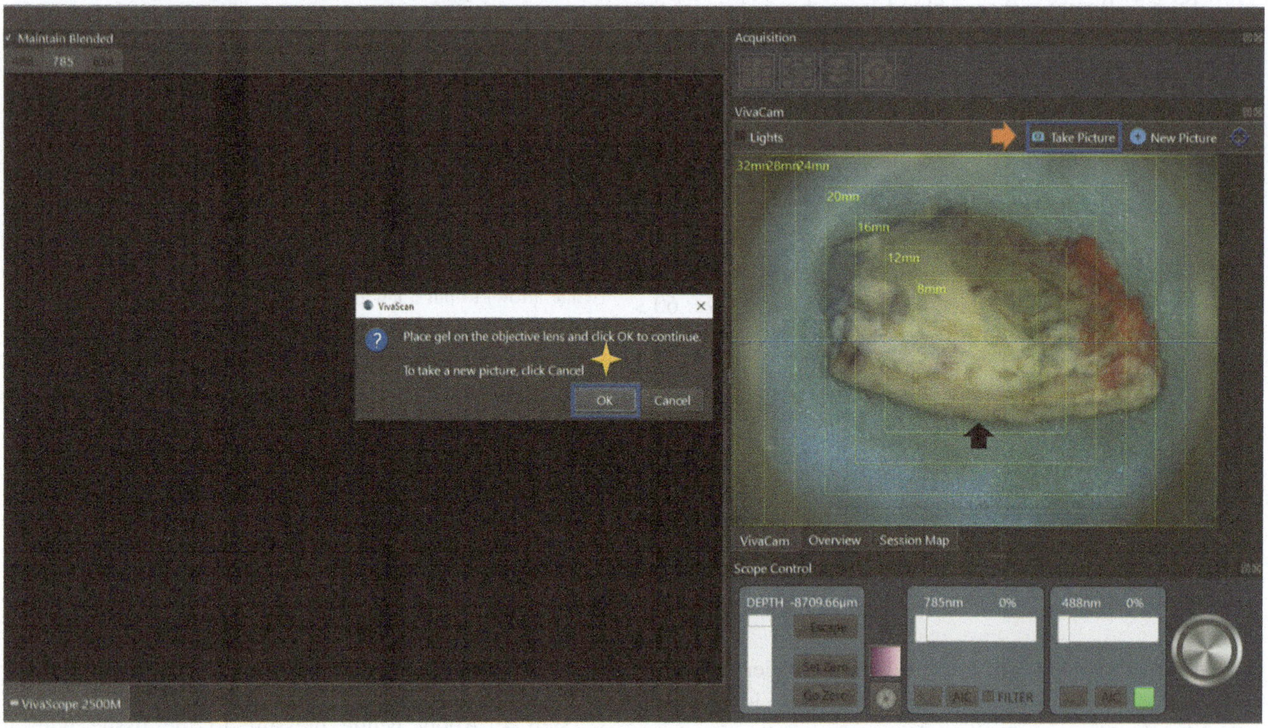

Fig. 3.5 A screenshot showing controls for acquiring macroscopic image of the Tissue. Image shows a fresh tissue piece (black arrow) centered in the imaging window on the right. Press "Take Picture" button (orange arrow) and then "OK" button (yellow star) in succession to acquire a macroscopic image of the tissue

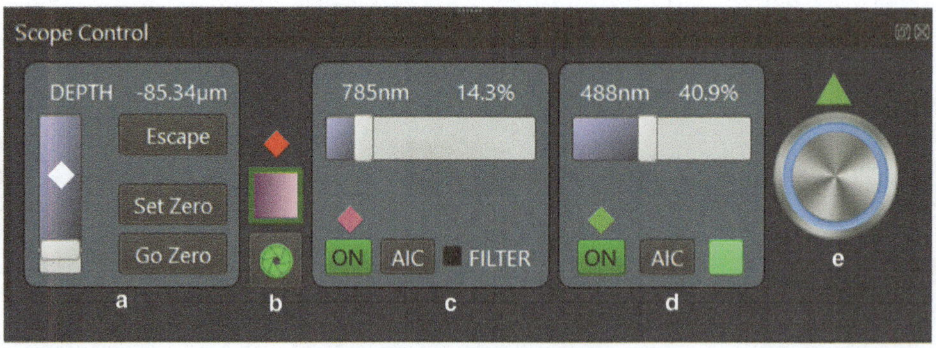

Fig. 3.6. A screenshot of controls for adjusting z-depth of imaging and intensity of Various laser channels. From left to right side of the image: **a** Sliding scale bar (white diamond) for bringing the tissue in focus and adjusting the z-depth during imaging. **b** Button (red diamond) to select pseudo-color (DHE) viewing mode. **c** Sliding scale for adjusting 785 nm (pink diamond) laser intensity. **d** Sliding scale for adjusting 488 nm (green diamond) laser intensity. **e** "ON" button (green triangle) to switch on and off both laser channels (488 nm and 785 nm) off simultaneously

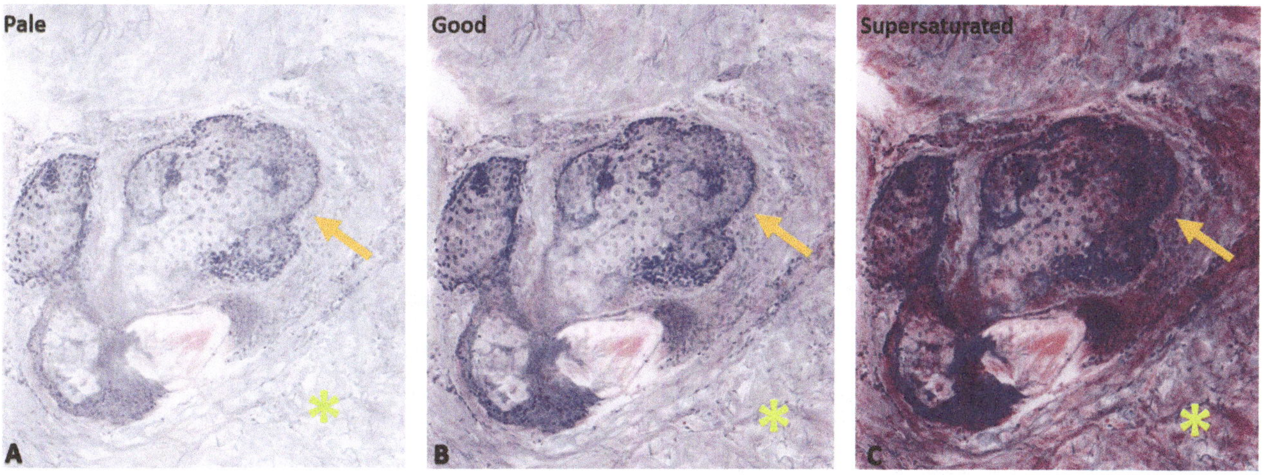

Fig. 3.7. Examples of pale, good quality, and supersaturated DHE (pseudo-color) images of a sebaceous gland acquired from a fresh skin tissue. **a** A pale DHE image shows less intense purple nuclei with a pale purplish cytoplasm in sebocytes (yellow arrow) and pale purplish surrounding collagen (yellow asterisk). **b** A good quality DHE image shows crisp and purple nuclei with purplish-pinkish cytoplasm in the sebocytes (yellow arrow) and purplish-pinkish dermis yellow asterisk). **c** A supersaturated DHE image shows very intense purple clumped nuclei and dark pink cytoplasm in the sebocytes (yellow arrow) and dark pink dermis (yellow asterisk)

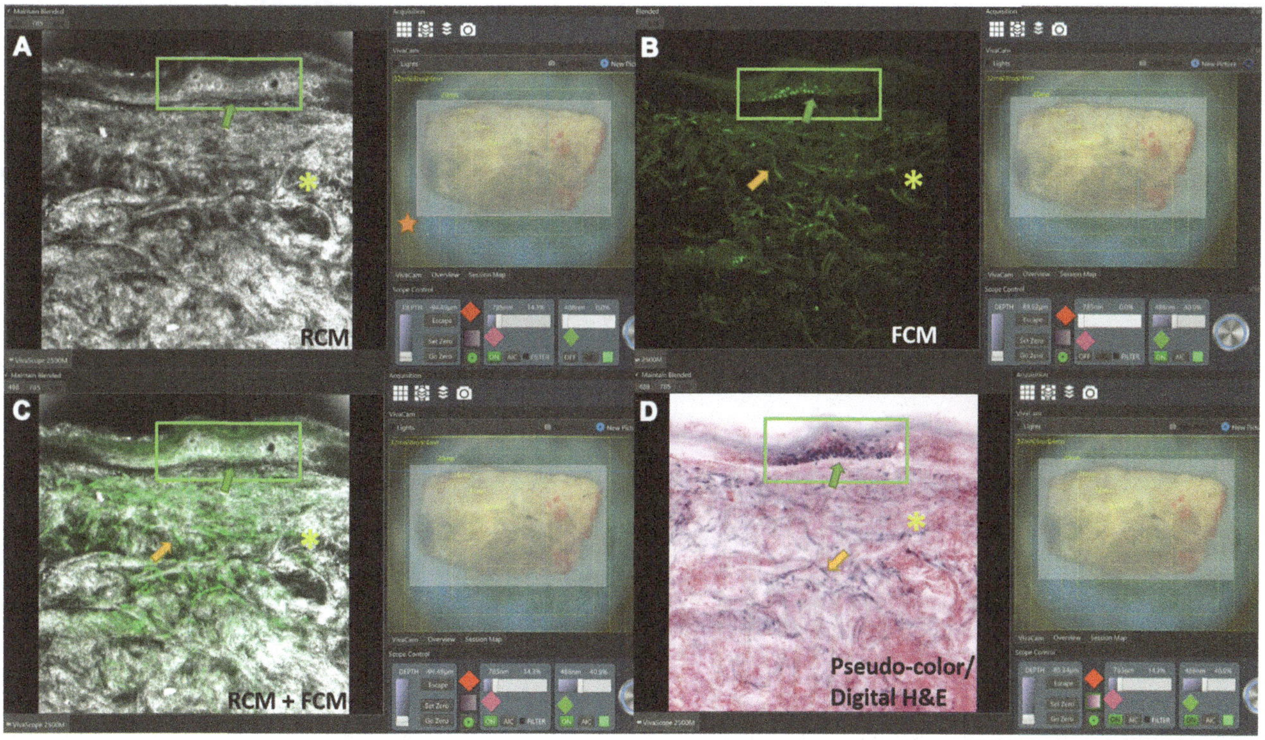

Fig. 3.8. Visualization of tissue in different imaging modes and creating a good quality DHE (pseudo-color) image. **a** RCM mode image shows a thin layer of epidermis composed of cells with dark nucleus and bright cytoplasm (green arrow; boxed area) and an underlying appears grayish-white dermis (yellow asterisk). **b** FCM mode image shows dark nucleus (in the RCM mode) as bright green structures (green arrow; boxed area) due to the uptake of fluorescent nuclear dye—acridine orange. Dermis has few bright green strands of connective tissue (orange arrow) admixed with less bright collagen bundles. **c** Combined RCM (grayscale color) and FCM (green color) mode image. **d** Digital H&E (DHE) pseudo-color image shows epidermis with purple nuclei (fluorescence signal converted to purple color) and dermis mainly pink in color (reflectance signal converted to pink color) with scattered purple connective tissue fibers (orange arrow). FOV of confocal images **a–d** = 550 µm × × 550 µm. ROI window is highlighted in panel A (orange star). Pseudo-color mode (red star) is switched off in panels a–c and is switched on in panel D. At the bottom of each image panel, slide bar for adjusting reflectance (785 nm channel) and fluorescence (488 nm channel) intensity is shown

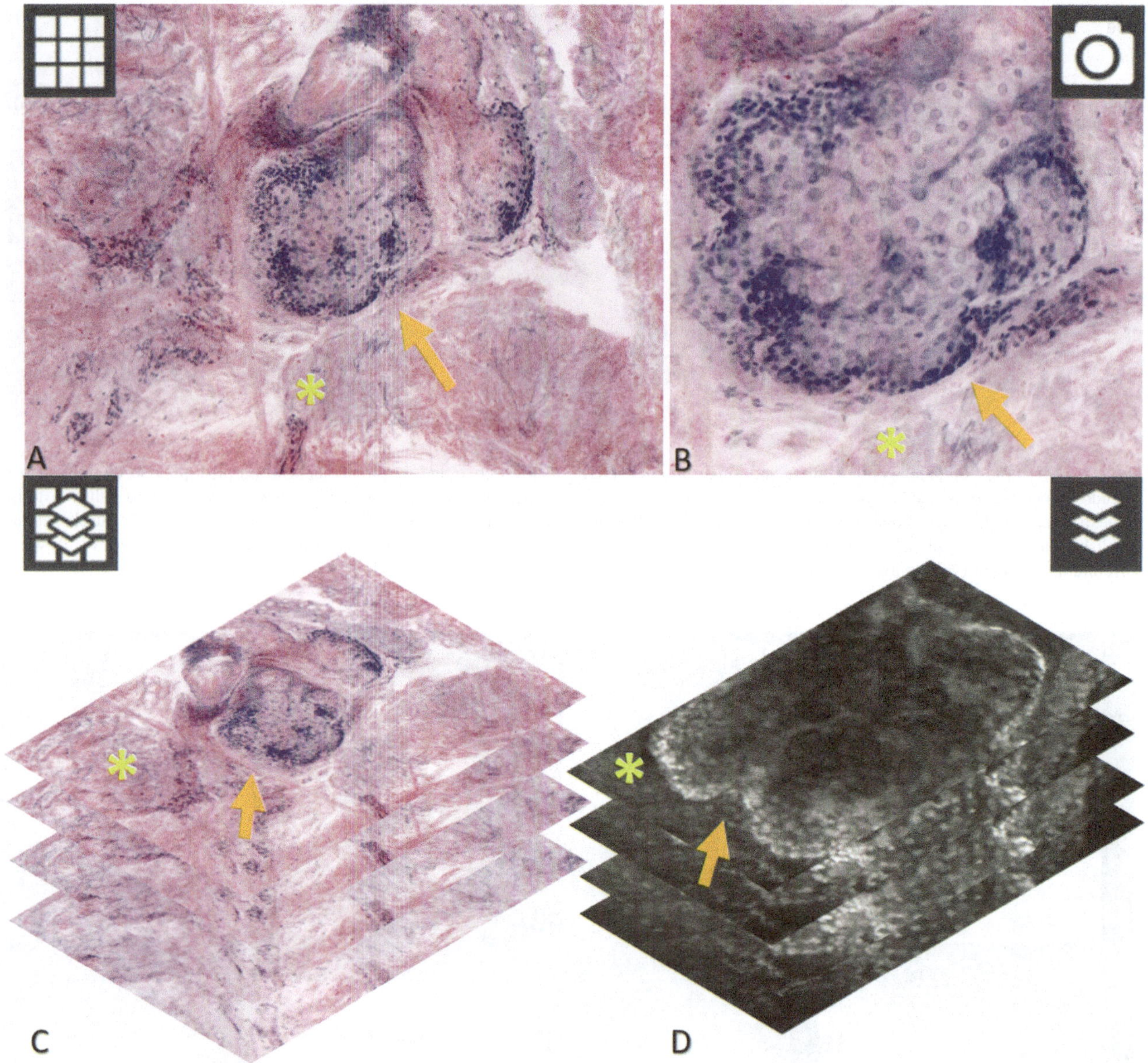

Fig. 3.9. Options for Image Acquisition. **a** A single mosaic acquired in DHE (pseudo-color) mode at a selected plane shows dermis (yellow asterisk) with a sebaceous gland (orange arrow). **b** A single-frame DHE image (550 μm × 550 μm) acquired by taking a snapshot in the mosaic (**a**) shows cellular details of the sebaceous gland at higher magnification (blue arrow). **c** A cube with a series of ∼ 1 mm mosaics acquired at increasing depths within the tissue. **d** A stack with series of single-frame images (550 μm × 550 μm) acquired at increasing imaging depths within the tissue. Note: Stack can only be visualized in grayscale. At the upper left corner of each panel, there is a symbol depicting each mode of imaging

Step 5: Once you have reached the tissue, finely adjust the z-depth with the ball of the computer mouse until you have identified your optimum imaging depth.

Step 6: The goal of EVCM imaging is to generate best-quality DHE (purple and pink) image. Take the following steps to acquire a good quality DHE image [3, 4] (Figs. 3.6, 3.7, 3.8, 3.9):

- First, focus on the epidermal nuclei by finely adjusting the z-depth until the nuclei (visible as purple color on DHE/pseudo-color mode) appear crisp and sharp. Then adjust 488 nm (FCM) channel to obtain a desirable purple color (neither too pale nor too saturated (Fig. 3.7).
- Next, focus on the collagen color (visible as pink color on DHE/pseudo-color mode) and sharpness in the underlying dermis by adjusting the 785 nm (RCM) channel.

Step 7: Once the tissue is in focus and both RCM and FCM channels are adjusted (Fig. 3.8), take a mosaic covering the entire tissue (if possible) to get a low magnification architectural view of the tissue. This mosaic would serve as a guide for acquiring images from the tissue.

Step 8: Select the image types that best suits the purpose of imaging: mosaic, single-frame images, cubes, or stacks as described in the "**Options for Image Acquisition**" section below (Fig. 3.9).

Step 9: Press "Finished Imaging" when the session is complete.

3.7 Cleaning the Equipment

Step 1: Remove the magnetic lid.

Step 2: Remove the slide with the tissue and set it aside.

Step 3: First, gently wipe off the excess gel from the objective lens with a clean optical lens paper, followed by another wipe with lens paper soaked in 70% ethanol to ensure that all gel has been removed.

3.8 Modes of Imaging (Fig. 3.8)

1. Reflectance Confocal Microscopy (RCM) mode: To visualize the tissue in reflectance mode, select 785 nm channel (infrared light). The images appear as grayscale in this channel (Fig. 3.8a). The reflectance signal mainly originates from the cell cytoplasm and connective tissue.
2. Fluorescence Confocal Microscopy (FCM) mode: To visualize tissue in the fluorescence mode, select 488 nm (blue light). The images appear in green color in this mode (Fig. 3.8b). The fluorescence signal mainly originates from the nuclei.
3. Combined RCM + FCM mode (fuCM): Images can be also visualized simultaneously in a combined reflectance and fluorescent mode. This is achieved by overlaying RCM and FCM images (Fig. 3.8c). The combined RCM + FCM mode can also be visualized as digitally pseudo-colored purple and pink images (Fig. 3.8d), where reflectance signal appears pink and fluorescence signal appears purple, simulating eosin and hematoxylin staining, respectively, of a conventional H&E-stained image.

3.9 Options For Image Acquistion (Fig. 3.9)

1. Mosaic (Fig. 3.9a): A mosaic is a wide field of view (FOV) image covering maximum up to 25 mm × 25 mm of the tissue, acquired at a given depth in the tissue. Mosaics provide architectural details of the tissue [5], which is akin to low magnifications (2 × and 4x) images on conventional histopathology.
2. Single-frame Image (Fig. 3.9b): A single-frame is a small FOV image (550 µm x 550 µm) be acquired using the snapshot button [5, 6].
3. Cube (Fig. 3.9c): A cube is a series of mosaics acquired at various z-depths within the tissue. Number of layers and their spacing from each other can be selected based on the imaging requirements [3, 4].
4. Stack (Fig. 3.9d): A stack is a collection of small FOV images (550 µm x 550 µm) acquired in the z-depth [3, 5] within the tissue. Similar to a 'cube', number of layers and their spacing from each other can be selected based on the imaging requirements.

3.10 Challenges Encountered During Ex Vivo Confocal Imaging (Figs. 3.10 and 3.11)

Tissue flattening is a major challenge while imaging with an EVCM device [2] and may cause the following issues:

1. Incomplete visualization of the epidermal margin (Fig. 3.10): This occurs when the entire tissue does not come into the contact with the glass slide [3, 4] during the tissue mounting procedure. This may hinder the visualization of superficial basal cell carcinoma (BCC) or other epidermal tumors. To better visualize the epidermis and flatten the tissue see steps described in Section D. Additionally, other mechanical [4, 7] devices can be used for this purpose. Recently, a digital tissue flattening approach has been described (see details in Chap. 18) [8] that maybe used in future. Nevertheless, nuclear details need always to be observed.
2. Saline bubbles (Fig. 3.11): Saline bubbles are often seen in the images, which may hinder the visualization of tissue. They are usually formed due to accumulation of excess saline within the natural folds of the tissue. Their appearance is described in RCM, FCM and DHE image in Fig. 3.11. During imaging, these bubbles can be minimized by gently pressing the tissue. If the bubbles are not expelled by pressure, de-mount the tissue and briefly wash for a few seconds in the acridine orange dye and then re-mount it and image [4–6].
3. Incomplete visualization of the dermal components: Apply an extra pressure with the use of a double and denser sponge and/or perform a stack, as previously described [5, 6].
4. Artifacts: Artifacts may hinder visualization of the tissue. To mitigate this issue, reduce the artifacts by keeping the scanning surface clean of any tissue fibers from gauze, excess blood or ink (usually applied at the surgical margins) before imaging the tissues.

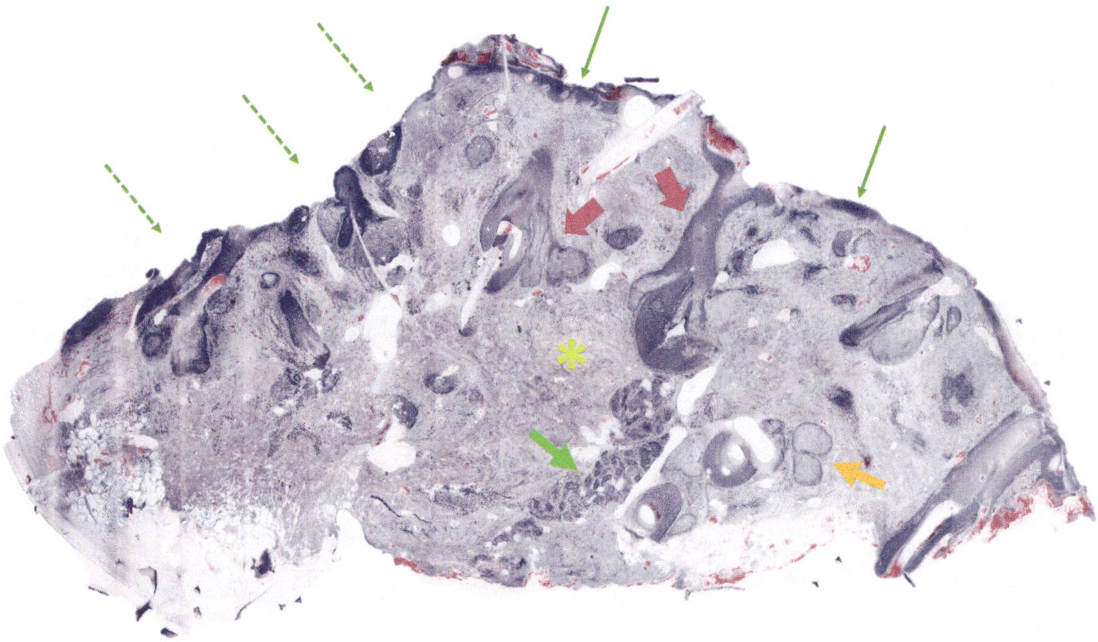

Fig. 3.10. Incomplete epidermal margin visualization in a Mohs surgical tissue margin imaged with an EVCM device. DHE image shows epidermis (dark green arrows) and anunderlying dermis (yellow asterisk) with sebaceous glands (orange arrow), pilosebaceous units (pink arrows) and eccrine units (light green arrow). Note, loss of epidermal lining (dashed-darkgreen arrows) at multiple area in the image

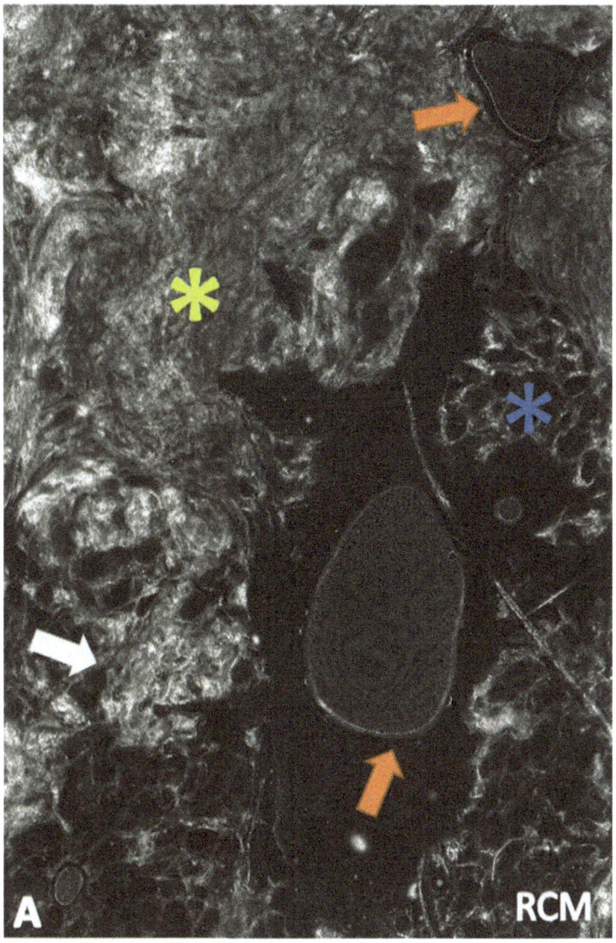

Fig. 3.11. Appearance of saline bubbles in the EVCM images. **a** An RCM image (785 nmchannel) shows bubbles (orange arrows) as oval darkish-gray sharply demarcated structureswith central fringes (fingerprint-like pattern) within the grayish-white dermis (yellow asterisk) and dark fatty tissue of subcutis (blue asterisk). **b** An FCM image (488 nm channel) of the same tissue shows bubbles (orange arrows) as intense dark but sharply demarcated oval structures(less clearly visible than in the RCM mode) within the dark grayish hypoflourescent dermis(yellow asterisk) and subcutis (blue asterisk). **c** In DHE image, the bubble (orange arrows)appears as oval sharply demarcated light pink structures with central fingerprint-like pattern-within a pinkish dermis (yellow asterisk) and white lobules of fatty tissue (blue asterisk)

References

1. Malvehy J, Pérez-Anker J, Toll A, Pigem R, Garcia A, Alos LL, Puig S. Ex vivo confocal microscopy: revolution in fast pathology in dermatology. Br J Dermatol. 2020;183(6):1011–25. https://doi.org/10.1111/bjd.19017 Epub 2020 May 10 PMID: 32134506.
2. Jain M, Rajadhyaksha M, Nehal K. Implementation of fluorescence confocal mosaicking microscopy by "early adopter" Mohs surgeons and dermatologists: recent progress. J Biomed Opt. 2017;22 (2):024002. https://doi.org/10.1117/1.JBO.22.2.024002.
3. Pérez-Anker J, Toll A, Puig S, Malvehy J. Six steps to reach optimal scanning in ex vivo confocal microscopy. J Am Acad Dermatol. 2021:S0190–9622(21)00193–6. https://doi.org/10.1016/j.jaad.2021.01.044. Epub ahead of print. PMID: 33476729
4. Pérez-Anker J, Puig S, Malvehy J. A fast and effective option for tissue flattening: optimizing time and efficacy in ex vivo confocal microscopy. J Am Acad Dermatol. 2020;82(5):e157–8. https://doi.org/10.1016/j.jaad.2019.06.041 Epub 2019 Jun 28 PMID: 31255748.
5. Pérez-Anker J, Ribero S, Yélamos O, García-Herrera A, Alos L, Alejo B, Combalia M, Moreno-Ramírez D, Malvehy J, Puig S. Basal cell carcinoma characterization using fusion ex vivo confocal microscopy: a promising change in conventional skin histopathology. Br J Dermatol. 2020;182(2):468–476. https://doi.org/10.1111/bjd.18239. Epub 2019 Oct 27. PMID: 31220341; PMCID: PMC6923630.
6. Vivascope Operations Manual, page 64
7. Cinotti E, Grivet D, Labeille B, et al. The "tissue press": a new device to flatten fresh tissue during ex vivo confocal microscopy examination. Skin Res Technol. 2017;23(1):121–4. https://doi.org/10.1111/srt.12293.
8. Sendín-Martín M, Kose K, Harris U, Rossi A, Lee E, Nehal K, Rajadhyaksha M, Jain M. Complete visualization of epidermal margin during ex vivo confocal microscopy of excised tissue with 3D-mosaicking and intensity projection. J Am Acad Dermatol. 2020:S0190–9622(20)30909–9. https://doi.org/10.1016/j.jaad.2020.05.044. Epub ahead of print. PMID: 32428611.

Part II
Normal Skin

Normal Skin on Ex Vivo Confocal Microscopy: Patterns and Histopathologic Correlation

4

Mercedes Sendín-Martín, Ucalene Harris, Banu Farabi, Matthew Moronta, Anthony Rossi, Erica Lee, Chih-Shan Jason Chen, Kishwer Nehal, Melissa Pulitzer, and Manu Jain

Abstract

In this chapter, we will describe the morphological features of normal layers of skin and its appendages using ex vivo confocal microscopy(EVCM). EVCM is now being integrated during Mohs surgery for margin assessment of the residual skin cancers, such as basal cell carcinoma (BCC) and squamous cell carcinoma (SCC). However, to identify skin cancers it is quintessential to first learn the features of normal skin on EVCME. Using this device, fresh or frozen tissues stained with an acridine orange dye (a fluorescent nuclear dye that enhances visualization of cell nuclei) can be rapidly imaged in reflectance mode (RCM; Reflectance confocal microscopy) or in fluorescence mode (FCM; Fluorescence confocal microscopy), separately or in a combined RCM and FCM mode. The nuclear signal originating from the fluorescentchannel is digitally converted to purple colour, resembling hematoxylin-stain on conventional H&E images. The reflectance signal, originating from cell cytoplasm and collagen, is digitally converted to pink colour, resembling eosin-stain on conventional H&E images. Thus, the resultant combined fluorescent and reflectance signal creates a digital H&E (DHE:Digital H&E), purple and pink-coloured images.

Skin is the largest organ of the human body, accounting for ~ 15% of the total body weight in adults [1]. Skin plays crucial functions such as temperature, sensory, and metabolic regulation, and protection from the external environment. There are three main layers of the skin: epidermis, dermis, and hypodermis. The overall thickness of skin and thickness of each layer varies at different anatomical locations in the body.

In this chapter, we will describe the morphological features of normal layers of skin and its appendages using ex vivo confocal microscopy (EVCM). EVCM is now being integrated during Mohs surgery for margin assessment of the residual skin cancers such as basal cell carcinoma (BCC) and squamous cell carcinoma (SCC). EVCM features of BCC and SCC are described in Chaps. 7 and 8, respectively. However, to identify skin cancers it is quintessential to first learn the features of normal skin on EVCM.

Using this device, fresh or frozen tissues stained with an acridine orange dye (a fluorescent nuclear dye that enhances visualization of cell nuclei) can be rapidly imaged in reflectance mode (RCM) or fluorescence mode (FCM), separately [2] or in a combined RCM and FCM mode. The nuclear signal originating from the fluorescent channel is digitally converted to purple color, resembling hematoxylin stain on conventional H&E images. The reflectance signal, originating from cell cytoplasm and collagen, is digitally converted to pink color, resembling eosin stain on conventional H&E images. Thus, the resultant combined fluorescent and reflectance signal creates a digital H&E (DHE), purple and pink-colored images [3]. FCM signals (hyperfluorescent, hypofluorescent, and nonfluorescent) and their color equivalence in FCM mode, DHE mode, and on corresponding

M. Sendín-Martín · U. Harris · M. Moronta · A. Rossi · E. Lee · C.-S.J. Chen · K. Nehal · M. Pulitzer
Department of Medicine, Memorial Sloan Kettering Cancer Center (MSKCC), New York, NY 10021, USA

M. Sendín-Martín
Dermatology Department, Hospital Universitario Virgen del Rocío, Sevilla, Spain

B. Farabi
Department of Internal Medicine, Saint Peter's University Hospital, New Brunswick, NJ, USA

M. Jain (✉)
Assistant Attending, Dermatology Service, Department of Dermatology, Memorial Sloan Kettering Cancer Center (MSKCC), New York, NY 10021, USA
e-mail: jainm@mskcc.org

M. Jain et al. (eds.), *Cutaneous Atlas of Ex Vivo Confocal Microscopy*,
https://doi.org/10.1007/978-3-030-89316-3_4

Table 4.1 FCM signal and their color equivalence on DHE mode of ex vivo confocal microscopy and corresponding conventional H&E-stained tissue sections

FCM signal	Color on FCM grayscale images	Color on FCM green-scale images	Color on DHE	Color on conventional H&E
Nonfluorescent (lack of signal)	Black (dark)	Black (dark)	White, whitish pink	White
Hypofluorescent (low signal)	Grayish dark to grayish	Dull green to dark green	Pinkish purple, purplish pink, and pink	Pink or purplish pink
Hyperfluorescent (high signal)	Bright, white	Bright green	Purple, dark purple	Purple

Table 4.2 Morphological features of normal skin on FCM, DHE, and corresponding conventional H&E-stained images

Fluorescence confocal microscopy (FCM) mode (grayscale image) (Figs. 4.1a, 4.2a, 4.3a, 4.4a, 4.5a, 4.6a, 4.7a, 4.8a, 4.9a, 4.10a, 4.11a, 4.12a, 4.13a, 4.14a, 4.15a, 4.16a and 4.17a)	Digital H&E (combined FCM and RCM modes) (Figs. 4.1b, 4.2b, 4.3b, 4.4b, 4.5b, 4.6b, 4.7b, 4.8b, 4.9b, 4.10b, 4.11b, 4.12b, 4.13b, 4.14b, 4.15b, 4.16b and 4.17b)	Conventional H&E-stained image (Figs. 4.1c, 4.2c, 4.3c, 4.4c, 4.5c, 4.6c, 4.7c, 4.8c, 4.9c, 4.10c, 4.11c, 4.12c, 4.13c, 4.14c, 4.15c, 4.16c and 4.17c)
Epidermis (Figs. 4.1 and 4.2)		
Low magnification (Fig. 4.1a): Thin bright (hyperfluorescent) layer High magnification (Fig. 4.2a): Multi-layered, bright (hyper-fluorescent) epidermis with the following layers (from top to bottom): *St. corneum:* Grayish (hypofluorescent) anucleate layer *St. granulosum:* 2–3 layers of flattened oval bright (hyperfluorescent) nuclei. Keratohyalin granules are not visible *St. spinosum:* 8–10 layers of polygonal keratinocytes with bright (hyperfluorescent) nuclei *St. basale:* Single layer of cuboidal bright (hyperfluorescent) nuclei. Melanocytes are not visible	Low magnification (Fig. 4.1b): Purple color thin layer High magnification (Fig. 4.2b): Multi-layered, purple color epidermis with the following layers (from top to bottom): *St. corneum:* Pinkish-purple anucleate layer *St. granulosum:* 2–3 layers of flattened oval purple nuclei. Keratohyalin granules appear as dark pink *St. spinosum:* 8–10 layers of polygonal keratinocytes with purple nuclei and scant pink cytoplasm *St. basale:* Single layer of cuboidal cells with dark purple nuclei and scant pink cytoplasm. Melanocytes appear pinkish and blurred (intense reflectance signal)	Low magnification (Fig. 4.1c): Thin layer of stratified squamous epithelium (bright on FCM; purple on DHE) High magnification (Fig. 4.2c): Stratified keratinized squamous epithelium (bright on FCM; purple on DHE) with following layers (from top to bottom): *St. corneum:* Flattened anucleate layer with keratin (grayish on FCM; pinkish-purple on DHE) *St. granulosum:* 2–3 layers of flattened nucleated cells (bright on FCM; purple on DHE) with keratohyalin granules (dark pink on DHE) *St. spinosum:* 8–10 cell thick layer of with polygonal keratinocytes (bright on FCM; purple on DHE) *St. basale:* Single layer of cuboidal basal keratinocytes (bright on FCM; purple on DHE) and few melanocytes (not visible on FCM; pinkish and blurred on DHE)
DEJ (Fig. 4.3)		
Undulating to flat interface between bright (hyperfluorescent) St. basale layer and grayish dark (hypofluorescent) papillary dermis	Undulating to flat interface between dark purple color St. basale layer and pinkish-purple color papillary dermis	Undulating to flat interface between St. basale layer (bright on FCM; dark purple on DHE) and papillary dermis (grayish-dark on FCM; pinkish-purple on DHE)
Dermis (Figs. 4.4, 4.5, and 4.6)		
Papillary dermis (Fig. 4.4a): Grayish (hypofluorescent) fibrillar loose collagen with scattered thin bright (hyperfluorescent) wavy elastic fibers (Fig. 4.5a) **Reticular dermis** (Fig. 4.4a): Grayish (hypofluorescent) dense and thicker collagen fibers **Inflammatory cells** (Fig. 4.6a): Small bright (hyperfluorescent) cells clustered or scattered throughout the grayish dermis (hypofluorescent)	**Papillary dermis** (Fig. 4.4b): Purplish pink fibrillar collagen with scattered thin purple wavy elastic fibers (Fig. 4.5b) **Reticular dermis** (Fig. 4.4b): Purplish pink dense and thicker collagen fibers **Inflammatory cells** (Fig. 4.6b): Small, round, dark purple color cells clustered or scattered within a pinkish purple dermis	**Papillary dermis** (Fig. 4.4c): Delicate reticulate collagen fibers (grayish on FCM; purplish pink on DHE). Elastic fibers (bright on FCM; purple on DHE) (Fig. 4.5c) are not readily visible on H&E (requires Verhoeff Van Gieson stain) **Reticular dermis** (Fig. 4.4c): Dense pink collagen bundles (grayish on FCM; purplish pink on DHE) **Inflammatory cells** (Fig. 4.6c): Scattered or clustered lymphocytic infiltrate (bright on FCM; dark purple on FCM) in the dermis
Hypodermis (Figs. 4.1 and 4.7)		
Low magnification (Fig. 4.1a): Sheet of dark (nonfluorescent) cells with intervening grayish (hypofluorescent) fibrous septae High magnification (Fig. 4.7a):	Low magnification (Fig. 4.1b): Sheet of whitish color cells with intervening purple fibrous septae High magnification (Fig. 4.7b):	Low magnification (Fig. 4.1c): Sheet of adipocytes with clear cytoplasm (dark on FCM; whitish DHE) and intervening pink fibrous septate (grayish on FCM; purple on DHE) High magnification (Fig. 4.7c):

(continued)

Table 4.2 (continued)

Fluorescence confocal microscopy (FCM) mode (grayscale image) (Figs. 4.1a, 4.2a, 4.3a, 4.4a, 4.5a, 4.6a, 4.7a, 4.8a, 4.9a, 4.10a, 4.11a, 4.12a, 4.13a, 4.14a, 4.15a, 4.16a and 4.17a)	Digital H&E (combined FCM and RCM modes) (Figs. 4.1b, 4.2b, 4.3b, 4.4b, 4.5b, 4.6b, 4.7b, 4.8b, 4.9b, 4.10b, 4.11b, 4.12b, 4.13b, 4.14b, 4.15b, 4.16b and 4.17b)	Conventional H&E-stained image (Figs. 4.1c, 4.2c, 4.3c, 4.4c, 4.5c, 4.6c, 4.7c, 4.8c, 4.9c, 4.10c, 4.11c, 4.12c, 4.13c, 4.14c, 4.15c, 4.16c and 4.17c)
Adipocytes: Dark (nonfluorescent) polygonal cells with a thin grayish border (hypofluorescent) and a small bright (hyperfluorescent) peripheral nucleus *Fibrous septae:* Grayish (hypofluorescent) fibrous strands between dark adipocyte lobules	*Adipocytes:* Whitish polygonal cells with a thin purplish border and small dark purple peripheral nuclei *Fibrous septae*: Purplish fibrous strands between whitish adipocyte lobules	*Adipocytes:* Fat cells (dark on FCM; whitish on DHE) with thin borders (grayish on FCM; purplish on DHE) and small peripheral nuclei (bright on FCM; dark purple on DHE) *Fibrous septae* (grayish on FCM; purplish on DHE) separating adipose lobules (dark on FCM; whitish on DHE)
Muscle, Vessels and Nervers (Figs. 4.8, 4.9 and4.10)		
Smooth muscle (Fig. 4.8a): Grayish (hypofluorescent) bundles of muscle fibers with elongated bright (hyperfluorescent) cigar-shaped nuclei **Vessels** (Fig. 4.9a): Round to oval structures with a dark (nonfluorescent) central lumen and grayish (hypofluorescent) vessel wall with the following distinct layers (luminal to outwards): *Tunica intima:* Wavy thin bright (hyperfluorescent) elastic fibers and lining of small bright plump endothelial cell nuclei *Tunica media:* Thick grayish (hypofluorescent) smooth muscle layer with bright (hyperfluorescent) blunt-ended nuclei *Tunica adventitia:* Grayish (hypofluorescent) fibrous layer with thin, bright (hyperfluorescent) sparse nuclei **Nerves** (Fig. 4.10a): Grayish (hypofluorescent) oval to round structure with wavy pointed bright (hyperfluorescent) nuclei enclosed in a grayish fibrous sheath of the perineurium	**Smooth muscle** (Fig. 4.8b): Pale purple muscle bundles with elongated cigar-shaped purple nuclei **Vessels** (Fig. 4.9b): Round to oval structures with a white central lumen and a thick purplish pink vessel wall with the following layers (luminal to outwards): *Tunica intima:* Wavy thin purple layer elastic fibers and small plump purple endothelial cells *Tunica media:* Thick purplish pink smooth muscle layer with thin, purple blunt-ended nuclei *Tunica adventitia:* Pinkish fibrous layer with thin, purple sparse nuclei **Nerves** (Fig. 4.10b): Purplish pink oval to round structure with wavy, thin purple nuclei enclosed in a purplish pink fibrous sheath of the perineurium	**Smooth muscle** (Fig. 4.8c): Smooth muscle bundles (grayish on FCM; pale purple on DHE) with cigar-shaped nuclei (bright on FCM; purple on DHE) **Vessels** (Fig. 4.9c): Medium-sized vessels with the following layers of the vessel walls (luminal to outwards): *Tunica intima:* Wavy elastic layer (bright on FCM, purple on DHE) and endothelial cells (bright cells on FCM, and purple cells on DHE) *Tunica media:* Smooth muscle layer (grayish on FCM; purplish pink on DHE) *Tunica adventitia*: Connective tissue fibrous layer (grayish on FCM; pinkish on DHE) **Nerves** (Fig. 4.10c): Oval to round structures with wavy fibers (grayish on FCM, purplish pink on DHE) enclosed in perineurium (grayish on FCM; purplish pink on DHE)
Adnexal structures (Figs. 4.11, 4.12, 4.13, 4.14, 4.15, 4.16 and 4.17)		
Hair follicles and hair shaft (Fig. 4.11a): Hair follicles are tubular to round bright (hyperfluorescent) structures that harbor a bright (hyperfluorescent) elongated hair shaft in the center <u>*Layers of hair follicles:*</u> <u>A. Infundibular portion:</u> Stratified layer of bright (hyperfluorescent) nuclei (Fig. 4.11a) B. Isthmus and lower segment (outermost to innermost layers) (Fig. 4.12a): - Connective tissue sheath: Grayish (hypofluorescent) layer with few thin, bright (hyperfluorescent) nuclei - Outer root sheath: Multiple layers of bright (hyperfluorescent) cuboidal nuclei - Inner root sheath: Grayish (hypofluorescent) layer with few bright (hyperfluorescent) nuclei C. Bulb (Fig. 4.12a): Multi-layer of intensely bright (hyperfluorescent) crowded nuclei of matrical cells surrounding the central grayish dark (hypofluorescent) dermal papillae with bright (hyperfluorescent) nuclei of mesenchymal cells *Layers of a hair shaft* (Fig. 4.13a): - Medulla: Grayish (hypofluorescent) material in the core of a terminal hair - Cortex: Grayish (hypo-fluorescent) middle layer with bright (hyperfluorescent) elongated spindle-shaped nuclei	**Hair follicles and hair shaft** (Fig. 4.11b): Hair follicles are tubular to round purple structures enclosing a dark pink color hair shaft <u>*Layers of hair follicles:*</u> <u>A. Infundibular portion:</u> Stratified layer with dark purple nuclei (Fig. 4.11b) B. Isthmus and lower segment (outermost to innermost) (Fig. 4.12b): - Connective tissue sheath: Pinkish layer with thin purple nuclei - Outer root sheath: Multiple layers of cells with dark purple nuclei and scant purplish cytoplasm - Inner root sheath: Purplish color layer with few purple cells C. Bulb (Fig. 4.12b): Matrical cells have dark purple crowded nuclei and surround central pinkish dermal papillae with purple nuclei of mesenchymal cells *Layers of a hair shaft* (Fig. 4.13b): - Medulla: Pinkish-purple material in the core of the terminal hair - Cortex: Purplish middle layer with purple spindle-shaped nuclei - Cuticle: Dark purple keratinized outermost layer of the hair shaft **Sebaceous glands:** <u>Low magnification</u> (Fig. 4.14b): Lobular pale purple structure in the dermis often seen attached to a hair follicle	**Hair follicles and hair shaft** (Fig. 4.11c): Hair follicles (bright on FCM; purple on DHE) are tubular to round structures enclosing a reflective hair shaft (bright on FCM, dark pink on DHE) <u>*Layers of hair follicles:*</u> <u>A. Infundibular portion:</u> Keratinized stratified squamous epithelium, resembling epidermis (bright on FCM; dark purple on DHE) (Fig. 4.11c) B. Isthmus and lower segment (outermost to innermost layers) (Fig. 4.12c): - Connective tissue sheath (grayish on FCM; pinkish on DHE) - Outer root sheath: Multi-layer of cuboidal cells with basophilic nuclei (bright on FCM; dark purple on DHE) and scant cytoplasm (purplish on DHE) - Inner root sheath: Amphophilic layer between the outer root sheath and the hair shaft (grayish on FCM; purplish on DHE) C. Bulb (Fig. 4.12c): Matrical cells are basophilic epithelial cells (bright on FCM; dark purple on DHE) surrounding the central dermal papillae (grayish dark on FCM; pinkish on DHE) *Layers of a hair shaft* (Fig. 4.13c): - Medulla: Partially keratinized amorphous material in the inner core of a terminal hair shaft (grayish on FCM; pinkish purple on DHE)

(continued)

Table 4.2 (continued)

Fluorescence confocal microscopy (FCM) mode (grayscale image) (Figs. 4.1a, 4.2a, 4.3a, 4.4a, 4.5a, 4.6a, 4.7a, 4.8a, 4.9a, 4.10a, 4.11a, 4.12a, 4.13a, 4.14a, 4.15a, 4.16a and 4.17a)	Digital H&E (combined FCM and RCM modes) (Figs. 4.1b, 4.2b, 4.3b, 4.4b, 4.5b, 4.6b, 4.7b, 4.8b, 4.9b, 4.10b, 4.11b, 4.12b, 4.13b, 4.14b, 4.15b, 4.16b and 4.17b)	Conventional H&E-stained image (Figs. 4.1c, 4.2c, 4.3c, 4.4c, 4.5c, 4.6c, 4.7c, 4.8c, 4.9c, 4.10c, 4.11c, 4.12c, 4.13c, 4.14c, 4.15c, 4.16c and 4.17c)
- Cuticle: Very bright (hyperfluorescent) keratinized outermost layer of the hair shaft **Sebaceous glands:** Low magnification (Fig. 4.14a): Grayish dark (hypofluorescent) lobular structure in the dermis often seen attached to a hair follicle High magnification (Fig. 4.15a): Sebocytes: Round, darkish gray (hypofluorescent) cells with a central small, bright (hyper-fluorescent) round nuclei Germinative layer: Outer bright (hyperfluorescent) layer of flat to cuboidal cells **Eccrine glands and ducts** (Fig. 4.16a): Low magnification: Coiled and round, bright (hyperfluorescent) structures are seen in grayish color (hypo-fluorescent) deeper dermis and dark (nonfluorescent) subcutis. Straight portions of ducts appear intensely bright ducts and may be seen opening in the epidermis High magnification (Fig. 4.17a): *Glands:* Composed of a single layer of bright hyperfluorescent) cuboidal nuclei *Ducts:* Composed of two layers of bright (hyperfluorescent) cuboidal nuclei *Lumen of glands and ducts:* Dark (nonfluorescent) to grayish (hypofluorescent) secretions	High magnification (Fig. 4.15a): Sebocytes: Round cells with pale purple to pale pink cytoplasm and small, purple round nuclei Germinative layer: Outer dark purple (hyperfluorescent) layer of flat to cuboidal cells with scant pink cytoplasm **Eccrine glands and ducts** (Fig. 4.16b): Low magnification: Coiled and round, purple structures are seen in purplish to pink deeper dermis and whitish subcutis. Straight portions of the ducts appear dark purple and may be seen opening in the epidermis High magnification (Fig. 4.17b): *Glands:* Composed of a single layer of cuboidal epithelium with purple nuclei and purplish pink cytoplasm *Ducts:* Composed of two layers of the cuboidal epithelium with purple nuclei and scant or no purplish cytoplasm *Lumen of glands and ducts:* Whitish to pinkish secretions	- Cortex: Middle layer composed of elongated spindle-shaped cells (grayish on FCM; purplish on DHE) - Cuticle: Keratinized outer layer of hair shaft (bright on FCM; dark purple on DHE) **Sebaceous glands:** Low magnification (Fig. 4.14c): Lobular structure (grayish on FCM, pale purple on DHE) in the dermis often seen attached to a hair follicle High magnification (Fig. 4.15c): Lobules of vacuolated round sebocytes (darkish gray on FCM, pale purple on DHE) with small central nuclei (bright on FCM, purple on DHE) surrounded by germinative layer (bright on FCM, dark purple on DHE) **Eccrine glands and ducts** (Fig. 4.16c): Low magnification: Coiled and round structures (bright on FCM, purple on DHE) seen in the deeper dermis and subcutis. Straight portions of ducts may be seen opening in the epidermis High magnification (Fig. 4.17c): *Glands:* Composed of a single layer of cuboidal epithelium (bright on FCM, purple on DHE) *Ducts:* Composed of two layers of the cuboidal epithelium (bright on FCM, purple on DHE) *Lumen of glands and ducts:* Empty luminal spaces

H&E-stained tissue sections are described in Tables 4.1 and 4.2 (Figs. 4.1, 4.2, 4.3, 4.4, 4.5, 4.6, 4.7, 4.8, 4.9, 4.10, 4.11, 4.12, 4.13, 4.14, 4.15, 4.16 and 4.17). For more information on the principles of EVCM imaging, please refer to Chap. 2.

The latest prototype of EVCM device, Vivascope 2500 (Caliber ID, Rochester, NY, USA), was utilized to acquire images for this chapter. Tissues, imaged to describe normal skin structures, were obtained from the frozen negative Mohs surgical margins and stained with the most commonly used fluorescent dye, acridine orange. Tissue handling and staining is described in detail in Chap. 3.

4.1 Epidermis

The epidermis is the most superficial layer of the skin. Keratinocytes are the major cell type of the epidermis comprising 95% of the cells [4]. The remaining 5% of the cells are melanocytes, Langerhans cells, and Merkel cells [5]. Epidermis consists of keratinized stratified squamous epithelium, which is composed of four major layers (top to bottom): *stratum corneum* or horny layer, *stratum granulosum* or granular layer, *stratum spinosum* or spinous layer, and *stratum basale*

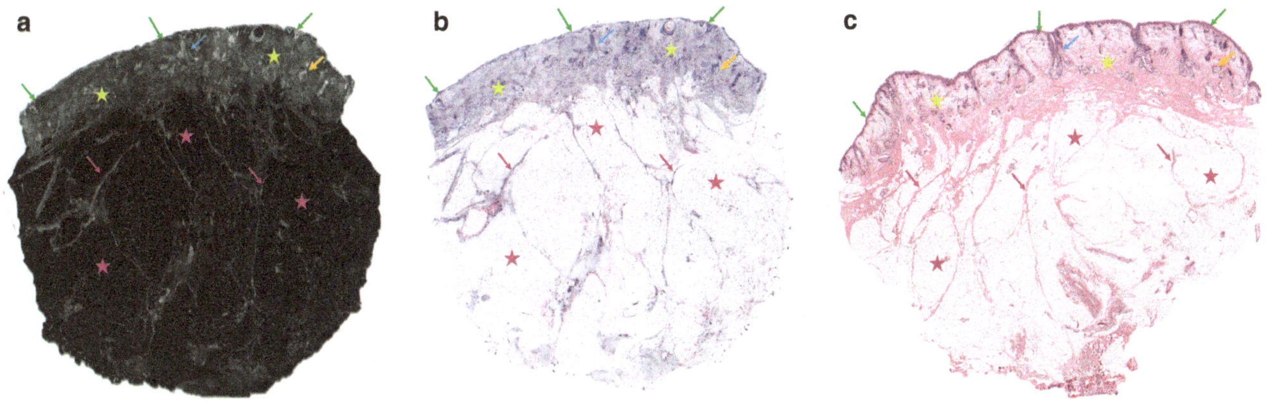

Fig. 4.1 Low magnification mosaic images of normal skin from right cheek of an 84-year-old female showing various skin layers. **a** FCM image shows a thin bright epidermis (green arrows), a grayish dermis (yellow stars), and a dark hypodermis (pink stars) with grayish fibrous septae (pink arrows). Hair follicles (blue arrow) and sebaceous glands (orange arrow) with variable brightness can be discerned in the grayish background of the dermis even at this low magnification. **b** On the corresponding digital H&E (DHE) image, the epidermis (green arrows) appears as a thin, purple color layer (bright on FCM), the dermis (yellow stars) appears as a purplish pink (grayish on FCM) layer, and the hypodermis (pink stars) as whitish color with thin purple (bright on FCM) septae (pink arrows). Hair follicles (blue arrow) and sebaceous glands (orange arrow) appear as purple structures (bright to grayish on FCM). **c** Corresponding conventional H&E-stained image shows an excellent correlation with FCM (**a**) and DHE (**b**) images for the epidermis (green arrows), dermis (yellow stars), hypodermis (pink stars) with thin fibrous septae (pink arrows), hair follicle (blue arrow), and sebaceous gland (orange arrow). H&E magnifications = 2X

or basal layer [4]. The topmost layer, the stratum corneum layer, is composed of dead (anucleate) cells and keratin. Next is the stratum granulosum layer, which is composed of 3–5 layers of flattened keratinocytes with distinct keratohyalin granules. Below is the stratified 8–10-cell thick stratum spinosum composed of polygonal-shaped keratinocytes held together by desmosomes [6]. Lastly, the lowest layer, the basal layer, is composed of a single layer of small cuboidal cells. On certain anatomical sites where the skin is thicker such as the skin of the palms and soles, a fifth layer is present, which is called the *stratum lucidum* or clear layer [7] (see figures 4.1 and 4.2).

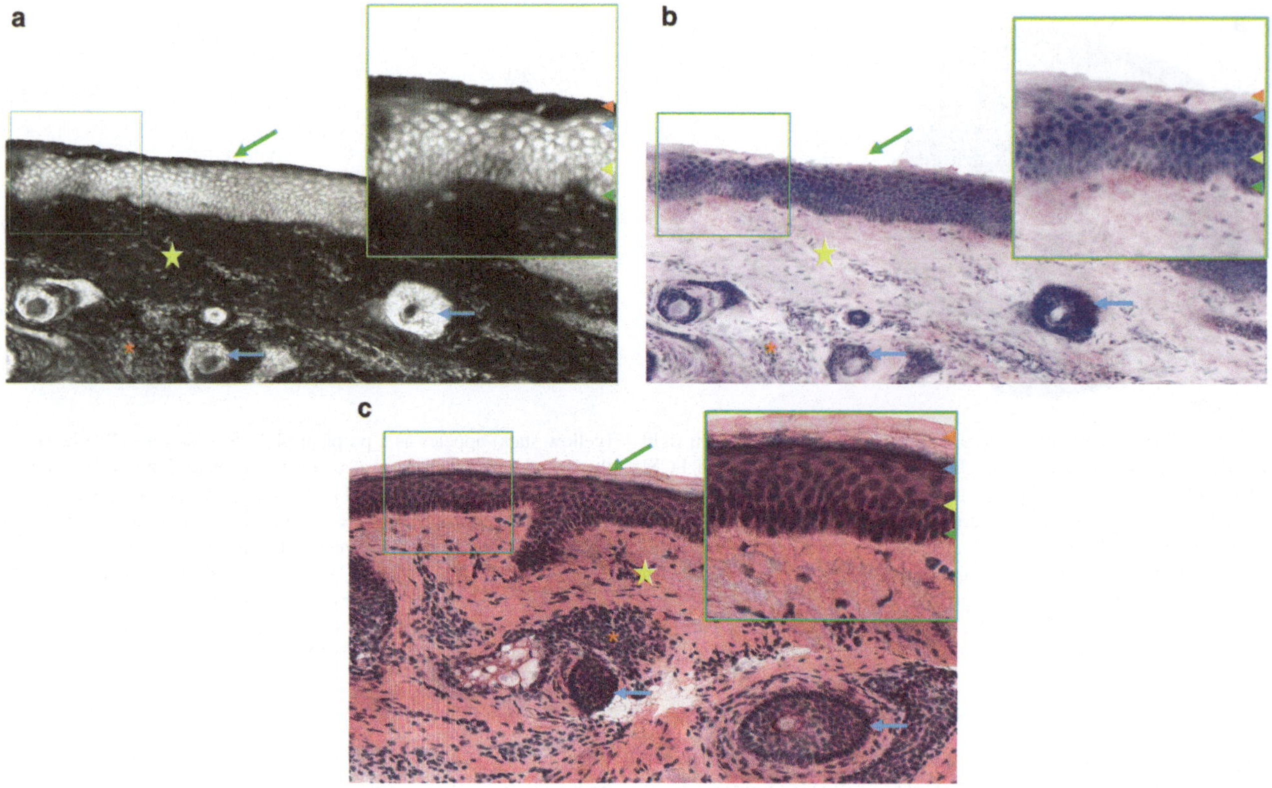

Fig. 4.2 Medium to high magnification submosaic images of skin from right forehead of a 59-year-old female skin showing various epidermal layers. **a** FCM image shows a bright multi-layered epidermis (green boxed area and green arrow) with four distinct layers (inset): a grayish anucleate *St. corneum* (orange arrowhead), a bright *St. granulosum* (bright blue arrowhead) with 2–3 layers of flattened nuclei, a bright *St. spinosum* (yellow arrowhead) with layered polygonal nuclei, and a bright single layer of cuboidal nuclei of *St. basale* layer (green arrowhead). Underlying papillary dermis (yellow star) appears grayish to dark with bright hair follicles (blue arrows), and bright inflammatory cells (orange asterisk). **b** On the corresponding DHE image, the four distinct layers (inset) of the stratified epidermis (boxed area) appear as: *St. corneum* (orange arrowhead) as a pinkish-purple (grayish on FCM) anucleate layer, *St. granulosum* (blue arrowhead) as a 2–3 cells thick layer with flattened to oval purple color (bright on FCM) nuclei and dark pink keratohyalin granules (not visible on FCM), *St. spinosum* (yellow arrowhead) as an 8–10 cell thick layer of polygonal keratinocytes with purple nuclei and scant pink cytoplasm, and *St. basale* (green arrowhead) as a single layer of dark purple (bright on FCM) cuboidal nuclei with scant pink cytoplasm. Underlying papillary dermis (yellow star) appears pinkish-purple (grayish on FCM) with purple color (bright on FCM) hair follicles (blue arrows) and inflammatory cells (orange asterisk). **c** Corresponding conventional H&E-stained image shows an excellent correlation with FCM (**a**) and DHE (**b**) images for all the layers of epidermis (green boxed area and inset): *St. corneum* (orange arrowhead), *St. granulosum* (bright blue arrowhead), *St. spinosum* (yellow arrowhead), and *St. basale* (green arrowhead), and for the underlying papillary dermis (yellow star), and inflammatory cells (orange asterisk). Insets in images A, B, and C are obtained by digitally zooming in the green boxed areas. H&E magnifications = 10X and 20X (inset)

4.2 Dermoepidermal Junction

The dermoepidermal junction (DEJ) is the interface that anchors the basal layer of the epidermis to the underlying papillary dermis with cell adhesion molecules [8, 9]. Normal DEJ shows an undulating pattern due to the protrusion of dermal papillae into the epidermis and the protrusion of epidermal rete ridges into the dermis. However, this undulated aspect is lost in aging skin and at facial sites, where the DEJ appears flattened as the rete ridges are effaced [10]. For example, for age-related changes in the DEJ, *see* Figs. 4.3 and 4.5.

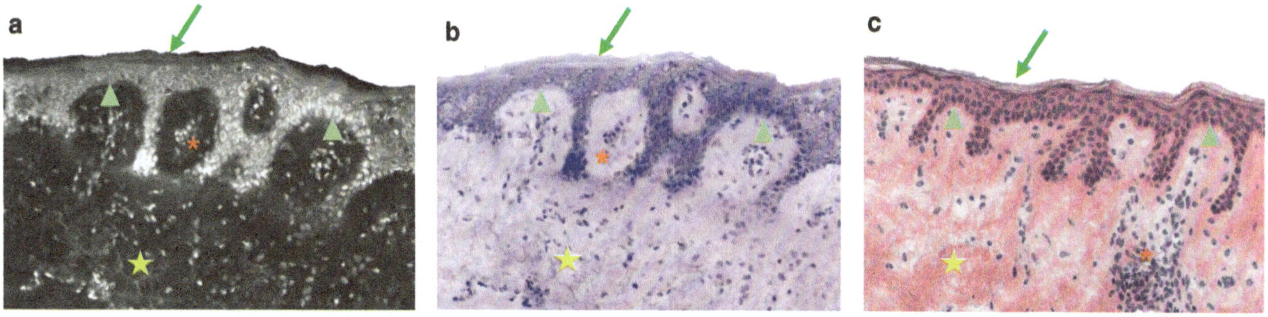

Fig. 4.3 High magnification submosaic images of a normal dermoepidermal junction (DEJ) from scapular skin of a 76-year-old male. **a** FCM image shows an undulating pattern of the DEJ (light green arrowheads) as an interface between the bright basal layer of epidermis (green arrow) and grayish dark papillary dermis (yellow stars) with scattered bright inflammatory cells (orange asterisk). **b** Corresponding DHE image shows this undulating DEJ pattern (light green arrowheads) as an interface between dark purple (bright on FCM) basal epidermal layer (green arrow) and pinkish-purple (grayish on FCM) papillary dermis (yellow stars) with scattered small purple (bright on FCM) inflammatory cells (orange asterisk). **c** Corresponding conventional H&E-stained image shows an excellent correlation with FCM (**a**) and DHE (**b**) images for the normal undulating pattern of DEJ at the interface of the basal layer of epidermis (green arrow) and papillary dermis (yellow stars) with scattered inflammatory cells (orange asterisk). H&E magnifications = 20X

4.3 Dermis

The dermis has two major components: papillary dermis and reticular dermis. Papillary dermis is the uppermost layer of the dermis. This layer is composed of a fine reticular network of collagen and elastic fibers. Just beneath this layer is the reticular dermis, which is characterized by the presence of thicker and denser collagen bundles [11]. Various adnexal structures (hair follicles, sebaceous, and eccrine glands), muscle, nerves, and vessels reside in the dermis. The thickness of the dermis is variable depending on the anatomical site (Figs. 4.4, 4.5, 4.6) [12].

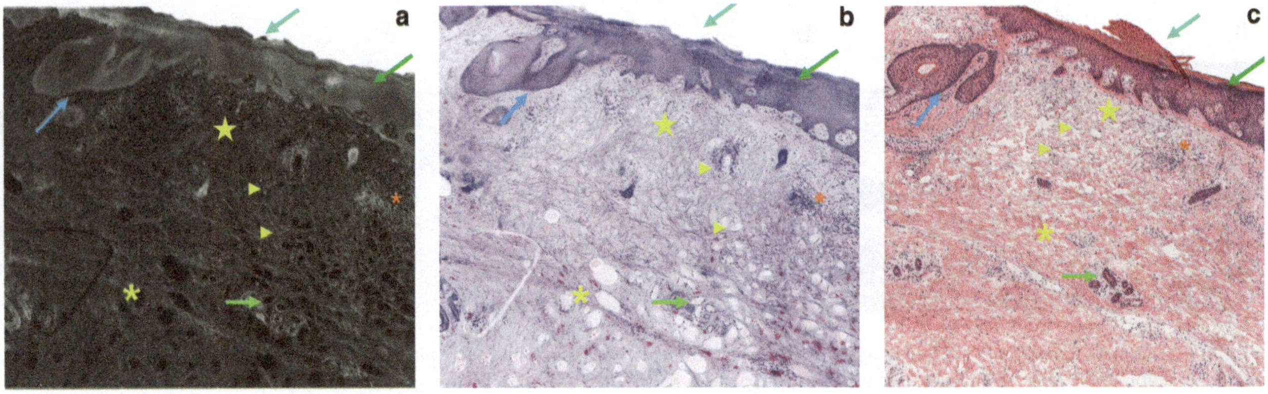

Fig. 4.4 Medium magnification submosaic images of skin from dorsal right hand of a 40-year-old female showing dermal layers. **a** FCM image shows a grayish white epidermis (green arrow) with a dark grayish parakeratotic scale (turquoise arrow) on top. Papillary dermis (yellow star) appears grayish with loose connective tissue, and bright wavy elastic fibers (yellow arrowheads) and has bright inflammatory cells (orange asterisk), and a grayish bright hair follicle (blue arrow). Beneath the papillary dermis, grayish reticular dermis (yellow asterisk) with denser collagen bundles and bright eccrine ducts (fluorescent green arrow) can be identified. **b** On the corresponding DHE image, epidermis (green arrow) appears purple in color (grayish white on FCM) and the parakeratotic scale (turquoise arrow) appears pale purple (dark grayish on FCM) with compressed purple nuclei. Papillary dermis (yellow star) appears purplish pink (grayish on FCM) with loose connective tissue and prominent wavy purple color (bright on FCM) elastic fibers (yellow arrowhead) and has small purple (bright on FCM) inflammatory cells (orange asterisk), and a purple-colored hair follicle (blue arrow). Reticular dermis (yellow asterisk) appears as purplish pink (grayish on FCM) bundles of dense collagen within which purple eccrine ducts (fluorescent green arrow) are identified. **c** Corresponding conventional H&E-stained image shows an excellent correlation with FCM (**a**) and DHE (**b**) images for the epidermis (green arrow), parakeratotic scale (turquoise arrow); papillary dermis (yellow star) with elastic fibers (yellow arrowhead), inflammatory cells (orange asterisk), and hair follicle (blue arrow); and reticular dermis (yellow asterisk) with eccrine ducts (fluorescent green arrow). H&E magnifications = 10X

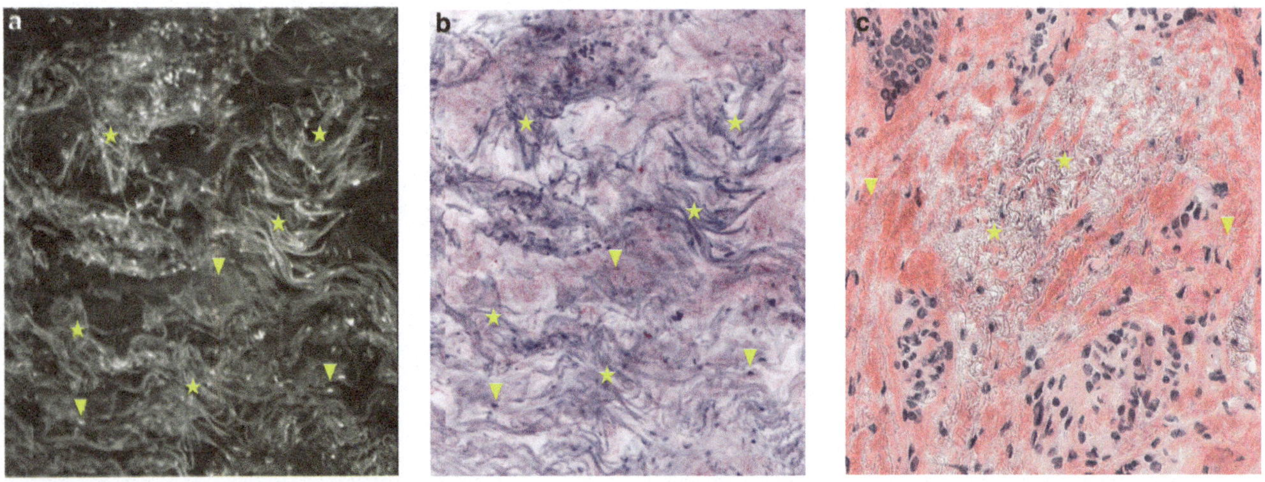

Fig. 4.5 High magnification submosaic images of elastic fibers in papillary dermis from the same sun-exposed skin as in Fig. 4.4. **a** FCM image shows thin bright wavy elastic fibers (yellow stars) admixed with grayish collagen bundles (yellow arrowheads). **b** On the corresponding DHE image, elastic fibers (yellow stars) appears as thin purple (bright on FCM) wavy strands admixed with pinkish-purple (grayish on FCM) collagen bundles (yellow arrowheads). **c** Corresponding conventional H&E-stained image shows an excellent correlation with FCM (**a**) and DHE (**b**) images for elastic fibers (yellow stars) and surrounding collagen bundles (yellow arrowheads). H&E magnifications = 20X

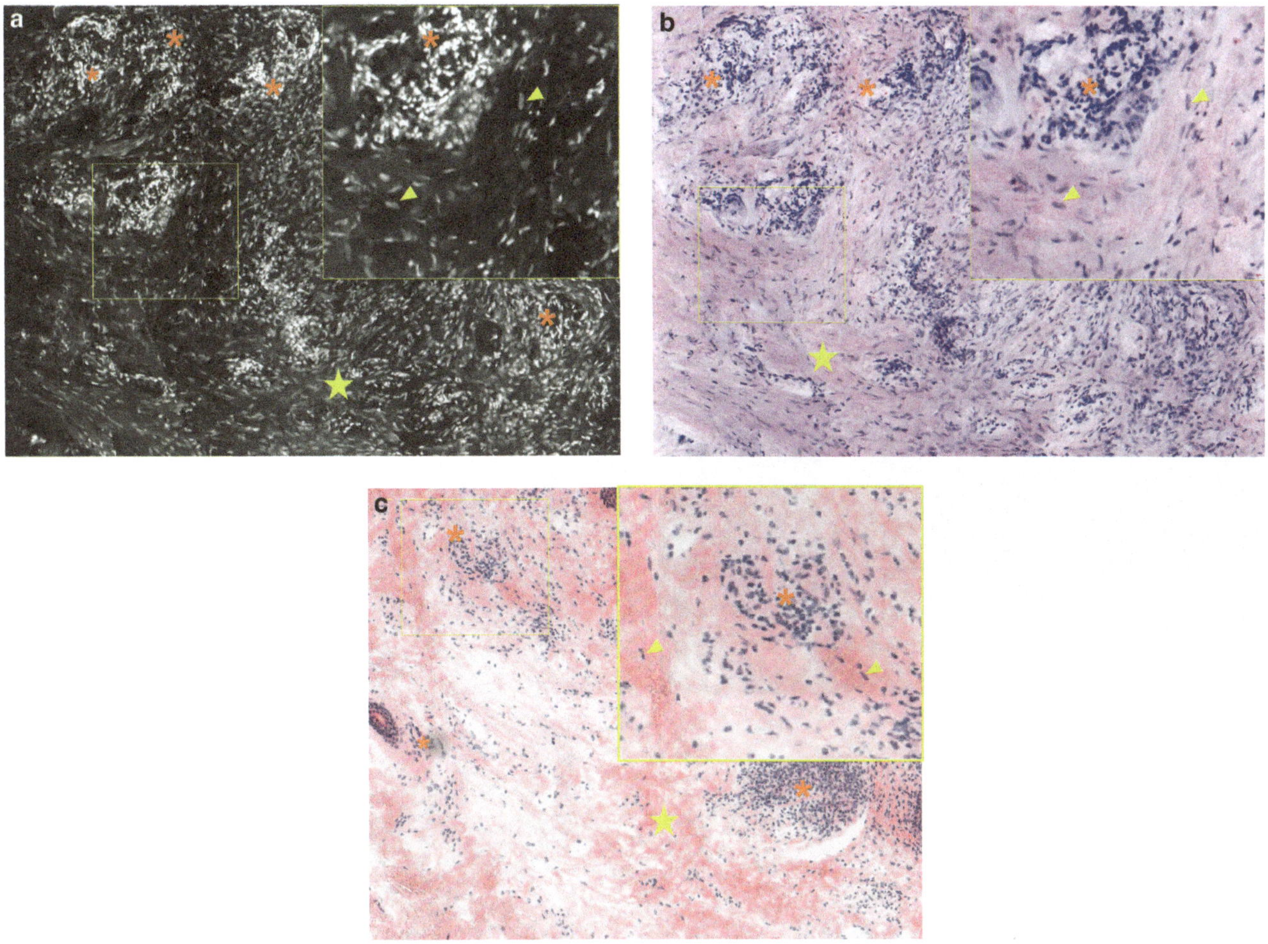

Fig. 4.6 Medium to high magnification submosaic images of inflammatory cells in dermis obtained from sun-damaged scapular skin of a 76-year-old male. **a** FCM image shows inflammatory cells (orange asterisks and inset) as small bright cells clustered and scattered throughout the grayish dermis (yellow star). These inflammatory cells can be readily distinguished from the surrounding thin bright elongated nuclei of fibroblasts (yellow arrowheads and inset). **b** On the corresponding DHE image, inflammatory cells (orange asterisks and inset) appears as dark purple (bright on FCM) small round cells within a pinkish dermis (yellow star); while the surrounding fibroblasts (yellow arrowheads and inset) have a thin purple (bright on FCM) elongated nuclei. **c** Corresponding conventional H&E stained image shows an excellent correlation with FCM (**a**) and DHE (**b**) images for an extensive lymphocytic infiltrate (orange asterisk) and surrounding fibroblasts (yellow arrowheads) within dermal collagen (yellow star). H&E magnifications = 10X and 20X (inset)

4.4 Hypodermis

Hypodermis or subcutis/subcutaneous layer is the lowermost layer of the skin. It is primarily composed of adipocytes, which are organized as lobules separated by thin connective tissue fibrous *septae*. Within these *septae,* neurovascular bundles are located. Deeper portions of eccrine lobules and terminal hair follicles can be seen in this layer (Fig. 4.7).

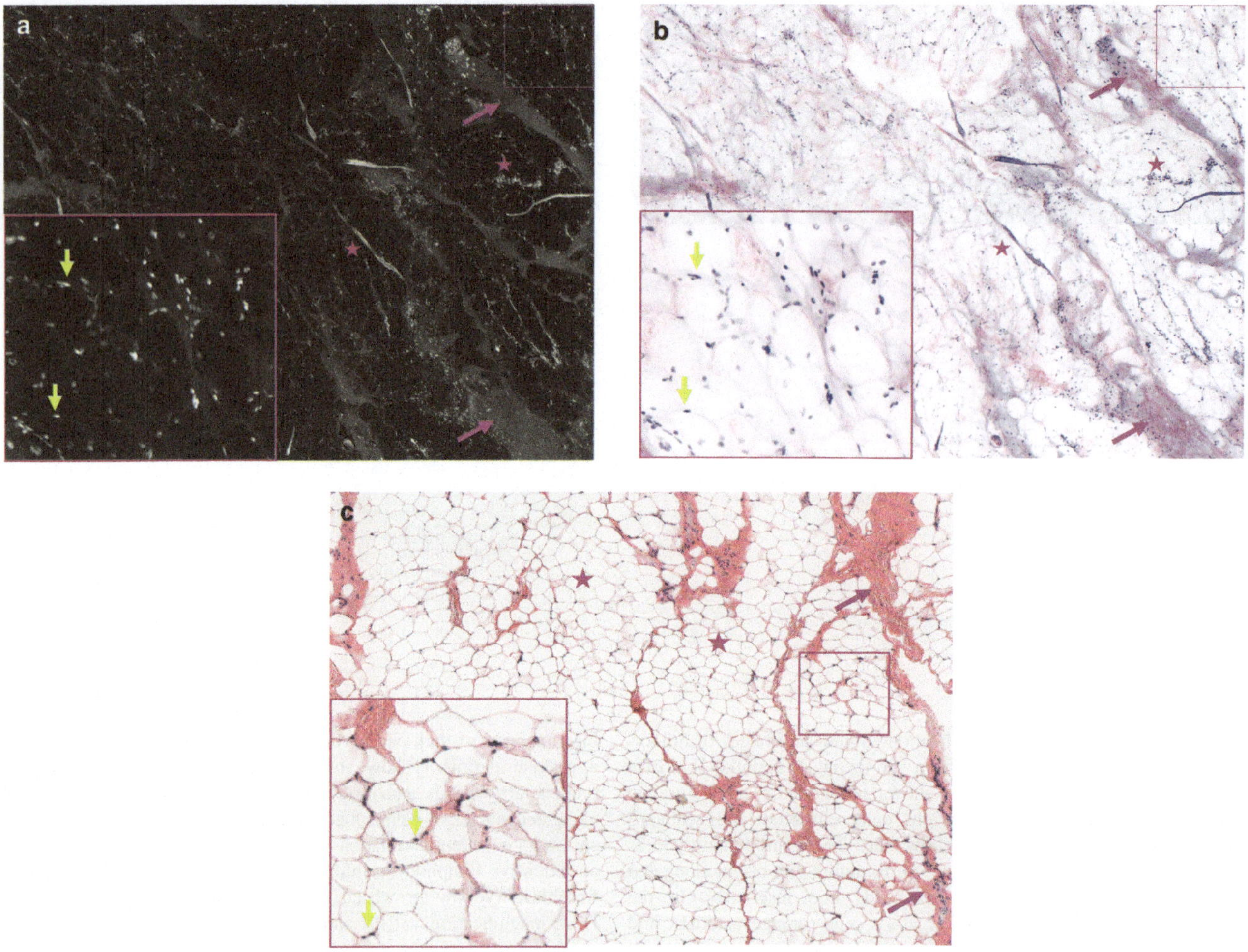

Fig. 4.7 Medium to high magnification submosaic images of hypodermis obtained from scalp skin of a 65-year-old male. **a** FCM image shows a sheet of dark adipocytes (pink stars) with intervening grayish fibrous septae (pink arrows). Adipocytes (inset) appear as dark polygonal cells with a thin grayish border and small bright peripheral nuclei (yellow arrows). **b** On the corresponding DHE image, a sheet of adipocytes (pink stars) appears whitish (dark on FCM) and the fibrous septae (pink arrows) appears purplish in color. In the inset, adipocytes appear as whitish (dark on FCM) polygonal cells with a thin purplish border (grayish on FCM) and a small dark purple (bright on FCM) peripheral nuclei (yellow arrows). **c** Corresponding conventional H&E-stained image shows an excellent correlation with FCM (**a**) and DHE (**b**) images for adipocytes (pink stars and inset) with small peripheral nuclei (yellow arrows) in the hypodermis, and intervening fibrous septae (pink arrows). H&E magnifications = 10X and 20X (inset from pink boxed areas)

4.5 Muscle, Vessels, and Nerves

Muscle, vessels, and nerves are present in the dermis and hypodermis.

- Muscle: Mostly composed of smooth muscle fibers that appear as pink bundles without striations or strips (unlike skeletal muscles) with cigar-shaped nuclei [13]. These are either attached to hair follicles (arrector pili muscle, causing piloerection) or are found in the vessel walls (Fig. 4.8).
- Vessels: Vessels range from small capillaries present in the superficial dermis to larger vessels in the deep dermis and subcutis. They have a central lumen lined by an endothelial layer. Medium- to large-sized vessels have three layers: the elastic *tunica intima*, the *tunica media* of smooth muscle fibers, and *tunica adventitia* of connective tissue (Fig. 4.9).
- Nerves: Small- to large-sized peripheral nerve fibers with wavy Schwann cell nuclei enclosed in connective tissue sheath of epineurium can be seen around hair follicles and eccrine or apocrine glands (Fig. 4.10). Sensory receptors such as Pacinian and Meisner's corpuscles can be seen at specific anatomic sites [13].

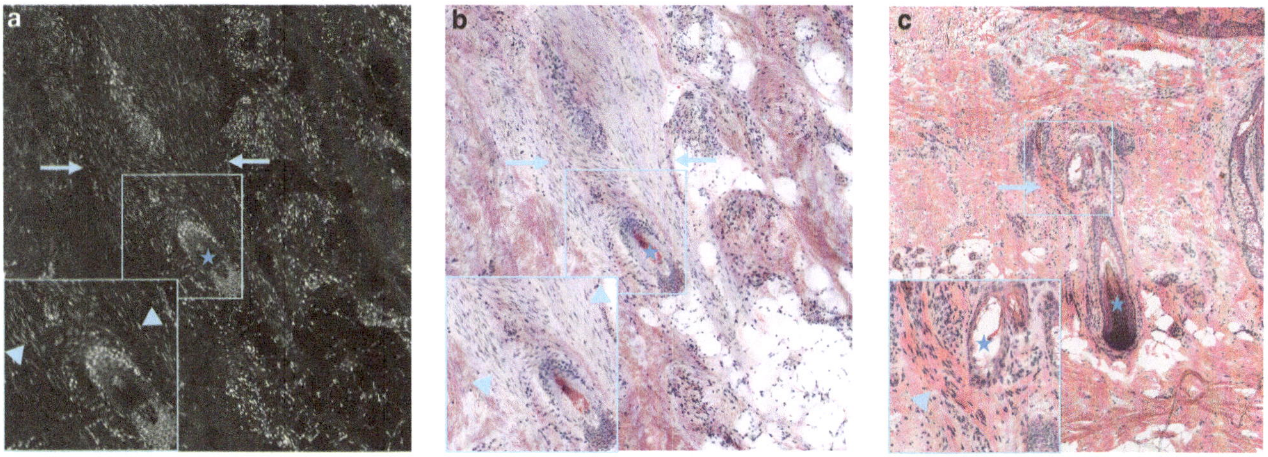

Fig. 4.8 Medium to high magnification submosaic images of an arrector pili muscle in a longitudinal section obtained from scalp skin of a 65-year-old male. **a** FCM image shows grayish fascicles of smooth muscles (light blue arrows) in a longitudinal section around a bright hair follicle (blue star). Within these grayish muscle bundles, elongated cigar-shaped bright nuclei (light blue arrowheads) are seen (inset and light blue boxed area). **b** On the corresponding DHE image, smooth muscle bundles (light blue arrows) appears pale purple in color (grayish on FCM) bundles with purple color (bright on FCM) elongated nuclei (light blue arrowheads and light blue boxed area) (inset). Hair follicle (blue star) appears as a purplish color (bright on FCM) structure. **c** Corresponding conventional H&E-stained image shows an excellent correlation with FCM (**a**) and DHE (**b**) images for smooth muscle bundles (light blue arrows) and hair follicle (blue star). H&E magnifications = 10X and 20X (inset)

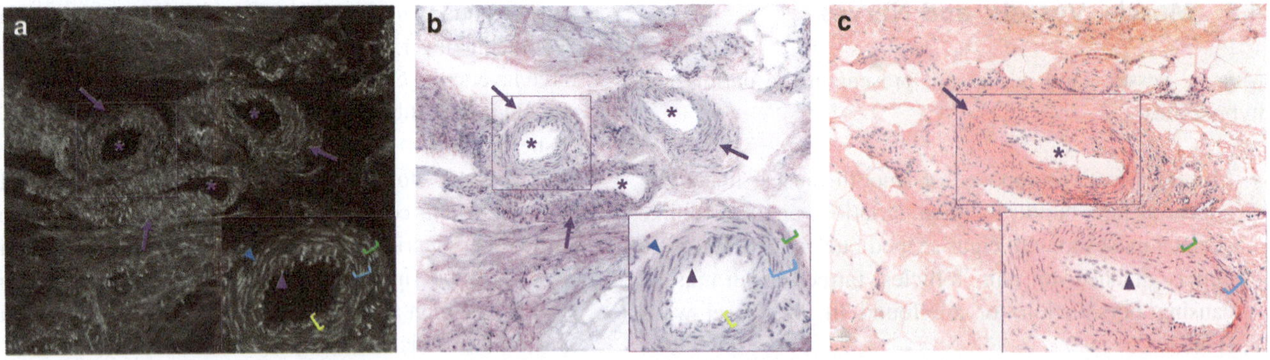

Fig. 4.9 Medium to high magnification submosaic images of dermal vessels in cross-sectional view obtained from right dorsal hand skin of a 57-year-old male. **a** FCM image shows three medium-sized blood vessels with a central dark lumen (purple asterisks) and grayish vessel wall (purple arrows). Inset shows three distinct layers of the vessel wall: *intima* with small bright plump endothelial cells (purple arrowhead) and thin wavy bright internal elastic membrane (yellow bracket); *media* (blue bracket) with thick grayish concentric layers of smooth muscle with bright blunt-ended nuclei (blue arrowhead); and grayish *adventitia* (green bracket) with sparse bright thin nuclei of fibroblasts. **b** Corresponding DHE image shows these vessels with white color (dark on FCM) central lumen (purple asterisks) and a thick purplish pink (grayish on FCM) vessel wall (purple arrows). Inset shows three distinct layers of the vessel wall: *intima* with small purple (bright on FCM) plump endothelial cells (purple arrowhead) and wavy thin purple (bright on FCM) internal elastic membrane (yellow bracket); *media* (blue bracket) with purplish pink (grayish on FCM) concentric layers of smooth muscle with purple (bright on FCM) blunt-ended nuclei (blue arrowhead); and a pinkish (grayish on FCM) *adventitia* (green bracket) with sparse thin purple nuclei of fibroblast. **c** Corresponding conventional H&E-stained image shows an excellent correlation with FCM (**a**) and DHE (**b**) images for the dermal vessels (purple arrows and purple asterisks) and all its three layers: *intima* with endothelial cells (purple arrowhead), media (blue bracket), and *adventitia* (green bracket). Note the internal elastic membrane is not prominent on the H&E image, in contrast to the FCM and DHE images. H&E magnifications = 10X and 20X (inset)

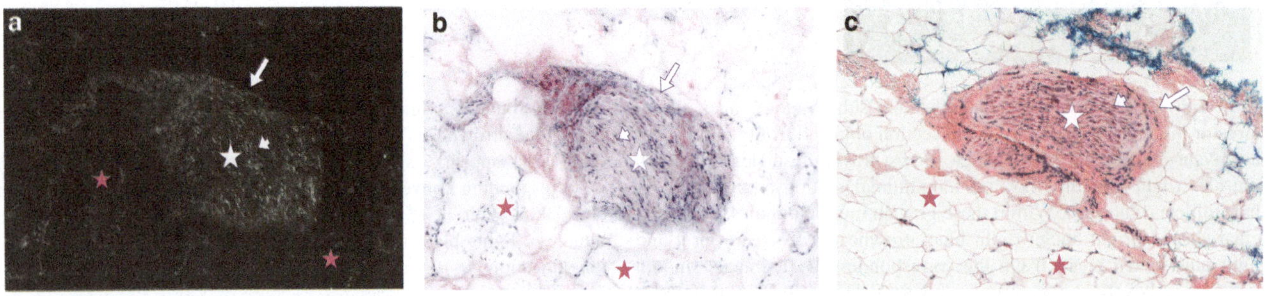

Fig. 4.10 Higher magnification submosaic images of a nerve in a longitudinal section obtained from right forehead skin of a 59-year-old female. **a** FCM image shows a grayish nerve (white star with purple borders) located within dark adipocytes of hypodermis (pink stars). The nerve fiber is composed of bright wavy pointed nuclei (white arrowhead with purple borders) and is encased in a thin grayish fibrous sheath of perineurium (white arrow with purple borders). **b** On the corresponding DHE image, the nerve (white star with purple borders) appears as a purplish pink oval structure (grayish on FCM) within the surrounding whitish (dark on FCM) adipocytes of hypodermis (pink stars). The wavy, pointed nerve nuclei (white arrowhead with purple borders) appear purple in color (bright on FCM) and the perineurium (white arrow with purple borders) appears as purplish pink (grayish on FCM) fibrous sheath. **c** Corresponding conventional H&E image shows an excellent correlation with FCM (**a**) and DHE (**b**) images for the nerve (white star with purple borders) with its wavy nuclei (white arrowhead with purple borders) and perineurium (white arrow with purple borders), and its surrounding hypodermis (pink stars). H&E magnifications = 20X

4.6 Adnexal Structures

Skin contains multiple adnexal structures including hair follicles, and exocrine glands such as sebaceous, eccrine, and apocrine glands.

- Hair and hair follicles: Hair follicles may appear as tubular, round, or oblique structures in the hair-bearing skin on histology sections. They are mostly seen in the dermis but may be seen in the hypodermis. They encompass the hair shaft, which has three layers: the innermost medulla, outer cortex, and outermost cuticle. Hair follicles have three main segments (from epidermis to dermis) [13, 14] (Figs. 4.11, 4.12, 4.13).

  Infundibulum. The uppermost portion extends from the follicular orifice in the epidermis to the entrance of the sebaceous duct. It is lined by a keratinized squamous epithelium (similar to the epidermis) and contains the hair shaft, secretions of sebaceous and/or apocrine glands, and commensal bacteria, yeast, and parasites such as Demodex folliculorum [15, 16].

  Isthmus. The middle portion extends between the entrance of the sebaceous duct to the insertion of the arrector pili muscle. It is lined by an internal root sheath of keratinized cells from the hair matrix and an external/outer root sheath from the tubular invagination of the epidermis.

  Inferior segment: Supra-bulb and Bulb portion: This is the lowest segment, extending from the insertion of the erector pili muscle to the base of the follicle. It is surrounded by a hair matrix that contains proliferating cells. The bulb has a central dermal papilla with blood vessels.
- Sebaceous glands and ducts: They appear as a lobular structure composed of round sebocytes filled with lipid vacuoles, which produce sebum [13]. They are attached to the hair follicle to form a pilosebaceous unit (Figs. 4.14 and 4.15) [15].
- Eccrine glands: They are sweat glands most commonly located not only on the palms and soles but also on other part of the body. They have three portions: a spiral intraepidermal duct (or acrosyringium), a straight intradermal duct, and a coiled secretory portion in the deeper dermis and subcutis. The duct is lined by two layers of cuboidal cells, whereas the secretory portions are lined by a single layer of cuboidal cells (Figs. 4.16 and 4.17) [13].

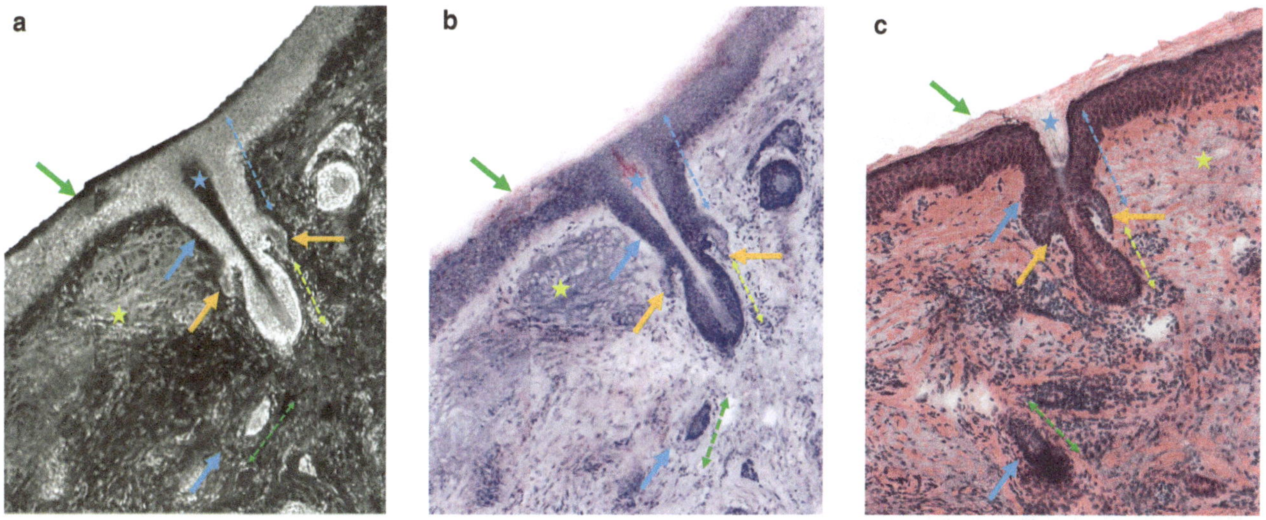

Figs. 4.11. Medium magnification submosaic images of a vellus hair follicle in longitudinal section obtained from left nasal tip skin of a 75-year old female. **a** FCM image shows a tubular vellus hair follicle (blue arrows) lined by a bright multi-nucleated layer seen in continuation from the overlying epidermis (green arrow) and extending up to the grayish papillary dermis with solar elastosis (yellow star). In the center of the follicle, there is a dark tubular area for the hair shaft (not visible). All three segments of the hair follicle can be identified: an infundibulum (blue dashed double-headed arrow), an isthmus (yellow dashed double-headed arrow), and a lower segment (green dashed double-headed). At the junction of infundibulum and isthmus, the sebaceous mantle (orange arrows) appears as a small grayish white bulbous structure. **b** On the corresponding DHE image, the lining of the vellus hair (blue arrows) appears as multi-layered dark purple (bright on FCM) nuclei and in its center there is a dark pink hair shaft (blue star; not visible on FCM) and pinkish keratin is identified. All three segments of the hair follicles (as described above) are seen. Sebaceous mantle (orange arrows) appears as a purplish (bright on FCM) bulbous structure. **c** Corresponding conventional H&E image shows an excellent correlation with FCM (**a**) and DHE (**b**) images for all segments of the vellus hair follicles (blue arrows and dashed double-headed), reflective hair shaft (blue star), and sebaceous mantle (orange arrows), and for papillary dermis with solar elastosis (yellow star). H&E magnifications = 10X

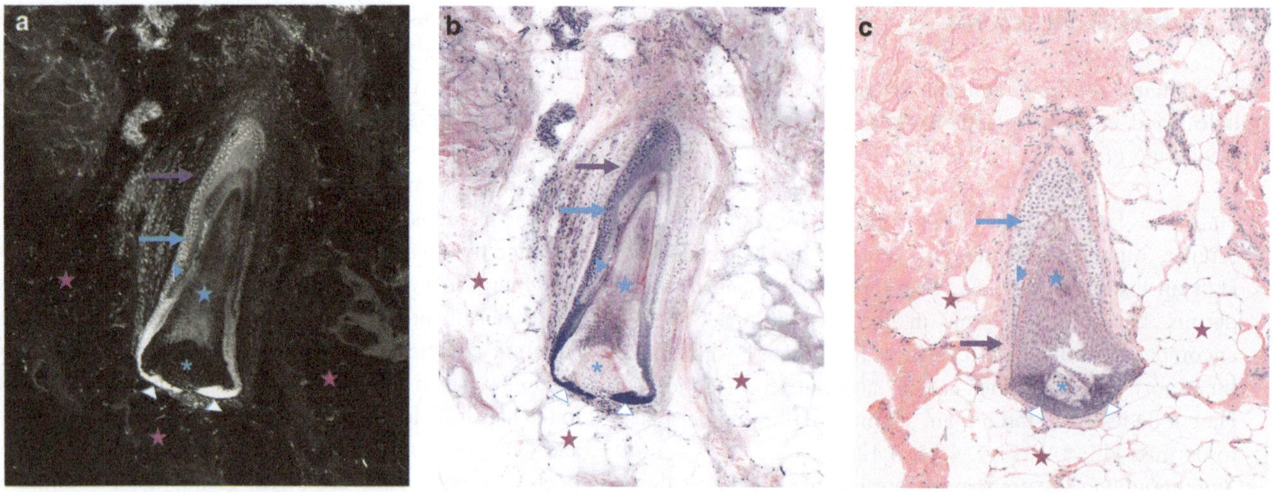

Figs. 4.12. Medium magnification submosaic images of a terminal hair follicle in longitudinal section obtained from scalp skin of a 65-year-old female. **a** FCM image shows a lower segment of a terminal hair follicle with an outer root sheath (ORS; blue arrow) composed of multi-layered bright nuclei, a grayish internal root sheath (IRS; blue arrowhead) with a few bright nuclei, and a grayish to whitish central hair shaft (blue star). In the hair bulb portion, an intensely bright (hyperfluorescent) crowded layer of matrical cells (white arrow with blue borders) surrounds the central grayish dark (hypofluorescent) dermal papillae (blue asterisk) with bright (hyperfluorescent) nuclei of mesenchymal cells. The hair follicle is surrounded by a grayish fibrous sheath (purple arrow) and is located within the dark hypodermis (pink stars). **b** On the corresponding DHE image, ORS (blue arrow) cells have dark purple nuclei (bright cells on FCM) and scant purplish cytoplasm, IRS (blue arrowhead) appears purplish color (grayish on FCM) with few purple nuclei, and the hair shaft (blue star) appears purplish pink. Fibrous sheath (purple arrow) surrounding hair follicle appears pinkish (grayish on FCM). Matrical cells (white arrow with blue borders) appear dark purple (bright on FCM) and the dermal papillae (blue asterisk) appear pinkish in color (grayish dark on FCM) with purple nuclei of mesenchymal cells. The hair follicle is seen within the whitish (dark on FCM) hypodermis (pink stars). **c** Corresponding conventional H&E-stained image shows an excellent correlation with FCM (**a**) and DHE (**b**) images for all the layers of hair follicles (fibrous sheath, ORS, IRS), the hair shaft (blue star), the hair bulb with matrical cells (white arrow with blue borders), and the dermal papilla (blue asterisk) seen within the hypodermis (pink stars). H&E magnifications = 10X

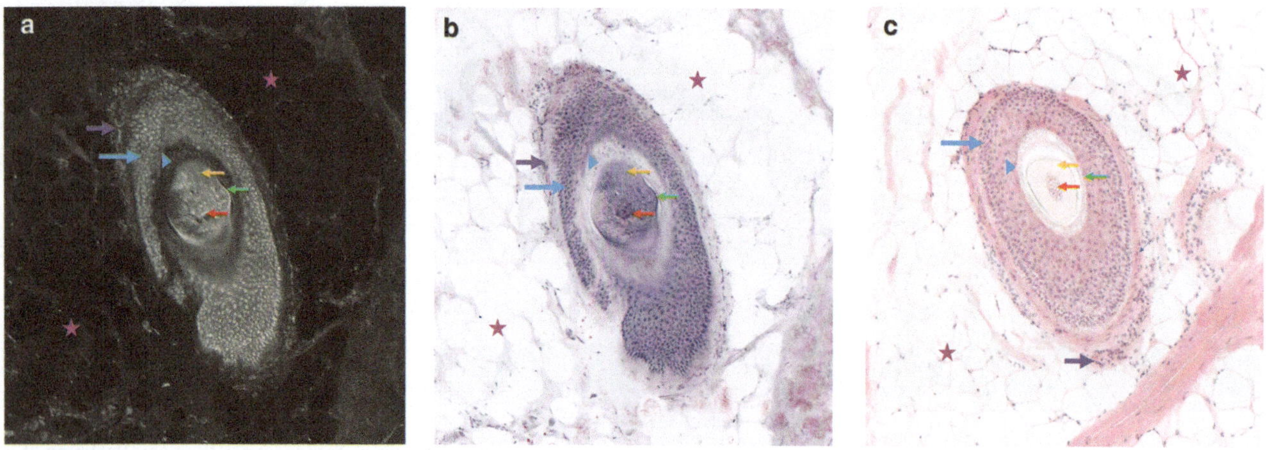

Figs. 4.13. Medium magnification submosaic images of a terminal hair follicle in a cross-section obtained from skin of the beard area in a 70-year-old man. **a** FCM image shows a terminal hair follicle with an outermost grayish fibrous sheath (purple arrow) with sparse bright elongated nuclei, an outer root sheath (ORS; blue arrow) composed of multiple layers of cuboidal bright cells, and a grayish internal root sheath (IRS; blue arrowhead). In the center of the hair follicle, a hair shaft is identified with a central grayish amorphous medulla (red arrow), a grayish layer of cortex (orange arrow) with elongated bright nuclei, and a thin very bright cuticle (green arrow) layer. Surrounding hypodermis appears dark (pink stars). **b** On the corresponding DHE image, fibrous sheath (purple arrow) of the hair follicle appears pinkish-purple (grayish on FCM), ORS (blue arrow) has multi-layered dark purple (bright cells on FCM) cuboidal cells, and IRS (blue arrowhead) appears purplish (grayish on FCM). The medulla (red arrow) of the hair shaft medulla appears pinkish-purple (grayish on FCM), the cortex (orange arrow) appears purplish (grayish on FCM) with purple spindle-shaped nuclei, and the cuticle (green arrow) appears as thin dark purple (bright on FCM) layer. The hypodermis (pink stars) appears whitish (dark on FCM). **c** Corresponding conventional H&E-stained image shows an excellent correlation with FCM (**a**) and DHE (**b**) images for all the layers of the hair follicle (fibrous sheath, ORS, IRS), and the hair shaft (medulla, cortex, and cuticle), and its surrounding hypodermis (pink stars). H&E magnifications = 10X

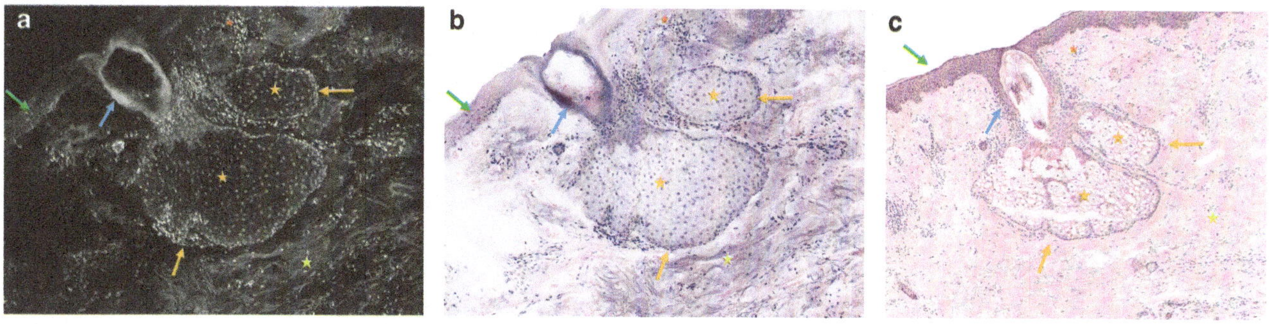

Fig. 4.14 Medium magnification submosaic images of a pilosebaceous unit obtained from skin in right parotideomasseteric area of a 70-year-old male. **a** FCM image of a pilosebaceous unit composed of grayish dark lobules of sebaceous glands (orange arrows) connected to a dark tubular infundibular portion of a hair follicle (blue arrow) with grayish keratin. The sebaceous lobules (orange arrows) are composed of sebocytes (orange stars) with darkish gray cytoplasm and central small bright round nuclei. The pilosebaceous unit is seen in continuation with the overlying bright epidermis (green arrow). The dermis (yellow star) appears grayish in color with scattered bright inflammatory cells (orange asterisk). **b** On the corresponding DHE image, sebaceous lobules (orange arrows) is composed of pale purple (grayish on FCM) sebocytes (orange stars) with small purple (bright on FCM) nuclei (orange stars). The infundibular portion of the hair follicle (blue arrow) appears dilated and white with purple to pink keratin material within. The overlying epidermis (green arrows) appears purple in color (bright on FCM). The dermis (yellow star) appears purplish (grayish on FCM) with scattered purple (bright on FCM) inflammatory cells (orange asterisk). **c** Corresponding conventional H&E stained image shows good correlation with FCM (**a**) and DHE (**b**) images for the pilosebaceous unit with sebaceous glands (orange arrows) with its sebocytes (orange asterisk), the follicular infundibulum (blue arrow), and for the overlying epidermis (green arrow), and the dermis (yellow arrow) and lymphocytic infiltrate (orange asterisk). H&E magnifications = 10X

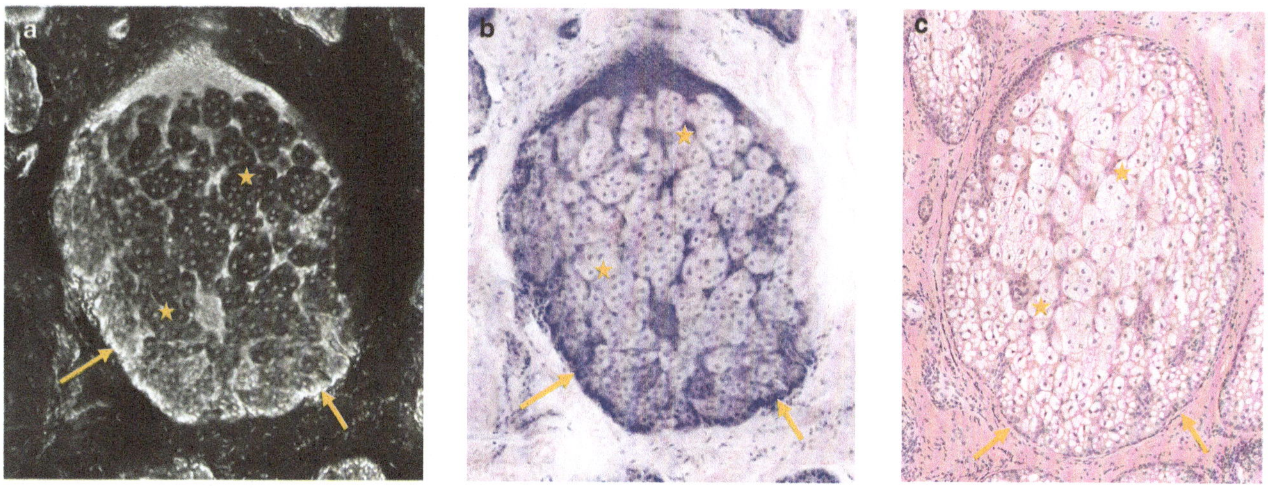

Fig. 4.15 High magnification submosaic images of a sebaceous lobule obtained from skin of left nasal sidewall of a 75-year-old male. **a** FCM image of a sebaceous gland lobule shows central sebocytes (orange stars) with darkish gray cytoplasm and small bright round nuclei and an outer bright germinative layer of flat to cuboidal nuclei (orange arrows). **b** On the corresponding DHE image, sebocytes (orange stars) appear as pale purple to pink (dark on FCM) with small purple (bright on FCM) central nuclei and the germinative layer appears to have flat to cuboidal dark purple nuclei with scant pink cytoplasm (orange arrows). **c** Corresponding conventional H&E-stained image shows an excellent correlation with FCM (**a**) and DHE (**b**) images for the sebocytes (orange stars) and germinative cells (orange arrows) of the sebaceous lobule. H&E magnifications = 20X

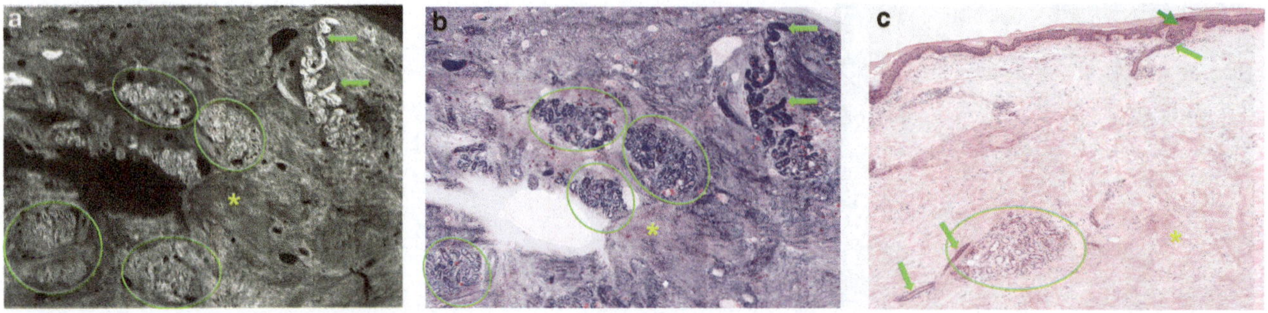

Fig. 4.16 Medium magnification submosaic images of eccrine ducts and glands obtained from left anterior tibial skin of an 80-year-old male. **a** FCM image shows bright coiled eccrine glands and bright oval to round ducts (fluorescent dashed green circles) within the dense grayish reticular dermis (yellow asterisk). Some of the ducts (fluorescent green arrows) appear intensely bright and are seen rising up in the dermis toward the epidermis (epidermis is not visible in this section due to tissue flattening issues). **b** On the corresponding DHE image, these clusters of coiled eccrine glands and ducts (fluorescent dashed green circles) appear purple in color (bright on FCM) within a dense purplish pink (grayish on FCM) reticular dermis (yellow asterisk). The straight portion of the eccrine ducts (fluorescent green arrows), seen rising up to the epidermis, appear dark purple in color (intensely bright on FCM). **c** Corresponding conventional H&E-stained image shows an excellent correlation with FCM (**a**) and DHE (**b**) images for the eccrine glands (fluorescent green circle), and eccrine ducts (fluorescent green arrows and fluorescent green circle) within the reticular dermis (yellow asterisk). In this tissue section, the eccrine ducts (fluorescence green arrows) are seen traversing the epidermis (green arrow). H&E magnifications = 10X

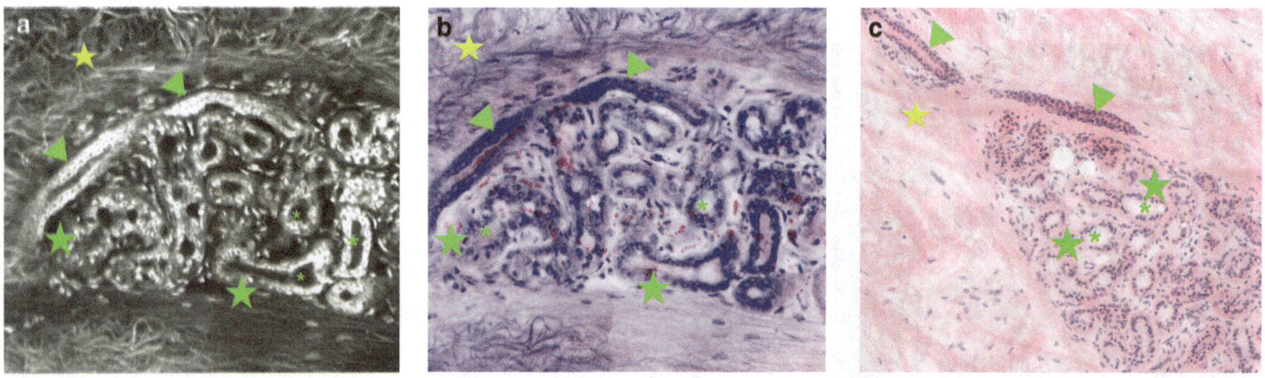

Fig. 4.17 Higher magnification submosaic images of the lining of eccrine ducts and glands. **a** FCM image shows eccrine glands (green stars) with a central dark lumen (green asterisk) lined by a single layer of bright cuboidal nuclei (green stars). Ducts are composed of two layers of bright cuboidal nuclei (green arrowheads). Dermis has grayish clumps of solar elastosis (yellow star). **b** On the corresponding DHE image, eccrine glands have a single layer of cuboidal cells with purple (bright on FCM) nuclei and purplish pink cytoplasm (green star), while the ducts have two distinct layers of cuboidal cells with purple nuclei (bright on FCM) and scant light purple cytoplasm (green arrowheads). Lumen (green asterisk) of the glands and ducts have a whitish to pinkish (dark on FCM) homogenous material. Solar elastosis appears as clumps of purple (grayish on FCM) fibers (yellow star). **c** Corresponding conventional H&E-stained image shows an excellent correlation with FCM (**a**) and DHE (**b**) images for the lining of eccrine glands (green stars) and eccrine ducts (green arrowheads). *Note:* Solar elastosis (yellow star) is less prominent on this frozen H&E section as compared to the FCM and DHE images. H&E magnifications = 20X

4.7 Regional Variations of Skin

The thickness of skin layers and the presence of adnexal structures may vary depending on the anatomical site, which facilitates the topographic identification of the skin [17]. For example, the highest density of sebaceous glands is found in the nasal skin, or the increased epidermal thickness with a prominent corneal layer is typically seen in the palmoplantar skin (Table 4.3). Examples of regional variations on EVCM can be found in Figs. 4.18, 4.19, 4.20, 4.21, 4.22, 4.23, 4.24 and 4.25.

Table 4.3 Morphological features of skin related to the anatomical location on FCM, DHE, and corresponding conventional H&E-stained images

Fluorescence confocal microscopy (FCM) mode (grayscale image) (Figs. 4.18a, 4.19a, 4.20a, 4.21a, 4.22a, 4.23a, 4.24a and 4.25a)	Digital H&E (combined FCM and RCM modes). Figs. 4.18b, 4.19b, 4.20b, 4.21b, 4.22b, 4.23b, 4.24b and 4.25b	Conventional H&E-stained image (Figs. 4.18c, 4.19c, 4.20c, 4.21c, 4.22c, 4.23c, 4.24c and 4.25c)
Cartilage: Low magnification (Fig. 4.18a): Grayish white (hypofluorescent) dense sheet below the dermis High magnification (Fig. 4.19a): Sheets of spherical-shaped chondrocytes with bright (hyperfluorescent) nuclei and grayish (hypofluorescent) cytoplasm seen within a dark (nonfluorescent) lacunae. The surrounding extracellular matrix (ECM) is grayish-white (hypofluorescent) with some thin bright (hyperfluorescent) fibers	**Cartilage**: Low magnification (Fig. 4.18b): Dense purple color sheet below the dermis High magnification (Fig. 4.19b): Sheets of spherical-shaped chondrocytes with purple nuclei and pinkish cytoplasm seen within whitish lacunae. The ECM appears purplish pink in color with some thin purple elastic fibers	**Cartilage**: Low magnification (Fig. 4.18c): Dense purplish (grayish on FCM; purple on DHE) sheet of chondrocytes below the dermis. Seen in the skin of the pinna and nose High magnification (Fig. 4.19c): Sheets of chondrocytes (bright on FCM; purple on DHE) within lacunae (dark on FCM; whitish on DHE). ECM (grayish on FCM; purplish pink on DHE) is rich in collagen II and elastic fibers (in elastic cartilage) (not always visible on H&E stain)
Thick skin: (Figs. 4.22a and 4.23a): - Thickened *St. corneum* with layers of grayish white (hypofluorescent) keratin - Prominent *St. granulosum* with flat bright (hyperfluorescent) nuclei - Bright eccrine glands and ducts - No hair follicles or sebaceous glands	**Thick skin**: (Figs. 4.22b, and 4.23b) - Thickened *St. corneum* with layers of purple color keratin - Prominent *st. granulosum* with flat purple nuclei and dark pink keratohyalin granules - Dark purple eccrine glands and ducts - No hair follicles or sebaceous glands	**Thick skin**: (Figs. 4.22C and 4.23c): Palm, sole, and fingertips - Thickened *St. corneum* with layered keratin (grayish white on FCM; purple on DHE) - Prominent *St. granulosum* with keratohyalin granules (dark pink on DHE) - Prominent eccrine glands and ducts (bright on FCM; dark purple on DHE) - No hair follicles or sebaceous glands
Sun-damaged skin (Fig. 4.24a): *Solar elastosis:* Clumps of grayish (hypofluorescent) fibers or material mostly in the papillary dermis	**Sun-damaged skin** (Fig. 4.24b): *Solar elastosis:* Clumps of purple fibers or material mostly in the papillary dermis	**Sun-damaged skin** (Fig. 4.24c): *Solar elastosis:* Fibrillary basophilic material (grayish on FCM; purple on DHE) in the upper dermis replacing normal pink collagen fibers. Seen in sun-exposed sites such as face
Skeletal muscle: (Fig. 4.25a): *Cross-sectional view:* Grayish (hypofluorescent) round structures with small round bright (hyperfluorescent) peripheral nuclei *Longitudinal section view:* Grayish (hypofluorescent), elongated cylindrical (unbranched) structures with bright (hyperfluorescent), flattened, elongated peripheral nuclei. With or without evident cross-striations	**Skeletal muscle**: (Fig. 4.25b): *Cross-sectional view:* Purplish round structures with small, round purple color peripheral nuclei *Longitudinal section view:* Purplish elongated cylindrical (unbranched) structures with elongated, flattened purple peripheral nuclei. With or without evident cross-striations	**Skeletal muscle**: (Fig. 4.25c): Unbranched muscle fascicles (grayish on FCM; purplish on DHE) with cross-striations and a flattened peripheral round to elongated nuclei (bright on FCM; purple on DHE)

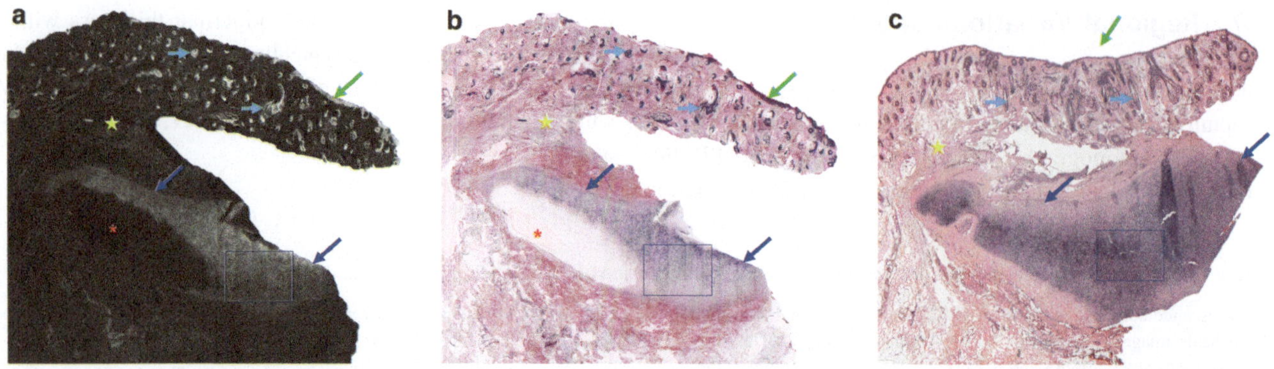

Fig. 4.18 Low magnification submosaic images of a cartilage obtained from skin of left tragus of a 60-year-old male. **a** FCM image shows bright epidermis (green arrow) and an underlying grayish dermis (yellow star) with several bright hair follicles (blue arrows). Below the dermis, cartilage is seen as a grayish white dense sheet (dark blue boxed area and dark blue arrows). **b** On the corresponding DHE image, cartilage (dark blue box and dark blue arrows) appears as a dense purple color (grayish white on FCM) sheet. The epidermis (green arrow) appears as thin purple color (bright on FCM) layer and the dermis (yellow star) as pink (grayish on FCM) layer with dark purple color hair follicles (blue arrows). *Note:* FCM and DHE image shows an artifact (red asterisk) caused by a saline bubble, which appears as a dark structure in the FCM image (**a**) and whitish to pinkish in the DHE image (**b**). **c** Corresponding conventional H&E-stained image shows an excellent correlation with FCM (**a**) and DHE (**b**) images for the epidermis (green arrow), the dermis (yellow star), and the cartilage (dark blue boxed area and dark blue arrows). H&E magnifications = 10X

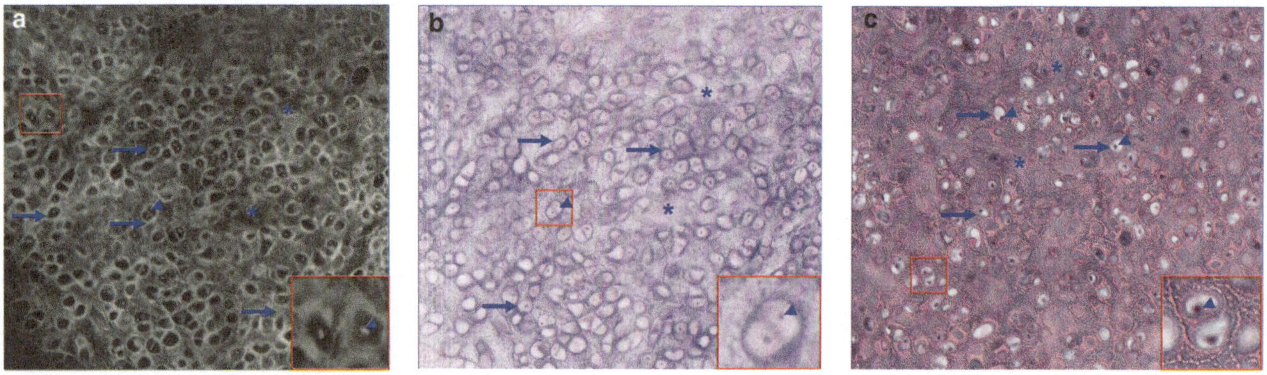

Fig. 4.19 High magnification submosaic images of chondrocytes obtained by digitally zooming in dark blue boxed area of Fig. 4.18. **a** FCM image shows the sheet of spherical-shaped chondrocytes (dark blue arrows and inset from the red boxed area). These chondrocytes have small round bright nuclei (dark blue arrowhead) and a grayish cytoplasm and are seen within dark lacunae. The surrounding extracellular matrix (blue asterisks) appears grayish white with scattered bright fibers (yellow star). **b** On the corresponding DHE image, the chondrocytes (dark blue arrows and inset) have small round purple nuclei (dark blue arrowhead; bright on FCM) and a pinkish cytoplasm (grayish on FCM) seen within whitish lacunae (dark on FCM). The surrounding extracellular matrix (blue asterisks) appears purplish pink in color (grayish on FCM) with a few thin purple elastic fibers (yellow star). **c** Corresponding conventional H&E-stained image shows high correlation with FCM image (**a**) and DHE image (**b**) for the chondrocytes (dark blue arrows) and its nuclei (dark blue arrowhead), and the surrounding matrix (blue asterisks). H&E magnifications = 20X and 40X (insets)

Fig. 4.20 Low magnification submosaic images showing normal skin layers of an eyelid from a 44-year-old male. **a** FCM image shows a thin bright epidermis (green arrow) and an underlying grayish dermis (yellow star) with a bright hair follicle (blue arrow) and a cluster of bright eccrine glands (bright green arrow). Beneath the dermis, grayish white band of orbicular muscle (light blue star) is seen. Toward the lower edge of the tissue, bright spherical structures of meibomian glands (orange boxed area) are seen. **b** On the corresponding FCM, thin purple color (bright on FCM) epidermis (green arrow) and pinkish purple (grayish on FCM) dermis (yellow star) with a dark purple hair follicle (blue arrow) and eccrine glands (bright green arrow) are seen. The orbicular muscle (light blue star) appears as light purple (grayish on FCM) bands, beneath which numerous purple meibomian glands (orange boxed area) can be identified. **c** Corresponding conventional H&E stained tissue section shows an excellent correlation with FCM image (**a**) and DHE image (**b**) for the epidermis (green arrow), the hair follicle (blue arrow), and the eccrine glands (bright green arrow) within the dermis (yellow star), and for the orbicular muscle (light blue star) and meibomian glands (orange boxed area). H&E magnifications = 2X

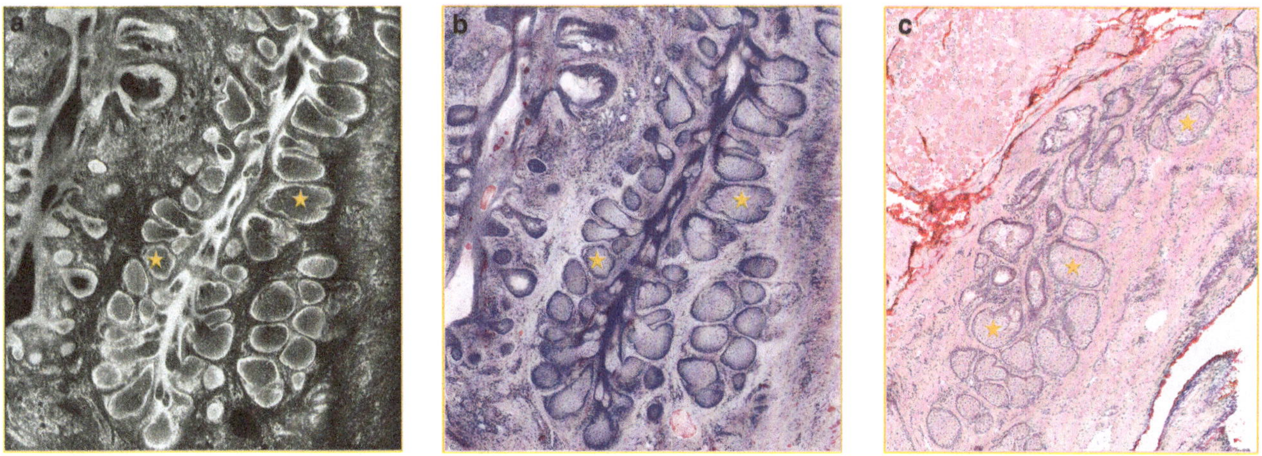

Fig. 4.21 Medium magnification submosaic images of meibomian glands obtained by digitally zooming in the orange boxed area of Fig. 4.20. **a** FCM image shows darkish to bright spherical lobules of meibomian glands (orange stars) with similar cellular details as sebaceous glands seen in Fig. 4.15a. **b** On the corresponding DHE image, these glands appear as pale purple to dark purple spherical structures (orange stars). **c** Corresponding conventional H&E-stained image shows an excellent correlation with FCM image (**a**) and DHE image (**b**) for the meibomian glands (orange stars). H&E magnifications = 10X

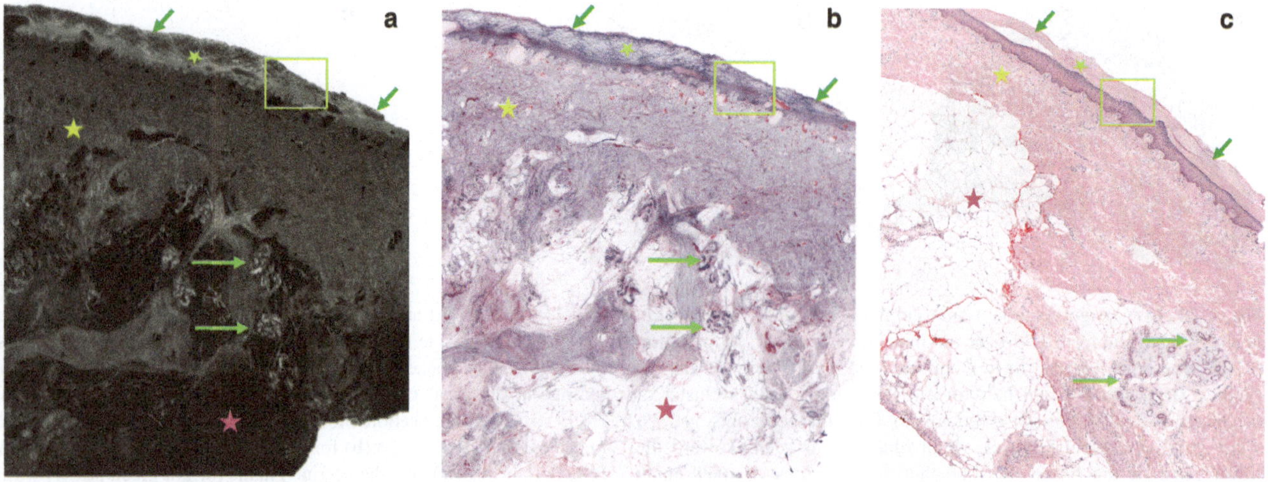

Fig. 4.22 Low magnification submosaic images of a thick skin from right foot of a 65-year-old female. **a** FCM image shows all layers of the skin: a bright layer of epidermis (green arrows), a dense grayish dermis (yellow star) with clusters of bright eccrine glands (bright green arrows), and a dark hypodermis (pink star). Stratum corneum (light green boxed area and light green star) is thickened and appears grayish. **b** On the corresponding DHE image, thickened stratum corneum (light green boxed area and light green star) appears purple (grayish on FCM) in color. The epidermis appears purple, dermis (yellow star) as pinkish purple with purple eccrine glands (bright green arrows), and fat lobules of hypodermis (pink star) whitish (dark on FCM) in color. **c** Corresponding conventional H&E-stained image shows good correlation with FCM image (**a**) and DHE image (**b**) for all the layers of thick skin including thickened stratum corneum (light green boxed area and light green star) of the epidermis, for the dermis (yellow star), for the eccrine glands (bright green arrows), and the hypodermis (pink star). H&E magnifications = 4X. *Note*: there is a lack of hair follicles and sebaceous glands in this thick skin

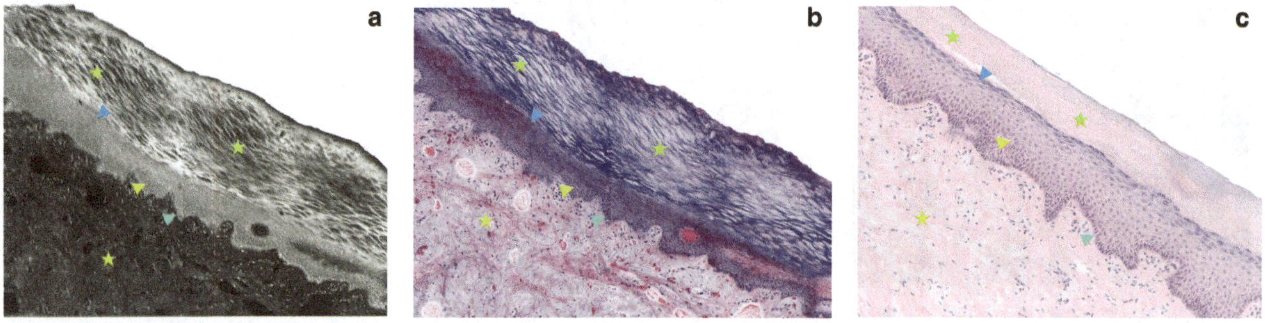

Fig. 4.23 Higher magnification submosaic images of the epidermis of thick skin obtained by digitally zooming in light green boxed area of Fig. 4.22. **a** FCM image shows a thickened stratum corneum (light green stars) with layers of grayish white keratin. Stratum granulosum (blue arrowhead) is prominent with flat bright nuclei, spinosum (yellow arrowhead) and basale (turquoise arrowhead) layers are bright with multi-layered polygonal cells and a single layer of cuboidal cells, respectively. Underlying dermis (yellow stars) has a dense grayish appearance. **b** On the corresponding DHE image, the thickened stratum corneum (light green stars) appears as layers of purple color (grayish white on FCM) keratin. Keratohyalin granules (blue arrowhead) appear dark pink in the stratum granulosum with flat purple (bright on FCM) nuclei. The *St. spinosum* (yellow arrowhead) has purple (bright on FCM) nuclei with purplish pink cytoplasm, while the basale layer (turquoise arrowhead) has dark purple nuclei with scant cytoplasm. The underlying dermis (yellow stars) appears pinkish purple with dense connective tissue. **c** Corresponding conventional H&E-stained image shows a good correlation with FCM image (**a**) and DHE image (**b**) for all the layers of epidermis with thickened stratum corneum (light green stars) and a prominent stratum granulosum (blue arrowhead), and for the dense underlying dermis (yellow stars). H&E magnifications = 20X

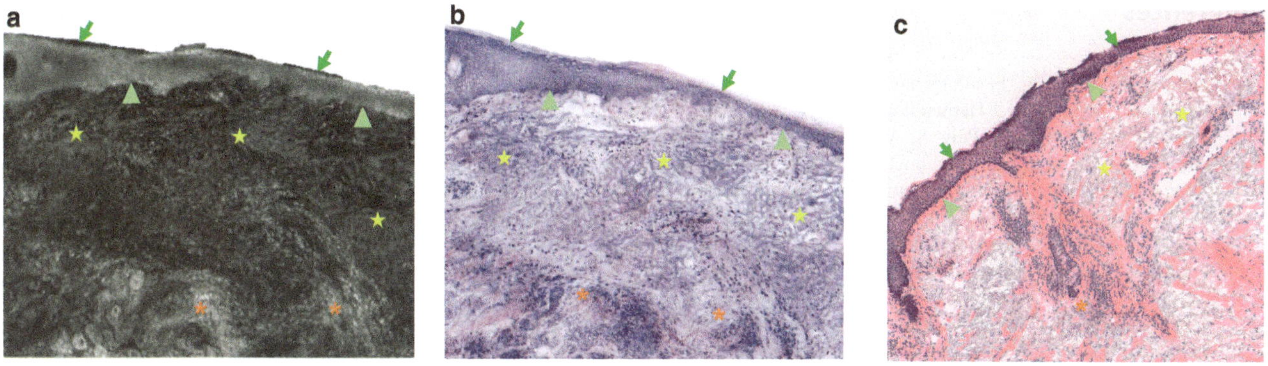

Fig. 4.24 Medium magnification submosaic images of sun-damaged skin from right cheek of an 84-year-old female. **a** FCM image shows a bright epidermis (green arrows) with a flattened or an effaced DEJ (green arrowheads). Underlying papillary dermis shows an extensive solar elastosis (yellow stars) that appears as clumps of grayish material. Foci of bright inflammatory cells (orange asterisks) can also be seen in the dermis. **b** On the corresponding DHE image, the epidermis (green arrows) appears purple (bright on FCM) in color with a flattened DEJ (green arrowheads). Solar elastosis (yellow stars) appears as clumps of purple (grayish on FCM) color material and small purple (bright on FCM) inflammatory cells (orange asterisks) are seen scattered throughout the dermis. **c** Corresponding conventional H&E-stained image shows a good correlation with FCM image (**a**) and DHE image (**b**) for the epidermis (green arrows), the flattened DEJ (green arrowheads), the extensive solar elastosis (yellow stars), and the lymphocytic inflammatory infiltrate (orange asterisks). H&E magnifications = 10X

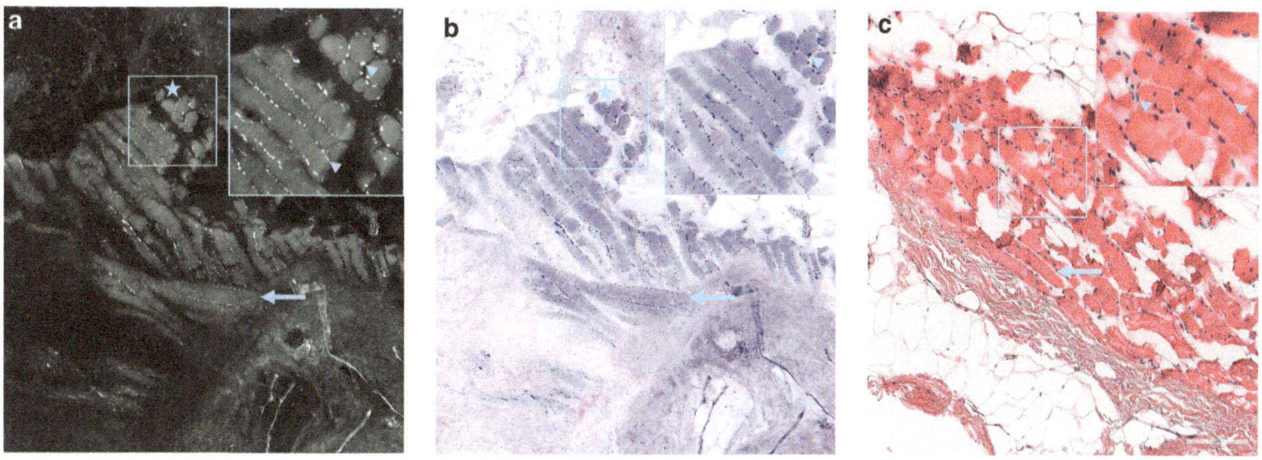

Fig. 4.25 Medium magnification submosaic images of skeletal muscles obtained from left cheek of an 83-year-old female. **a** FCM image shows grayish thick unbranched bundles of skeletal muscles in cross-section (light blue star) and longitudinal section (light blue arrow). Inset: In the cross-section, muscle bundles appear round grayish structures with bright small round peripheral nuclei (light blue arrowheads), while in the longitudinal section the muscle bundles appear as thick cylindrical grayish bands lined by elongated bright peripheral nuclei (light blue arrowheads). **b** On the corresponding DHE image, skeletal muscle bundles (light blue star and light blue arrow) appear purplish (grayish on FCM) in color with dark purple peripheral nuclei (light blue arrowheads and inset). **c** Corresponding conventional H&E-stained image shows a good correlation with FCM image (**a**) and DHE image (**b**) for the skeletal muscles bundles in longitudinal (light blue arrow) and cross-section (light blue star) with their nuclei (light blue arrowhead; inset). H&E magnifications = 10X and 20X (inset)

References

1. Kanitakis J. Anatomy, histology and immunohistochemistry of normal human skin. Eur J Dermatology. 2002;12:390–401.
2. Bennàssar A, Carrera C, Puig S, Vilalta A, Malvehy J. Fast evaluation of 69 basal cell carcinomas with ex vivo fluorescence confocal microscopy. JAMA Dermatol. 2013;149:839. https://doi.org/10.1001/jamadermatol.2013.459.
3. Mu EW, Lewin JM, Stevenson ML, Meehan SA, Carucci JA, Gareau DS. Use of digitally stained multimodal confocal mosaic images to screen for nonmelanoma skin cancer. JAMA Dermatol. 2016;152:1335–41. https://doi.org/10.1001/jamadermatol.2016.2997.
4. Khavkin J, Ellis DAF. Aging skin: histology, physiology, and pathology skin histology skin physiology skin aging skin architecture photoaging. Facial Plast Surg Clin N Am. 2011;19:229–34. https://doi.org/10.1016/j.fsc.2011.04.003.
5. Patterson JW. Weedon's Skin Pathology. 5th ed. Philadelphia: Elsevier; 2021.
6. Barbieri JS, Wanat K, Seykora J. Skin: Basic Structure and Function. Pathobiol Hum Dis Dyn Encycl Dis Mech. Elsevier Inc.; 2014, p. 1134–44. https://doi.org/10.1016/B978-0-12-386456-7.03501-2.
7. Zaidi Z, Lanigan SW. Skin: structure and function. Dermatology Clin Pract. London: Springer London; 2010, p. 1–15. https://doi.org/10.1007/978-1-84882-862-9_1.
8. Fenner J, Clark RAF. Anatomy, physiology, histology, and immunohistochemistry of human skin. Ski. Tissue Eng. Regen. Med., Elsevier Inc.; 2016, p. 1–17. https://doi.org/10.1016/B978-0-12-801654-1.00001-2.
9. Ciarletta P, Ben AM. Papillary networks in the dermal-epidermal junction of skin: A biomechanical model. Mech Res Commun. 2012;42:68–76. https://doi.org/10.1016/j.mechrescom.2011.12.001.
10. Langton AK, Halai P, Griffiths CEM, Sherratt MJ, Watson REB. The impact of intrinsic ageing on the protein composition of the dermal-epidermal junction. Mech Ageing Dev. 2016;156:14–6. https://doi.org/10.1016/j.mad.2016.03.006.
11. Rippa AL, Kalabusheva EP, Vorotelyak EA. Regeneration of dermis: scarring and cells involved. Cells. 2019;8:607. https://doi.org/10.3390/cells8060607.
12. Aasi SZ, Leffell DJ, Lazova RZ. Normal Skin. Atlas Pract. Mohs Histopathol., Springer New York; 2013, p. 3–26. https://doi.org/10.1007/978-1-4614-5161-7_2.
13. Paxton S, Peckham M, Knibbs A, Paxton S, Knibbs A, Peckham M. The Leeds Histology Guide 2003.
14. Brown TM, Krishnamurthy K. Histology. Hair and Follicle: StatPearls Publishing; 2020.
15. Requena L, Sangüeza O. Embryology, anatomy, histology, and physiology of the sebaceous glands. Cutan. Adnexal Neoplasms, Springer International Publishing; 2017, p. 755–64. https://doi.org/10.1007/978-3-319-45704-8_58.
16. Requena L, Sangüeza O. Embryology, Histology, and Physiology of the Hair Follicle. Cutan. Adnexal Neoplasms, Springer International Publishing; 2017, p. 409–39. https://doi.org/10.1007/978-3-319-45704-8_36.
17. Fernandez-Flores A. Regional variations in the histology of the skin. Am J Dermatopathol. 2015;37:737–54. https://doi.org/10.1097/DAD.0000000000000353.

Part III

Benign Skin Tumours

Features of Benign Epidermal Nonmelanocytic Lesions on Ex Vivo Confocal Microscopy and Histopathologic Correlation

5

Banu Farabi, Ucalene Harris, Daniela Hartmann, Babar K. Rao, and Manu Jain

5.1 Introduction

External stimuli such as ultraviolet radiation (UVR) or viruses can cause benign or malignant skin lesions. Solar lentigo (SL) and seborrheic keratosis (SK) are benign nonmelanocytic epidermal lesions that are commonly seen on sun-damaged skin of patients undergoing Mohs surgery; thus, it is crucial to identify them on frozen sections or freshly excised tissues and differentiate them from skin cancers such as basal cell carcinoma (BCC), squamous cell carcinoma (SCC), and melanoma.

Verruca vulgaris (VV) and molluscum contagiosum (MC) are viral-induced epidermal skin lesions, which are commonly seen in all age groups and can mimic other benign or malignant solitary lesions of the skin.

EVCM may rapidly evaluate these lesions and differentiate them from their malignant counterpart. Although EVCM features MC [1] and Herpes simplex virus [2] have been described before, features of VV, SL, and SK have never been described in the literature. In this chapter, we will first provide a brief introduction to clinical and pathologic features of SL, SK, VV, and MC and then define morphological features of these lesions on EVCM (fluorescence and digital H&E modes) and compare them with their histopathological findings.

All the EVCM images (Figs. 5.1, 5.2, 5.3, 5.4, 5.5, 5.6, 5.7, 5.8, 5.9, 5.10, 5.11 and 5.12) demonstrated in this chapter have been acquired with the new-generation Vivascope 2500 (CaliberID, Rochester, NY, USA) using fresh or frozen tissues. Standard staining procedure (described in Chap. 3) using acridine orange dye was employed. Note fluorescence images, in this chapter, are either seen in grayscale or green-scale mode.

5.2 Solar Lentigo

Solar lentigines are benign pigmented lesions, usually located on the sun-exposed sites of face, forearm, and extensor surfaces of extremities, and trunk. Their number and size increase with age and chronic sun exposure. SL is frequently seen in white elderly population as a part of skin senescence and solar damage ($\sim$95% white population over 60 years of age) [3].

Clinically, SL presents as multiple light brown pigmented, oval, and sharply demarcated lesions with irregular borders. Dermatoscopic features include moth-eaten borders, homogenous light-brown pigmentation, delicate reticulated pigment network, fingerprinting, symmetrical perifollicular pigmentation, and pseudo network on facial skin [4]. In the 'Ink spot' lentigo subtype, heavily pigmented thick reticular lines that abruptly ends at the edges of the lesion is seen [5].

Histologically, SL shows small buddings from the basal epidermis and elongation of rete ridges with hyperpigmentation of the basal layer, giving a 'dirty feet' appearance [6]. Papillary dermis often has solar elastosis. Solar lentigines can evolve to seborrheic keratosis (SK) and histologic sections commonly include both lesions.

Table 5.1 summarizes the clinical and histopathological features of a solar lentigo. Table 5.2 describes the morphological features of SL on EVCM (FCM and digital H&E modes) and compare them with their histopathological features (Figs. 5.1 and 5.2).

B. Farabi
Department of Internal Medicine, Saint Peter's University Hospital, New Brunswick, NJ, USA

U. Harris · M. Jain (✉)
Dermatology Service, Department of Medicine, Memorial Sloan Kettering Cancer Center (MSKCC), 530 E. 74th Street , New York, NY 10021, USA
e-mail: jainm@mskcc.org

D. Hartmann
Department of Dermatology and Allergy, University Hospital, LMU Munich, Munich, Germany

B. K. Rao
Department of Dermatology, Rutgers Center, Somerset, NJ, USA

B. K. Rao
Weil Cornell Medicine, New York, NY, USA

M. Jain et al. (eds.), *Cutaneous Atlas of Ex Vivo Confocal Microscopy*,
https://doi.org/10.1007/978-3-030-89316-3_5

Table 5.1 Basics of solar lentigo

Epidemiology and patient demographics: Common in white elderly population; However, individuals with sun exposure/sunburn during early childhood develops SL in earlier stages of life
Etiology: UV damage, sunburn in early childhood, chronic sun exposure
Clinical presentation: multiple light brown pigmented, oval, and sharply demarcated lesions with irregular borders, located on sun-exposed areas (face, forearms, extensor surfaces of the extremities, upper back)
Dermatoscopy: Moth-eaten borders, delicate reticular network, perifollicular uniform light brown pigmentation, homogenous light brown patch, pseudo network on facial skin
Histopathology: Increased basal keratinocyte pigmentation, mildly increased number of melanocytes in the basal layer, variable elongation of rete ridges, solar elastosis

Table 5.2 Morphological features of a solar lentigo on FCM, and DHE modes and their comparison with histopathology

Fluorescence confocal microscopy (FCM) mode (grayscale image) (Figs. 5.1a and 5.2a)	Digital H&E (combined FCM and RCM modes (Figs. 5.1b and 5.2b)	Conventional H&E-stained image (Figs. 5.1c, and 5.2c)
Pigmented basal keratinocytes: Basal keratinocyte nuclei appear brighter (hyperfluorescent), more compact, and smaller compared to the nuclei of spinous and granulosum layers	**Pigmented basal keratinocytes**: Basal keratinocytes appear dark purple, more compact, and smaller compared to keratinocytes of spinous and granulosum layers	**Pigmented basal keratinocytes**: Pigmented basal keratinocytes (brighter in FCM; dark purple in DHE)
Elongated rete ridges and bulbous projections: Elongated and interconnected thin, bright (hyperfluorescent) cords of epidermis forming bright round bulb-like projections	**Elongated rete ridges and bulbous projections**: Elongated and interconnected, thin dark purple colored cords of epidermis forming round bulb-like projections	**Elongated rete ridges and bulbous projections**: Elongation of the rete ridges forming bulbous projections on sun damaged skin with prominent solar elastosis in the superficial dermis. Pigmented bulbous projections give typical dirty feet appearance (bright on FCM; purple on DHE)
Solar elastosis: Grayish (hypofluorescent) clusters of curled fibers in the papillary dermis	**Solar elastosis**: Aggregates of curled light purple to dark purple fibers in the dermis	**Solar elastosis**: Purplish-colored clusters of irregularly thickened elastic fibers (grayish on FCM; purple on DHE) in the papillary dermis
Inflammatory cells: Scattered, bright (hyperfluorescent) small dots in the dermis	**Inflammatory cells**: Scattered dense, purple, small dots in the dermis	**Inflammatory cells**: Inflammatory cell infiltrate (bright on FCM; purple on DHE), mostly lymphocytes in the dermis

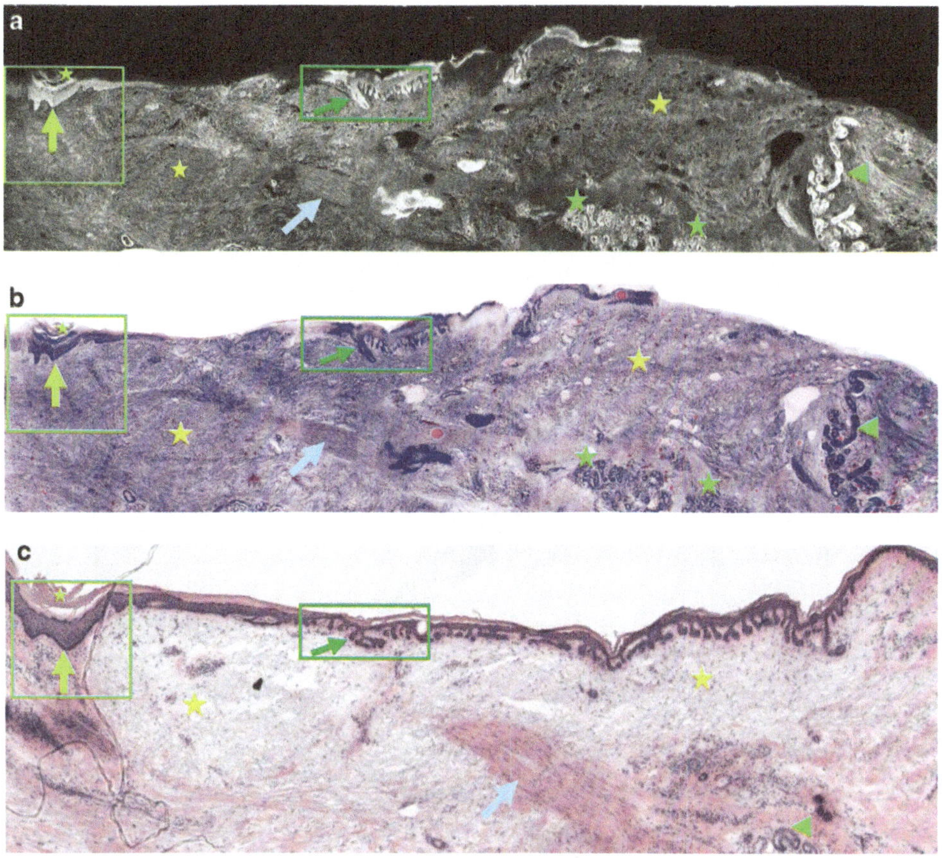

Fig. 5.1 Low magnification mosaic images of a solar lentigo (SL) and a seborrheic keratosis (SK) from face of a 70-year-old male. **a** FCM (grayscale) image shows an area of SK (light green boxed area) with bright acanthotic epidermis (light green arrow) and an overlying bright acellular scale (light green star). In the center of the image, SL with reticulated pattern (dark green box) can be seen with bright elongated rete ridges (dark green arrow). Underlying, papillary dermis shows grayish to bright clumps elastic fibers of solar elastosis (yellow stars). Eccrine glands (green stars) and eccrine ducts (green arrowhead) appear as bright structures, and arrector pili muscle (light blue arrow) appears as a grayish band with parallelly arranged bright elongated nuclei. **b** On the corresponding DHE image, SK (light green boxed area) appears with a dark purple (bright on FCM) color acanthotic epidermis (light green arrow) and a purplish pink overlying scale (light green star; bright on FCM). In the SL foci (dark green boxed area), elongated rete ridges (dark green arrow) appear dark purple (bright on FCM). Solar elastosis (yellow stars) appears as purplish clumps (grayish to bright on FCM) of elastic fibers. Eccrine glands (green stars) and ducts (green arrowhead) appears dark purple (bright on FCM), while arrector pili muscle (light blue arrow) appears as a pinkish purple band (grayish on FCM) with parallelly arranged, purple color (bright on FCM) elongated nuclei (bright on FCM). **c** Corresponding conventional H&E-stained image shows an excellent correlation with FCM image (**a**) and DHE image (**b**) for the features of SK (light green boxed area) and SL (light green boxed area) including acanthotic epidermis (light green arrow) with scale (light green star), elongated rete ridges (dark green arrow), and solar elastosis (yellow stars), and for an eccrine duct (green arrowhead), and arrector pili muscle (light blue arrow). H&E magnifications = 2x

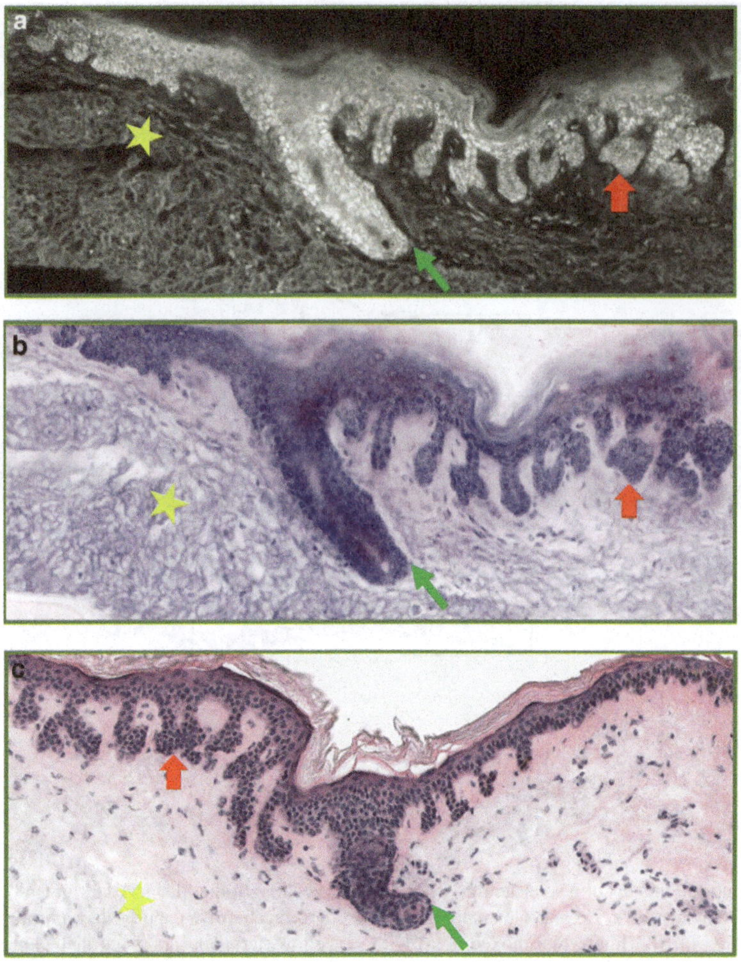

Fig 5.2 Medium magnification sub-mosaic images of a solar lentigo with reticulated pattern obtained by digitally zooming in the dark green boxed area of mosaics in Fig. 5.1. **a** FCM image shows elongated rete ridges (dark green arrow) lined by intensely bright and crowded basal cell nuclei and bulbous projections (red arrow). Solar elastosis (yellow star) appears as grayish clumps of elastic fibers in the papillary dermis. **b** On the corresponding DHE image, these elongated rete ridges (dark green arrow; bright on FCM) are seen lined by dark purple nuclei of basal keratinocytes, forming round bulbous projections (red arrow). Solar elastosis (yellow star) appears as purple (grayish to bright on FCM) crumpled material in the papillary dermis. **c** Corresponding conventional H&E-stained image shows a perfect correlation with FCM (**a**) and DHE (**b**) images for features of a reticulated solar lentigo including pigmented basal keratinocytes along the elongated rete ridges (dark green arrow), bulbous projection (red arrow), and solar elastosis (yellow star). H&E magnifications = 10x

5.3 Seborrheic Keratosis

SK are benign verrucous papules or plaque lesions that originate from keratinocytes. SK can be found anywhere in the skin except mucosal surfaces and palmoplantar areas. They are acquired and usually seen after 30 years of age. The pathogenesis is not well understood; however, the role of cumulative UV exposure, HPV infections [7], somatic activating mutations of FGFR3 and PIK3CA oncogenes [8, 9], underlying adenocarcinomas and chemotherapeutic exposure have been described as precipitating factors [10].

SK clinically present as light to dark brown oval, well-demarcated macule with a characteristic ‘stuck on’ appearance. The size of SK is variable and tends to enlarge as the lesion progresses. Follicular plugs can be seen both clinically and dermatoscopically, which are assuring features of this benign lesion [11].

There are different histologic variants of SK, including acanthotic, hyperkeratotic, reticulated (adenoid), clonal, irritated, and pigmented [12]. The most common variant is the acanthotic variant, which is characterized by acanthotic epidermis consisting of bland keratinocytes with scattered pseudo horn cysts. Pseudo horn cysts are epidermal invaginations that open to skin surface and are filled with orthokeratin. Hyperkeratotic variant is an exophytic lesion with significant epidermal hyperkeratosis and papillomatosis, mild degree of acanthosis, and has fewer pseudo horn cysts.

Table 5.3 Basics of a seborrheic keratosis

Epidemiology and patient demographics: Commonest benign skin tumor affects over 80 million Americans [22]. Usually seen after 30 years of age, number tends to increase as age advances
Etiology: Genetic mutations (FGFR3, PIK3CA), underlying malignancy, UV exposure, HPV coinfection, or chemotherapy, uncommonly as a sign of internal malignancy secondary to increased serum VEGF levels (Lesser-Trelat sign)
Clinical Presentation: Light to dark brown oval, well-demarcated macule with a 'stuck on' appearance seen anywhere except mucosal surfaces and palmoplantar skin
Dermatoscopy: Fingerprint appearance, milia-like cysts, comedo-like openings with a stuck-on appearance, and sharp demarcation
Histopathology: Acanthotic variant (most common) has dome-shaped epidermal proliferation with acanthotic epidermis, epidermal invaginations filled with keratin (pseudo horn cysts), slight hyperkeratosis, and papillomatosis. Other features include clonal proliferation of basaloid keratinocytes (clonal SK), heavy pigmentation of keratinocytes (melanoacanthoma), prominent hyperkeratosis (hyperkeratotic SK), thin anastomosing strands of basaloid cells arising from epidermis (adenoid/reticulated SK)

Table 5.4 Morphological features of a seborrheic keratosis on FCM, and DHE modes and their comparison with histopathology

Fluorescence confocal microscopy (FCM) mode (grayscale images) (Figs. 5.1a, 5.2a, 5.3a, 5.4a, 5.5a, 5.6a, 5.7a, 5.8a and 5.9a)	Digital H&E (combined FCM and RCM modes) (Figs. 5.1b, 5.2b, 5.3b, 5.4b, 5.5b, 5.6b, 5.7b, 5.8b and 5.9b)	Conventional H&E-stained image. (Figs. 5.1c, 5.2c, 5.3c, 5.4c, 5.5c, 5.6c, 5.7c, 5.8c and 5.9c)
Hyperkeratosis: Grayish-dark (non- to hypofluorescent) thickened material over bright (hyperfluorescent) epidermis	**Hyperkeratosis**: Purplish to pinkish, thickened anucleate material overlying purple epidermis	**Hyperkeratosis**: Thickened stratum corneum (grayish dark on FCM; purplish to pinkish on DHE)
Acanthosis: Thickened bright (hyperfluorescent) epidermis composed of multi-layered bright nucleated cells	**Acanthosis**: Thickened purple epidermis composed of multi-layered cells with purple nuclei and a variable amount of purplish pink cytoplasm	**Acanthosis**: Thickening of the epidermis (bright on FCM; purple on DHE)
Pseudo horn cysts: Round to oval structures with grayish loose (hypofluorescent) concentric material seen within a bright (hyperfluorescent) acanthotic epidermis	**Pseudo horn cysts**: Round to oval structures with dark pink loose concentric material seen within an acanthotic purple color epidermis	**Pseudo horn cysts**: Round to oval structures with loose concentric keratin material (grayish on FCM; dark pink on DHE) within an acanthotic epidermis (bright on FCM; purple on DHE)
Flattened (table-top) dermal–epidermal junction: Fused bright (hyperfluorescent) elongated rete ridges forming a flattened DEJ	**Flattened (table-top) dermal–epidermal junction**: Fused purple color elongated rete ridges forming a flattened DEJ	**Flattened (table-top) dermal–epidermal junction**: Fused elongated rete-ridges (bright on FCM; purple on DHE) forming a flattened DEJ

Reticulated variant is seen as interlacing thin strands of basaloid keratinocytes with pseudo horn cysts. Clonal variant consists of intraepidermal well-defined immature basaloid cells with a uniform appearance. Irritated SK has typical squamous eddies which are seen as whorls of eosinophilic keratinocytes with perivascular or diffuse lichenoid infiltrate in the dermis. And lastly, the heavily pigmented acanthotic SK is termed as 'melanoacanthoma' variant [12].

SK has typical clinical and dermatoscopic features; however, it may be difficult to diagnose SK when they are heavily pigmented, irritated, or inflamed [13, 14].

Table 5.3 summarizes the clinical and histopathological features of a seborrheic keratosis. Table 5.4 describes the morphological features of SK on EVCM (FCM and digital H&E modes) and compares them with their histopathological features (Figs. 5.3, 5.4, 5.5, 5.6, 5.7, 5.8 and 5.9).

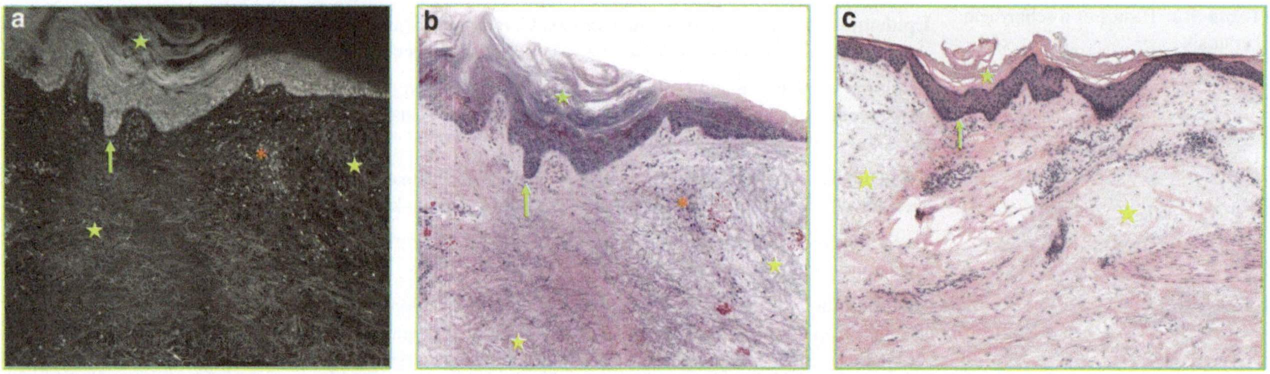

Fig 5.3 Medium magnification sub-mosaic images of a macular seborrheic keratosis (SK) obtained by digitally zooming in the light green boxed area of mosaics in Fig. 5.1. **a** FCM image shows an acanthotic epidermis composed of multi-layered bright nuclei (light green arrow) and an overlying grayish-white hyperkeratosis (light green star). Papillary dermis shows grayish clumps of curly elastic fibers of solar elastosis (yellow stars) and scattered small bright dots of inflammatory cells (orange asterisk). **b** On the corresponding DHE image, the acanthotic epidermis is composed of dark purple (bright on FCM) cells (light green arrow) and an overlying purplish (grayish white on FCM) hyperkeratosis (light green star). Clumps of solar elastosis (yellow star) appear purplish fibers (bright on FCM) and inflammatory cells (orange asterisk) appear as small purple (bright on FCM) dots. **c** Corresponding conventional H&E-stained image shows an excellent correlation with FCM (**a**) and DHE (**b**) images for the features of a macular SK including acanthosis with pigmented basal cells (light green arrow), hyperkeratosis (light green star), solar elastosis (yellow star), and inflammatory cells (orange asterisk). H&E magnifications = 10x

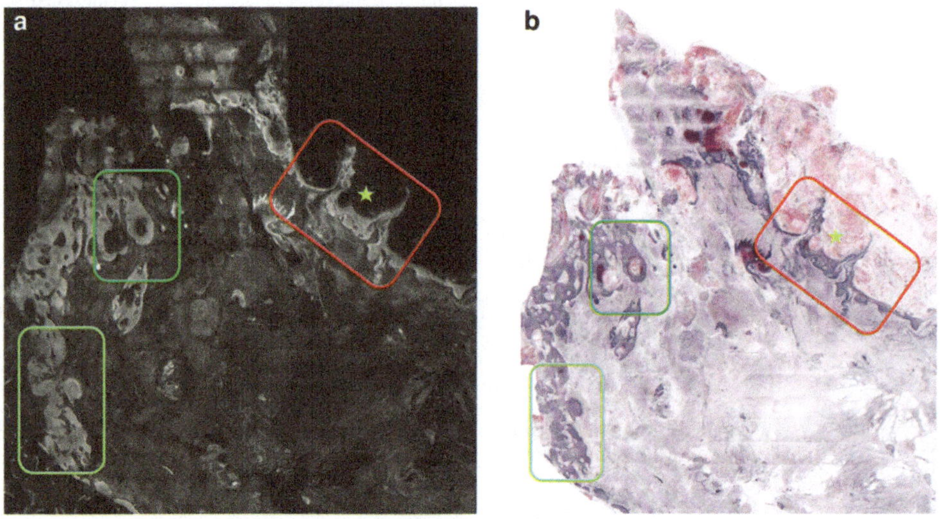

Fig 5.4 Low magnification (~2x) mosaic images of a seborrheic keratosis (SK) from left cheek of a 60-year-old male. **a** FCM (grayscale) image shows a bright acanthotic epidermis (light green boxed area) with dark, round pseudo horn cysts (dark green boxed area), and an area (red boxed) with prominent grayish-dark hyperkeratotic scale (light green star and red boxed area). **b** On the corresponding DHE image, the acanthotic epidermis (light green boxed area) appears as multi-layered dark purple color (bright on FCM) layer, round pseudo horn cyst (dark green boxed area) appears dark pink (dark on FCM), and the hyperkeratotic scale (red boxed area and light green star) appears as mounds of pinkish (grayish dark on FCM) anucleated material.

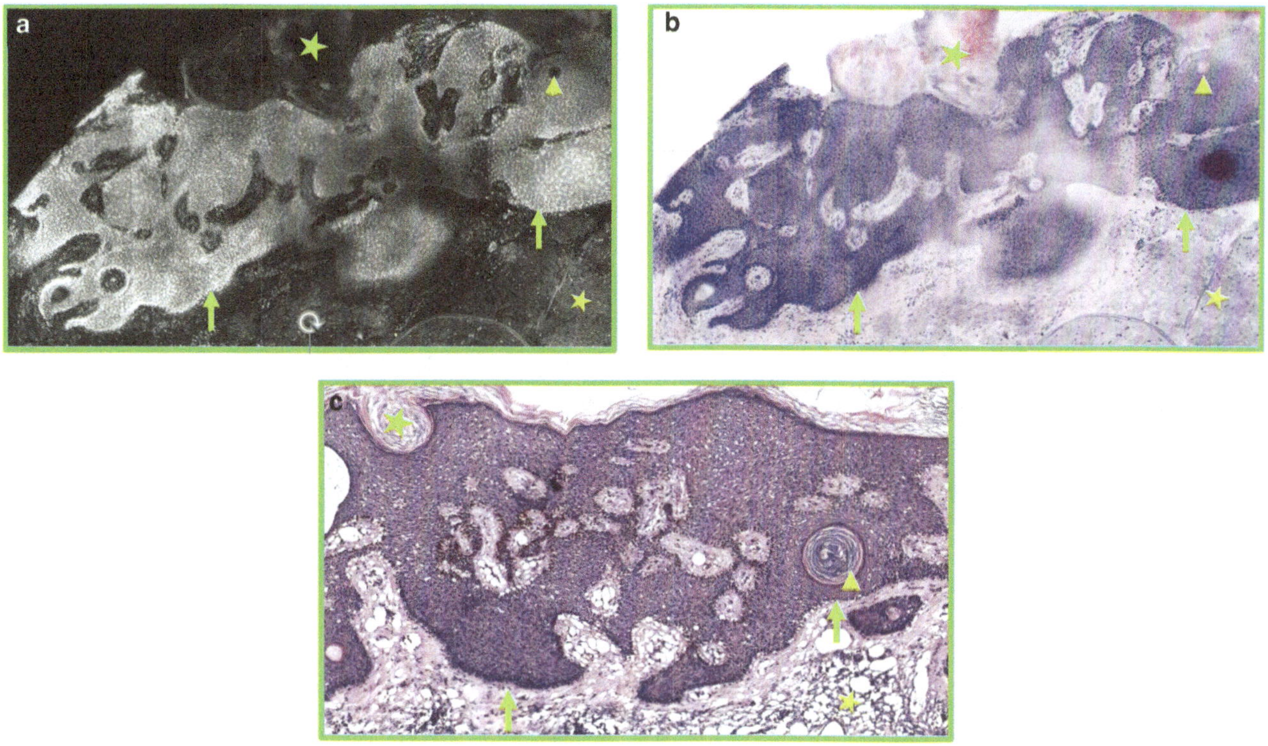

Fig 5.5 Medium magnification sub-mosaic images of an acanthotic seborrheic keratosis (SK) obtained by digitally zooming in light, green-boxed area of mosaics in Fig. 5.4. **a** FCM image shows an acanthotic epidermis composed of bright small monomorphic nuclei and a pigmented basal cell layer (light green arrows). Within the acanthotic epidermis, pseudo horn cyst (yellow arrowhead) appears as a round, dark structure. Hyperkeratosis (light green star) appears as grayish-dark acellular material on top of the bright acanthotic epidermis. Solar elastosis (yellow star) is identified as grayish crumpled material in the papillary dermis. **b** On the corresponding DHE image, the monomorphic keratinocytes of acanthotic epidermis have purple nuclei (bright on FCM) with light purple cytoplasm, and the pigmented basal cells (light green arrows) appear as dark purple layer. The pseudo horn cyst (yellow arrowhead) appears as pale pinkish (dark on FCM) round structure. Hyperkeratosis (light green star) appears as pinkish purple (grayish dark on FCM) mounds of acellular material on top of the acanthotic epidermis. Solar elastosis (yellow star) appears as purplish (grayish on FCM) clumps of elastic fibers in the papillary dermis. **c** Corresponding conventional H&E-stained image shows an exact correlation with FCM (**a**) and DHE (**b**) images for features of an acanthotic SK including an acanthotic epidermis with proliferation of small homogenous keratinocytes and pigmented basal layer (light green arrows), pseudo horn cyst (yellow arrowhead), hyperkeratosis (light green star), and solar elastosis (yellow star). H&E magnifications = 10x

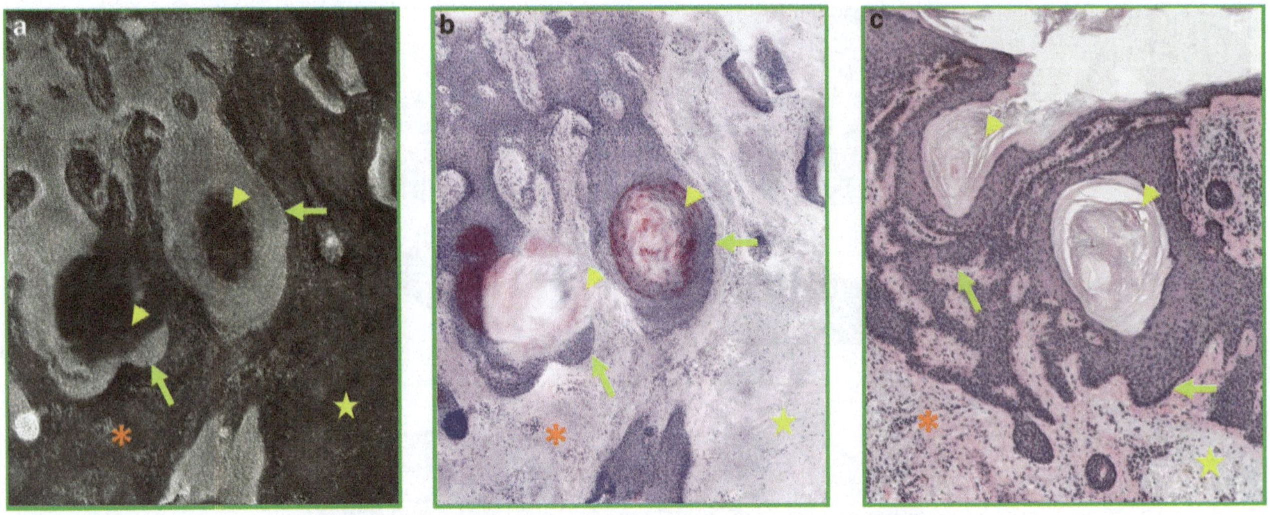

Fig 5.6 Medium magnification of sub-mosaic images of pseudo horn cysts in an acanthotic seborrheic keratosis obtained by digitally zooming in the dark green boxed area in mosaics in Fig. 5.4. **a** FCM image shows an acanthotic epidermis (light green arrows) composed of bright small monomorphic nuclei, within which dark round pseudo horn cysts with grayish keratin (yellow arrowheads) is seen. Grayish clumped material of solar elastosis (yellow star) and small bright dots of inflammatory cells (orange asterisk) are seen in the papillary dermis. **b** On the corresponding DHE image, pseudo horn cysts appear as round structures with dark pink (dark on FCM) central keratin (yellow arrowheads) within a purple (bright on FCM) acanthotic epidermis (light green arrows). Solar elastosis (yellow star) appears as purplish (grayish on FCM) clumped material and inflammatory cells (orange asterisk) appears as small purple (bright on FCM) dots in the dermis. **c** Corresponding conventional H&E-stained image shows a perfect correlation with FCM (**a**) and DHE (**b**) images for the pseudo horn cysts (yellow arrowheads) within acanthotic epidermis (light green arrows), solar elastosis (yellow star), and inflammation (orange asterisk) in the dermis. H&E magnifications = 10x

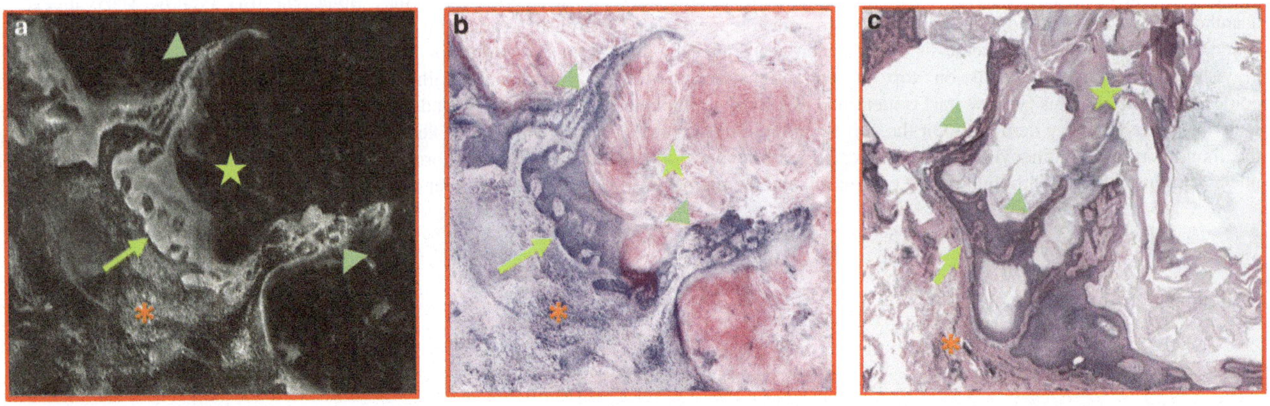

Fig 5.7 Medium magnification sub-mosaic images of an acanthotic and hyperkeratotic seborrheic keratosis (SK) obtained by digitally zooming in red boxed area of the mosaics in Fig. 5.4. **a** FCM image shows a cup-shaped invagination filled with an amorphous dark grayish material (light green star) and lined by a bright acanthotic epidermis (light green arrow), which is composed of small uniform bright nuclei and forms outward projections (spikes; light green arrowheads). A bright intense inflammatory infiltrate (orange asterisk) is seen in the papillary dermis. **b** On the corresponding DHE image, this cup-shaped invagination is filled with pinkish (dark grayish on FCM) keratin material (light green star) within a purple color (bright on FCM) acanthotic epidermis (light green arrow) forming spikes (light green arrowheads). Inflammatory infiltrate (orange asterisk) appears as clusters of small purple (bright on FCM) dots in the dermis. **c** Corresponding conventional H&E-stained image shows an excellent correlation with FCM (**a**) and DHE (**b**) images for features of a hyperkeratotic and acanthotic SK including acanthotic epidermis (light green arrow) with epidermal spicules (light green arrowheads), marked hyperkeratosis (light green star), and intense inflammation (orange asterisk) in the papillary dermis. H&E magnifications = 10x

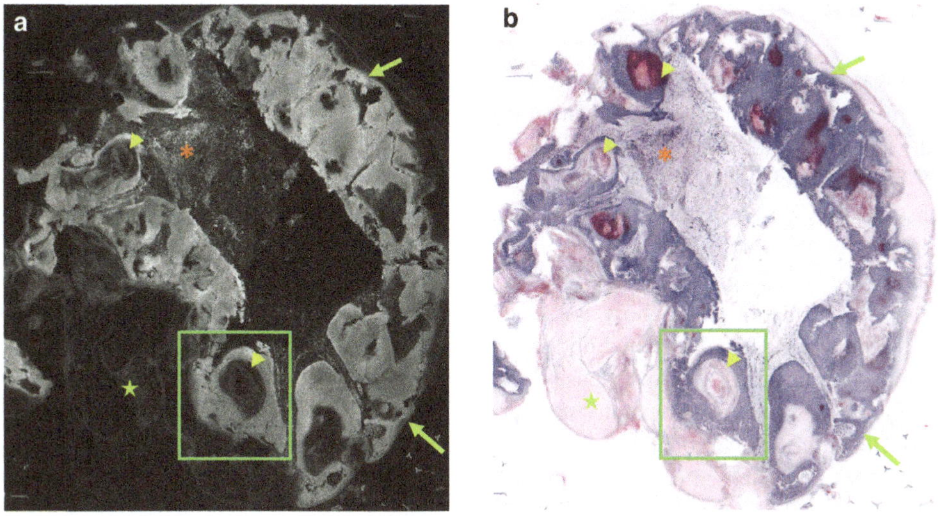

Fig 5.8 Low magnification (~2x) mosaic images of another acanthotic and hyperkeratotic seborrheic keratosis (SK) with a polypoid shape from right earlobe of a 70-year-old female. **a** FCM (grayscale) image shows a bright acanthotic epidermis (light green arrows), within which multiple dark round to oval pseudo horn cysts (yellow arrowheads and light green boxed area) with central grayish keratin is seen. Hyperkeratosis (light green star) appears as dark grayish loose material overlying the bright acanthotic epidermis (light green arrows). There is a dense bright inflammatory infiltrate (orange asterisk) in the dermis. **b** On the corresponding DHE image, round to oval pseudo horn cysts (yellow arrowheads and light green boxed area) have a pink (grayish on FCM) whorled keratin and are seen within purple color (bright on FCM) acanthotic epidermis (light green arrows). Overlying marked hyperkeratosis (light green star) appears pale pink (dark grayish on FCM) in color. Dermal inflammatory cells are evident as dark purple (bright on FCM) small cells (orange asterisk)

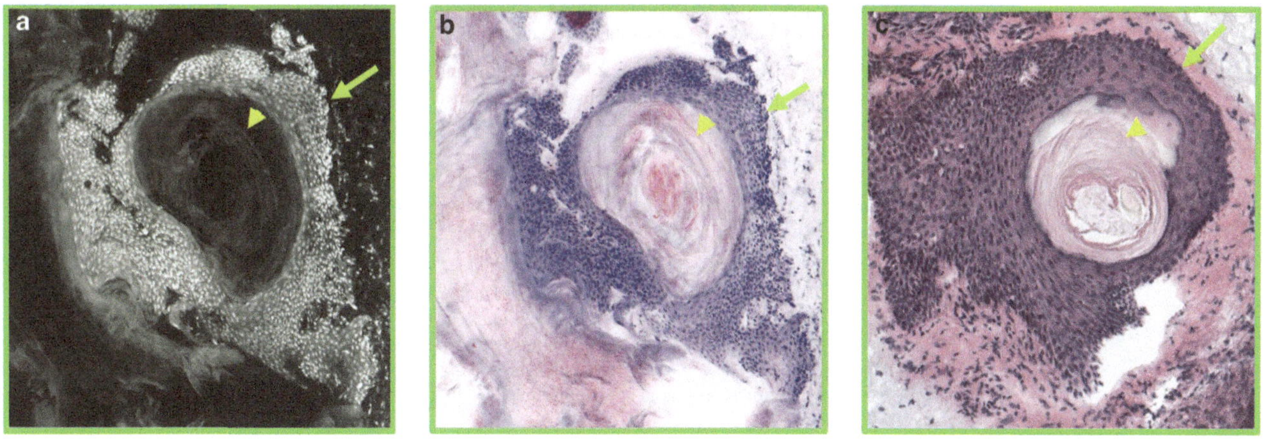

Fig. 5.9 High magnification mosaic images of a pseudo horn cyst in the acanthotic seborrheic keratosis (SK) obtained by digitally zooming in light green boxed area of the mosaics in Fig. 5.8. **a** FCM image of an oval dark pseudo horn cyst (yellow arrowhead) filled with grayish keratin and surrounded by a bright acanthotic epidermis (light green arrow). **b** On the corresponding DHE image, this pseudo horn cyst (yellow arrowhead) has pinkish (grayish on FCM) whorled keratin in the center surrounded by a purple color acanthotic epidermis (light green arrow). **c** Corresponding conventional H&E-stained image shows a great correlation with FCM (**a**) and DHE (**b**) images for this pseudo horn cyst with whorled keratin (yellow arrowhead) within an acanthotic epidermis (light green arrow). H&E magnifications = 20x

5.4 Verruca Vulgaris

Verruca vulgaris (VV) or common wart is a benign lesion human papilloma viruses (HPV-6 and 11). It is common in children, especially on palms and soles disrupting the dermatoglyphics. VV can appear clinically as vegetative, flat, or endophytic lesions and may be challenging to differentiate from malignant lesions of SCCs, Bowen's disease, Bowenoid papulosis, cutaneous horn, etc.

Dermoscopy can be very helpful to diagnose VV. Facial plane warts typically show regularly distributed dotted vessels on a brown-pinkish background [15]. As lesions become hyperkeratotic, dermoscopy shows papillomatous structures including looped or dotted vessels surrounded by a whitish halo. In a traumatized VV, thrombosed capillaries are seen, also called 'Auspitz sign' [16].

On histopathology, VV epidermis shows marked papillomatosis, hyperkeratosis, tiers of parakeratosis (over the tips of papillomatosis), and hypergranulosis. The elongated rete ridges are bend inward toward the center of the lesion, so-called 'toeing-in'. There are large pink inclusion bodies within enlarged epidermal cells with a clear cytoplasm and hyperchromatic nuclei also known as 'koilocytes'. Papillary dermis has dilated vessels [17]. Histologic differential diagnoses of VV include SK, epidermal nevus, verrucous carcinoma, acanthosis nigricans, and condyloma lata [18].

Table 5.5 summarizes the basic clinical–histopathological features of VV. Table 5.6 summarizes the morphological features of VV on FCM, and DHE modes and their comparison with histopathology (Figs. 5.10 and 5.11).

Table 5.5 Basics of a verruca vulgaris

Epidemiology and patient demographics: 7–12% of population [22]. Incubation period is ~ 2–6 months [23]. Common in all age groups, especially in children and young adults. More common in handlers of meat, poultry, and fish [24]
Etiology: Caused by HPV virus, transmitted through direct skin contact or sexually. Atopic dermatitis, immunosuppression (AIDS, chemotherapy, etc.) cause predisposition
Clinical presentation: Vegetative, flat, or endophytic lesions on facial and acral skin in children and on genitalia in adults
Dermoscopy: Plane wart shows regularly distributed dotted vessels on a brown to flesh-colored background; hyperkeratotic papillomatous lesions show linear or coiled vessels surrounded by whitish halo. Auspitz sign is visible due to thrombosed capillaries when stratum corneum is removed
Histopathology: Hyperkeratosis, acanthosis, papillomatosis with spiral of parakeratosis, Koilocytes in granular layer with a basophilic nucleus and perinuclear halo. Intracytoplasmic viral inclusion bodies and papillary dilated vessels are present

Table 5.6 Morphological features of a verruca vulgaris on FCM, and DHE modes and their comparison with histopathology

Fluorescence confocal microscopy (FCM) mode (green scale images) (Figs. 5.10a and 5.11a)	Digital H&E (combined FCM and RCM modes) (Figs. 5.10b and 5.11b)	Conventional H&E-stained image (Figs. 5.10c and 5.11c)
Acanthosis: Thickened epidermis with grayish-dark (nonfluorescent to hypofluorescent) superficial keratinocytes and bright (hyperfluorescent) basal layer	**Acanthosis**: Thickened epidermis with (nonfluorescent to hypofluorescent) superficial keratinocytes having abundant pink cytoplasm and light purple nuclei	**Acanthosis**: Thickened epidermis (bright on FCM; purple on DHE)
Koilocytes: Round cells with dark (nonfluorescent) peripheral halo and central bright (hyperfluorescent) round to oval nuclei, seen within the upper layers of epidermis	**Koilocytes**: Round cells with pale pink vacuole surrounding central oval to round purple nuclei within the upper layers of epidermis	**Koilocytes**: Vacuolated (dark on FCM; pale pink on DHE) keratinocytes with pyknotic raisin-like nuclei (bright on FCM; purple on DHE) in the superficial epidermis
Toeing-in of rete ridges: Inward bend of bright (hyperfluorescent) elongated rete ridges toward the center of the lesion	**Toeing-in of rete ridges**: Elongated rete ridges lined by purple basal keratinocytes bends inwards	**Toeing-in of rete ridges**: Elongated rete ridges (bright on FCM; purple on DHE) bending inwardly to the center of the lesion
Tiers of parakeratosis: Stacked-on layers of bright (hyperfluorescent) flattened nuclei within dull green to dark (nonfluorescent to hypofluorescent) thickened stratum corneum	**Tiers of parakeratosis**: Stacked-on layers of purple, flattened nuclei within pinkish thickened stratum corneum	**Tiers of parakeratosis**: Stacked-on layers of nuclei (bright on FCM; purple on DHE) within thickened stratum corneum (grayish-dark on FCM; pinkish white on DHE) over projecting dermal papillae

(continued)

Table 5.6 (continued)

Fluorescence confocal microscopy (FCM) mode (green scale images) (Figs. 5.10a and 5.11a)	Digital H&E (combined FCM and RCM modes) (Figs. 5.10b and 5.11b)	Conventional H&E-stained image (Figs. 5.10c and 5.11c)
Papillomatosis: Outward projections of bright to dark (hyperfluorescent to nonfluorescent) epidermis with elongated dark (nonfluorescent) dermal papillae. Dermal papillae appear as darkish (nonfluorescent to hypofluorescent) and have bright (hyperfluorescent) inflammatory cells and fibroblast and dark (nonfluorescent) small round to oval dilated capillaries	**Papillomatosis**: Outward projections of pinkish to purple epidermis with elongated pinkish purple dermal papillae. Dermal papillae appear as bluish pink and have purple inflammatory cells and fibroblast and small whitish round to oval dilated capillaries	**Papillomatosis**: Epidermal outward projections (bright to dark on FCM; pinkish to purple on DHE) with elongated dermal papillae. Dermal papillae (grayish dark on FCM; bluish pink on DHE) inflammatory cells, fibroblast, and round to oval (dark on FCM; whitish on DHE) dilated capillaries

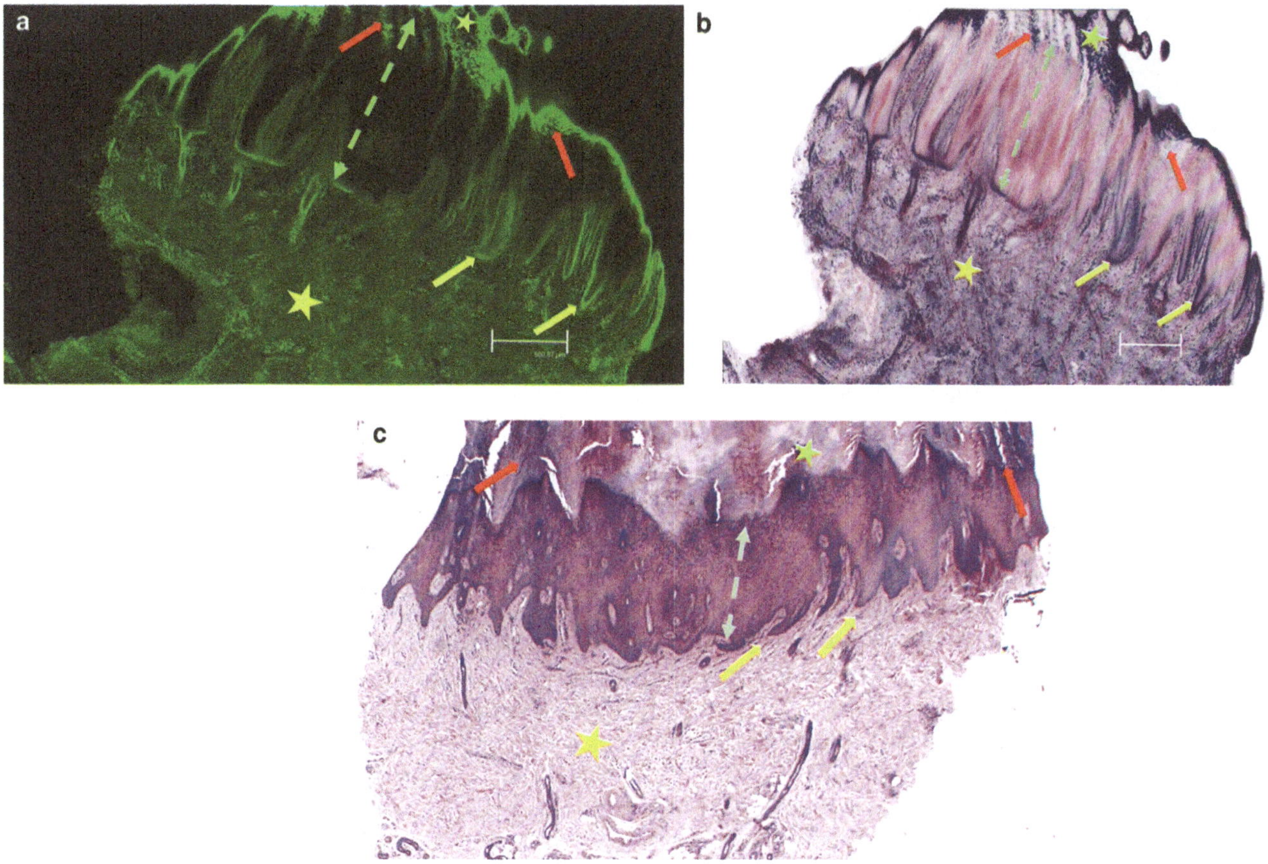

Fig 5.10 Low magnification mosaic images of a verruca vulgaris (VV) from right finger of a 12-year-old female. **a** FCM (green scale) image shows a bright to centrally darkish thickened epidermis with marked papillomatosis (green double-headed dashed arrow). Bright elongated rete ridges are turned in (yellow arrows) toward the center of the lesion ("toeing-in"). There is a marked, darkish hyperkeratosis (light green star) with tiers of bright compressed nuclei of parakeratosis (red arrows). Underlying dermis (yellow star) has greenish collagen bundles. **b** On the corresponding DHE image, the acanthotic epidermis with papillomatosis (green double-headed dashed arrow) appears pinkish purple (bright to darkish on FCM). 'Toeing-in' of elongated rete ridges (yellow arrows) appears purple in color (bright on FCM). Hyperkeratosis (light green star) appears pinkish (darkish green on FCM) acellular material with stacks of purple (bright green on FCM) nuclei of parakeratosis (red arrows) giving a 'church spire' appearance. Dermis (yellow star) appears pinkish purple (greenish on FCM). **c** Corresponding conventional H&E-stained image shows an excellent correlation with the corresponding FCM (**a**) and DHE (**b**) images for classic features of VV including an acanthotic epidermis with papillomatosis (green double-headed dashed arrow), marked hyperkeratosis (light green star), stacked parakeratosis (red arrows), and 'toeing-in' of rete ridges (yellow arrows). H&E magnifications = 2x

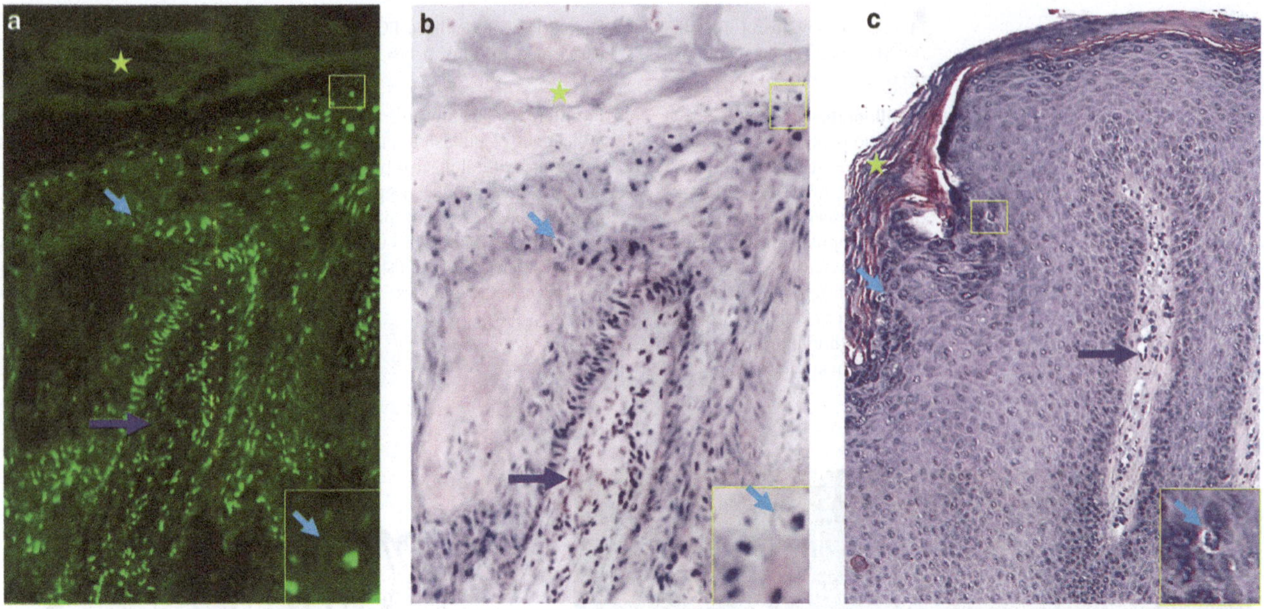

Fig 5.11 High magnification sub mosaic images of koilocytes in a verruca vulgaris from perianal area of a 50-year-old male. **a** FCM (green scale) image shows koilocytes (blue arrows, yellow boxed area, and inset) as round cell with dark peripheral halo and central bright round to oval nuclei within the upper layers of epidermis. Hyperkeratosis (light green star) appears as dull green in color. In the papillary dermis, prominent thin dark linear vessels are seen (purple arrow). **b** On the corresponding DHE image, koilocytes (blue arrows, yellow boxed area, and inset) show a pale pink (dark on FCM) vacuole surrounding central oval to round purple (bright green on FCM) nuclei. Hyperkeratosis (light green star) appears pale purple (dull green on FCM) in color. Blood vessel (purple arrow) appears prominent in the papillary dermis. **c** Corresponding conventional H&E-stained image shows an excellent correlation with the corresponding FCM (**a**) and DHE (**b**) images for koilocytes (blue arrows, yellow boxed area, and inset), hyperkeratosis (light green star), and prominent vessels (purple arrow) in the papillary dermis. H&E magnifications = 20x

5.5 Molluscum Contagiosum

Molluscum contagiosum (MC) is a chronic, localized, infectious, viral-induced disease caused by poxvirus. It is usually seen in childhood but can also be seen in adults. The main transmission route is direct skin-to-skin contact but may be transmitted sexually as.

Clinically, MC lesions appear as flesh-colored-pink, dome-shaped lesions on the skin. These lesions may show Koebnerization on the affected area. In immunocompromised individuals (such as HIV patients) atypical giant lesions can be seen. Clinical differential diagnoses include skin lesions due to fungal agents (cryptococcosis, histoplasmosis, penicillosis), VV, condyloma acuminata, condyloma lata, pyogenic granuloma, adnexal tumors, adnexal tumors, Langerhans cell histiocytosis, BCC, and amelanotic melanoma [19].

Although MC diagnosis can be made clinically, specific dermatoscopic features have been described, including central pore associated with poly-lobular white-yellow amorphous structures surrounded by linear fine vessels [20].

On histopathology, MC shows a cup-shaped lesion with inverted lobules of hyperplastic epithelium which expands to underlying epidermis. There are intracytoplasmic eosinophilic inclusion bodies (measuring up to 35 microns) within the cells, which are called 'Henderson-Patterson bodies'. The diameter of these bodies gradually increases through the upper layers of the epidermis and in the granular layer [21].

Table 5.7 summarizes the basic clinical–histopathological features of MC. Table 5.8 summarizes the morphological features of MC on FCM, digital H&E modes, and their corresponding histopathology features (Fig. 5.12).

Table 5.7 Basics of a molluscum contagiosum

Epidemiology and patient demographics: All ages, including children, with lesions on acral skin and extremities. Recent meta-analysis suggests a prevalence of 5.1% and 11.5% in children aged 0–16 years [25]. MC is common in HIV infected individuals and the presence of MC may signify advanced immunosuppression [26]
Etiology: Caused by Poxvirus and transmitted via direct skin contact and sexually. Four different genotypes have been defined; genotype 1 (MCV-1) represents the causative agent in 90% of the cases in the United States [27]
Clinical presentation: Lesions are flesh-colored to pink papules/dome-shaped on acral skin and extremities. Koebnerization may occur. Giant, atypical variants seen in immunocompromised individuals
Dermoscopy: Central pore surrounded by yellow-whitish amorphous material and fine linear vessels
Histopathology: A cup-shaped invagination surrounded by an acanthotic epidermis with a central crater. Intranuclear eosinophilic viral inclusion bodies (Henderson-Patterson bodies) are evident, which may occupy the entire granular layer

Table 5.8 Morphological features of molluscum contagiosum on FCM, digital H&E modes and their comparison with histopathology

Fluorescent confocal microscopy (FCM) mode (green-scale images) (Fig. 5.12a)	Digital H&E (combined FCM and RCM modes) (Fig. 5.12b)	Conventional H&E-stained image. (Fig. 5.12c)
Epidermal hyperplasia: Lobulated hyperplastic bright (hyperfluorescent) epidermis	**Epidermis hyperplasia** Lobulated hyperplastic epidermis with purple uniform nuclei and pinkish cytoplasm	**Epidermis hyperplasia** Lobulated hyperplastic epidermis (bright on FCM; purple on DHE)
Molluscum bodies: Large, round-oval, scattered or clustered, bright (hyper-fluorescent) bodies within small bright (hyperfluorescent) epidermal nuclei	**Molluscum bodies**: Large, round to oval, scattered or clustered dark pink to dark purple bodies within the purple uniform nuclei of epidermis	**Molluscum bodies**: Large intracytoplasmic eosinophilic inclusion bodies (bright on FCM; dark pink to dark purple on DHE)

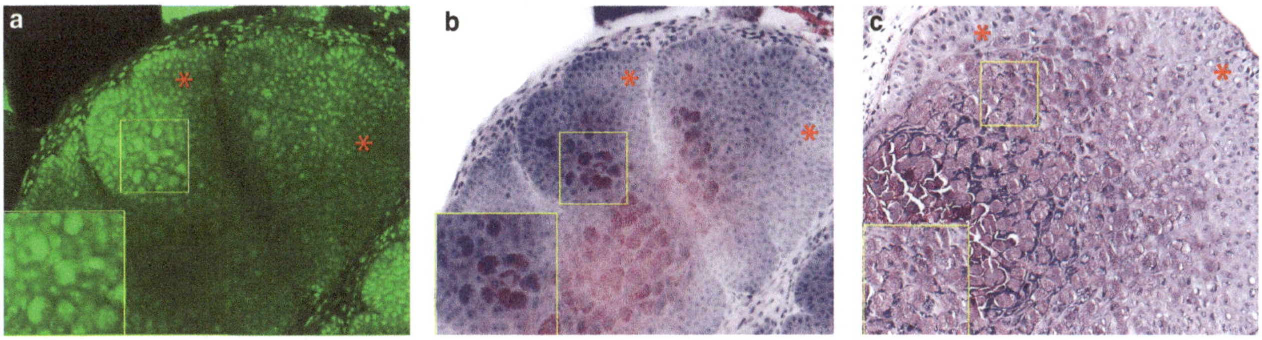

Fig 5.12 High magnification mosaic images of a molluscum contagiosum (MC) from trunk of a 4-year-old female. **a** FCM (green scale) image shows a well-demarcated, lobulated epidermis with multiple, large round bright globules of molluscum bodies (yellow boxed area and inset) surrounded by small bright keratinocyte nuclei (red asterisks). **b** On the corresponding DHE image, molluscum bodies (yellow boxed area and inset) appear as (bright green on FCM) dark pink round uniform globules with granulated appearance (bright green on FCM, left bottom inset) surrounded by (red asterisks) the epidermis. **c** Corresponding conventional H&E-stained image shows an excellent correlation with FCM (**a**) and DHE (**b**) images for the molluscum bodies (yellow boxed area and inset) surrounded by vacuolated keratinocytes within lobulated epidermal structures (red asterisks). H&E magnifications = 20x

References

1. Cinotti E, et al. Role of in vivo and ex vivo confocal microscopy and of optical coherence tomography as aids in the diagnosis of molluscum contagiosum. Ann Dermatol Venereol. 2016;143(8–9):564–6.
2. Cinotti E, et al. First identification of the herpes simplex virus by skin-dedicated ex vivo fluorescence confocal microscopy during herpetic skin infections. Clin Exp Dermatol. 2015;40(4):421–5.
3. Bastiaens M, et al. Solar Lentigines are Strongly Related to Sun Exposure in Contrast to Ephelides. Pigment Cell Res. 2004;17 (3):225–9.
4. dermoscopedia contributors–Ralph Braun, S.N. Solar lentigines. 2019 3 June 2019 06:10 UTC 14 May 2021 10:07 UTC]. https://dermoscopedia.org/w/index.php?title=Solar_lentigines&oldid=16362.
5. Argenziano G. Dermoscopy of melanocytic hyperplasias: subpatterns of lentigines (ink spot). Arch Dermatol. 2004;140(6):776.
6. Epidermis, in Atlas of Dermatopathology. 2018. p. 1–83.
7. Li YH, et al. Detection of epidermodysplasia verruciformis-associated human papillomavirus DNA in nongenital seborrhoeic keratosis. Br J Dermatol. 2004;151(5):1060–5.
8. Hafner C, et al. Spectrum of FGFR3 mutations in multiple intraindividual seborrheic keratoses. J Invest Dermatol. 2007;127 (8):1883–5.
9. Hafner C, et al. Oncogenic PIK3CA mutations occur in epidermal nevi and seborrheic keratoses with a characteristic mutation pattern. Proc Natl Acad Sci U S A. 2007;104(33):13450–4.
10. Hafner C, Vogt T. Seborrheic keratosis. J Dtsch Dermatol Ges. 2008;6(8):664–77.
11. Minagawa A. Dermoscopy–pathology relationship in seborrheic keratosis. J Dermatol. 2017;44(5):518–24.
12. Roh NK, et al. Clinical and Histopathological Investigation of Seborrheic Keratosis. Ann Dermatol. 2016;28(2):152–8.
13. Moscarella E, et al. Differential Diagnosis and Management on Seborrheic Keratosis in Elderly Patients. Clin Cosmet Investig Dermatol. 2021;14:395–406.
14. Longo C, et al. Not all lesions with a verrucous surface are seborrheic keratoses. J Am Acad Dermatol. 2014;70(6):e121–3.
15. Zalaudek I, et al. Entodermoscopy: a new tool for diagnosing skin infections and infestations. Dermatology. 2008;216(1):14–23.
16. Kaliyadan F. The Dermoscopic Auspitz Sign. Indian Dermatol Online J. 2018;9(4):290–1.
17. Bacaj P, Burch D. Human Papillomavirus Infection of the Skin. Arch Pathol Lab Med. 2018;142(6):700–5.
18. Heba Ahmed Abdelkader AE. Pathology of Verrucae. 2021 18 Feb 2021 [cited 2021 4th June]. https://www.pathologyoutlines.com/topic/skinnontumorwarts.html.
19. Isaacs SN. Molluscum Contagiosum. 2021 [cited 2021 4th June]. https://www.uptodate.com/contents/molluscum-contagiosum#H565299209.
20. Morales A, et al. Dermoscopy of molluscum contagiosum. Arch Dermatol. 2005;141(12):1644.
21. Badri T, Gandhi GR. Molluscum contagiosum. In StatPearls. 2021, StatPearls Publishing. Copyright © 2021, StatPearls Publishing LLC., Treasure Island (FL)
22. Braun RP, Ludwig S, Marghoob AA. Differential Diagnosis of Seborrheic Keratosis: Clinical and Dermoscopic Features. J Drugs Dermatol. 2017;16(9):835–42.
23. Bruggink SC, et al. Natural course of cutaneous warts among primary schoolchildren: a prospective cohort study. Ann Fam Med. 2013;11(5):437–41.
24. Karagounis TK, Pomeranz MK. Viral Venereal Diseases of the Skin. Am J Clin Dermatol. 2021;22(4):523–40.
25. Kilkenny M, Marks R. The descriptive epidemiology of warts in the community. Australas J Dermatol. 1996;37(2):80–6.
26. Olsen JR, et al. Epidemiology of molluscum contagiosum in children: a systematic review. Fam Pract. 2014;31(2):130–6.
27. Gur I. The epidemiology of Molluscum contagiosum in HIV-seropositive patients: a unique entity or insignificant finding? Int J STD AIDS. 2008;19(8):503–6.

Features of Benign Nonmelanocytic Dermal and Subcutaneous Lesions on Ex Vivo Confocal Microscopy and Histopathologic Correlation

6

Banu Farabi, Ucalene Harris, Daniela Hartmann, Babar K. Rao, and Manu Jain

6.1 Introduction

Benign nonmelanocytic dermal and subcutaneous lesions can have vascular (hemangioma, angiokeratoma, pyogenic granuloma), neural (neuroma, neurofibroma, schwannoma), fibrous (dermatofibroma, acrochordon), adnexal (sebaceous hyperplasia), or mesenchymal (lipoma, angiolipoma) origin. Although benign in nature, these lesions often clinically mimic malignant lesions such as BCC or other adnexal tumors, requiring biopsy for histopathology confirmation. Histopathology is the gold standard but requires tissue processing that delays diagnosis and management of the lesions.

Reflectance confocal microscopy (RCM) can readily image some of these lesions that are located in the superficial dermis including hemangioma, sebaceous hyperplasia, and dermatofibroma [1]; however, its role is limited for visualization of the deeper lesions such as lipoma and neurofibroma due to limited depth of imaging (~200 microns). Ex vivo confocal microscopy (EVCM) allows rapid evaluation of freshly excised tissues from various organ systems (e.g., esophagus, trachea, lung, breast tissues, and skin) [2] and is being explored as an alternative to a frozen section or during grossing surgical specimens [2]. EVCM could be useful to diagnose these deeper benign lesions such as lipoma and neurofibroma at the bedside, aiding in immediate management and alleviating patient's anxiety.

To the best of our knowledge, EVCM features of benign dermal nonmelanocytic lesions (NMLs) such as DF, EIC, lipoma, FEP, and NF have never been reported in the literature. In this chapter, we will give a brief introduction to clinical and pathologic features of dermatofibroma, lipoma, epidermal inclusion cyst (EIC), FEP, and NF, define their EVCM features, and correlate them with their histopathological findings. Knowing the ECVM features of these benign lesions is crucial not only for rapid bedside diagnosis to differentiate them from malignant lesions but also to identify them in Mohs surgical margins, where some of them (epidermal inclusion cyst) can be present as an incidentaly.

All EVCM images demonstrated in this chapter have been acquired with the new generation Vivascope 2500 (CaliberID, Rochester, NY, USA) using fresh or frozen tissues (Figs. 6.1, 6.2, 6.3, 6.4, 6.5, 6.6 and 6.7). Standard staining procedure was followed as described in Chapter 3 acridine orange dye was employed. Note fluorescence images in this chapter are either seen in grayscale or green scale.

B. Farabi
Department of Internal Medicine, Saint Peter's University Hospital, New Brunswick, NJ, USA

U. Harris · M. Jain (✉)
Dermatology Service, Department of Medicine, Memorial Sloan Kettering Cancer Center (MSKCC), 530 E. 74th Street , New York, NY 10021, USA
e-mail: jainm@mskcc.org

D. Hartmann
Department of Dermatology and Allergy, University Hospital, LMU Munich, Munich, Germany

B. K. Rao
Department of Dermatology, Rutgers Center, Somerset, NJ, USA

B. K. Rao
Weil Cornell Medicine, New York, NY, USA

M. Jain et al. (eds.), *Cutaneous Atlas of Ex Vivo Confocal Microscopy*,
https://doi.org/10.1007/978-3-030-89316-3_6

Table 6.1 Basics of a dermatofibroma

Epidemiology and patient demographics: DF occurs commonly between 3rd and 5th decades of life with a predominance in females (F:M; 2:1), and consists of 3% of all dermatopathology specimens [4]
Etiology: Controversial, inciting local trauma in 1/5th of the cases, spontaneous development, reactive fibrohistiocytic proliferation [4]
Clinical presentation: Usually located on the lower extremities as papule or nodule with a 'dimple sign' when laterally compressed
Dermatoscopy Features: Typical features are central white scar-like area and peripheral brown thin reticular lines and brown circles; atypical features are white shiny lines, crystalline structures, polymorphic vessels, scales and pigmentation [5]
Histopathology Features: Spindle-shaped cell proliferation in the dermis, peripheral collagen entrapment, epidermal acanthosis, elongated rete ridges and fusion of rete ridges gives a blunt appearance to the epidermis, increased pigmentation in basal keratinocytes

Table 6.2 Morphological features of dermatofibroma on FCM, DHE modes, and their comparison with histopathology

Fluorescence confocal microscopy (FCM) mode (grayscale images) (Fig. 6.1a)	Digital H&E (combined FCM and RCM modes) images (Fig.6.1b)	Conventional H&E-stained images(Fig. 6.1c)
Acanthosis: Multiple layers of bright (hyperfluorescent) epidermal cells nuclei	**Acanthosis**: Thickened intense purple color epidermal layer	**Acanthosis**: Epidermal hyperplasia (bright nucleated keratinocytes on FCM; dark purple on DHE)
Elongated rete ridges with basal cell hyperpigmentation: Elongated structures lined by bright (hyperfluorescent) basal cells. At an oblique angle, these structures appear as rings lined by bright (hyperfluorescent) basal cells with central grayish-dark (hypofluorescent to non-fluorescent) dermal papillae	**Elongated rete ridges with basal cell hyperpigmentation**: Elongated structures lined by dark purple basal cells. Rings with purplish-pink central core and rim of dark purple nuclei can be identified in an image acquired at an oblique angle	**Elongated rete ridges with basal cell hyperpigmentation** (bright nucleated keratinocytes on FCM; intense purple on DHE)
Proliferation of spindle-shaped cell in dermis: Proliferation of spindle-shaped bright (hyperfluorescent) wavy cells admixed with grayish dark collagen fibers forming a whorled pattern in the dermis. At higher magnification, these cells appear as plump bright (hyperfluorescent) cells of varied sizes	**Proliferation of spindle-shaped cell in dermis**: Proliferation of purplish-blue color spindle-shaped cells admixed with a pinkish purple collagen fiber forming a storiform pattern in the dermis. At higher magnification, these cells appear as plump purple cells of varied sizes	**Proliferation of spindle-shaped cell in dermis**: Fibroblastic proliferation of spindle cells (bright on FCM; purple on DHE) in the dermis forming a storiform to whorled pattern with entrapped collagen bundles (grayish dark on FCM; pinkish purple on DHE) at the periphery of the lesion

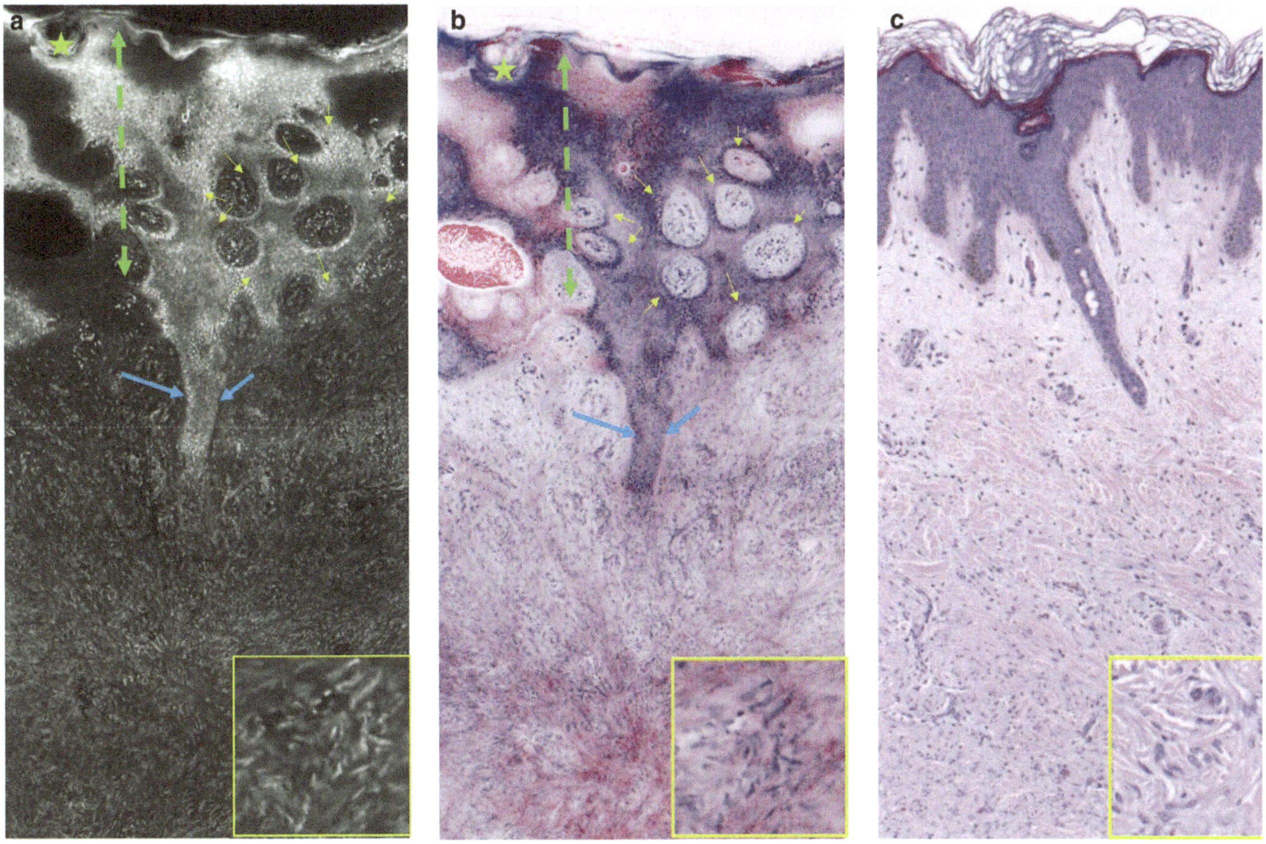

Fig. 6.1 Low magnification mosaic images of a dermatofibroma from anterior shin of a 40-year-old female. **a** FCM (grayscale) image shows an acanthotic epidermis (green double-headed arrow) composed of bright cells (green double-headed arrow) and an overlying grayish hyperkeratosis (light green star). Within the epidermis, round rings (yellow arrows) lined by a single bright layer of cells with central darkish-gray papillary dermis are seen. Elongated rete ridges are lined by bright cells (blue arrows). The underlying grayish papillary dermis shows proliferation of plump spindle-shaped, bright cells (inset) creating a storiform pattern. **b** Corresponding digital H&E image shows a purple-colored (bright on FCM) acanthotic epidermis (green double arrowhead) with an overlying purplish pink (darkish gray) hyperkeratosis (light green star). The lining cells in the round rings (yellow arrows) and elongated rete ridges (blue arrows) appear purple (bright on FCM). Dermal papillae within the rings appear purplish pink (darkish gray). In the purplish pink (grayish on FCM) dermis proliferation of purple-colored (bright on FCM) spindle-shaped cells (inset) are seen. **c** Corresponding conventional H&E-stained image shows an excellent correlation with FCM image (**a**) and DHE image (**b**) for an acanthotic epidermis (green double-headed arrow), hyperkeratosis (light green star), elongated rete ridges, pigmented basal keratinocytes (blue arrows), and storiform proliferation of spindle-shaped cell (inset) with focal collagen entrapment. H&E magnification = 4x

6.2 Dermatofibroma

Dermatofibromas (DF) are benign fibrohistiocytic proliferations of the skin, which usually occur on the lower extremities with a predisposition for young, middle-aged, female patients. DF clinically manifests as solitary or multiple pink to light-brown papules or nodules, measuring less than 10 mm in diameter. These lesions usually exhibit a dimple when laterally compressed, so-called the 'dimple sign'. They are usually asymptomatic. Although they are benign neoplasms, there have been reported cases of local recurrence, and rarely distant metastasis [3]. Etiopathogenesis is not well understood; however, a prior history of trauma to the affected area is present in one-fifth of the cases. More often, DF develops spontaneously, but a spontaneous resolution is not common. Underlying thyroid disorders create a predisposition for developing multiple DF lesions, especially hypothyroidism, lupus erythematosus, atopic dermatitis, and immunosuppression [4].

DF can clinically mimic seborrheic keratosis, dysplastic nevus, and melanoma. Dermatoscopic examination typically shows a central scar-like area with peripheral light brown thin reticular lines. Additionally, radial lines, brown dots and globules, crystalline-like structures, and yellow–brown scales can be seen. Especially, atypical forms of DF can mimic melanoma dermatoscopically and biopsy for diagnostic purposes may be required [5].

On histopathology, DF is characterized by the proliferation of spindle-shaped cells in the dermis forming a storiform pattern with entrapped peripheral pink collagen at low magnification. On higher magnification, plump spindle-shaped cells with pleomorphism in the center of the lesion are appreciated. Additionally, epidermal induction with adnexal differentiation (sebaceous and follicular differentiation) can be seen. There are different histopathological variants of DF including atrophic, cellular, xanthomatous, hemosiderotic, epithelioid, clear cell, and sclerosing hemangioma type [6].

Table 6.1 summarizes the basic clinical–histopathological features of a dermatofibroma.

Table 6.2 summarizes the morphological features of dermatofibroma features on fluorescence confocal microscopy (FCM), and pseudo-color or digital H&E (DHE) modes, and their comparison with histopathology (Fig. 6.1).

6.3 Lipoma

Lipomas are the most common benign soft-tissue neoplasms. They are usually seen in the skin; however, they can be incidentally found in the tracheobronchial system [7] or the musculoskeletal system [8].

Clinically, they present as solitary or multiple soft, painless subcutaneous nodules of varied sizes, mostly located on the trunk and upper extremity. Diagnosis is made clinically, and no treatment is necessary unless requested by the patient for cosmetic or other concerns. However, the presence of multiple lipomas or having lipoma located other than on skin may herald an underlying genetic mutation such as Cowden syndrome [9, 10].

Ultrasonographic examination of the skin may be required in rare conditions if suspicion for other benign (epidermoid cyst, ganglion cyst) or malignant (rapid enlargement, pain, restriction of movement) lesions are located in the subcutis [11].

Histologically, they consist of mature lipocytes with thin fibrous strands containing small vessels. The lesion may be enclosed by a thin fibrous capsule and located in the dermis and the subcutis.

Table 6.3 summarizes the basic clinical–histopathological features of a lipoma. Table 6.4 summarizes morphological features of lipoma on FCM, DHE, and comparison with their histopathology (Figs. 6.2 and 6.3).

Table 6.3 Basics of a Lipoma

Epidemiology and patient demographics: Affects 1% of the population. Occurs commonly in 4th–6th decades of life, and no gender bias is present
Etiology: Unclear, some studies revealed genetic link, trauma [26], obesity, alcohol abuse, liver disease and glucose intolerance have been proposed as etiologic factors [27]
Clinical presentation: Solitary or multiple lesions [27] located anywhere on the body but mostly on the trunk and upper extremities. A benign subcutaneous lesion, no risk for malignant transformation [28]
Histopathology: Proliferation of mature adipose tissue located in the dermis and the subcutis enclosed by a thin fibrous capsule

Table 6.4 Morphological features of lipoma on FCM, DHE, and comparison with their histopathology

Fluorescence confocal microscopy (FCM) mode (grayscale images)(Figs.6.2a and 6.3a)	Digital H&E (combined FCM and RCM modes) (Figs. 6.2b and 6.3b)	Conventional H&E-stained images (Figs. 6.2c and 6.3c)
Lipocytes: Sheet of large dark polygonal cells (nonfluorescent) with thin grayish (hypofluorescent) rim and tiny bright (hyperfluorescent), flat to round peripheral nuclei	**Lipocytes**: Sheet of pale pinkish polygonal cells with thin purple colored borders and small dark purple peripheral nuclei	**Lipocytes**: Mature lipocytes proliferation (dark cells on FCM; pinkish cells on DHE) with small peripheral nuclei (bright on FCM; dark purple on DHE)
Fibrous septate: Thick and brighter (hyperfluorescent) fibrous strands showing ramifications within the dark sheets of lipocytes	**Fibrous septate:** Dark purple thicker fibrous band with ramifications within fatty lobules	**Fibrous septate:** Intervening fibrous septate (bright strands on FCM; purple strands on DHE)

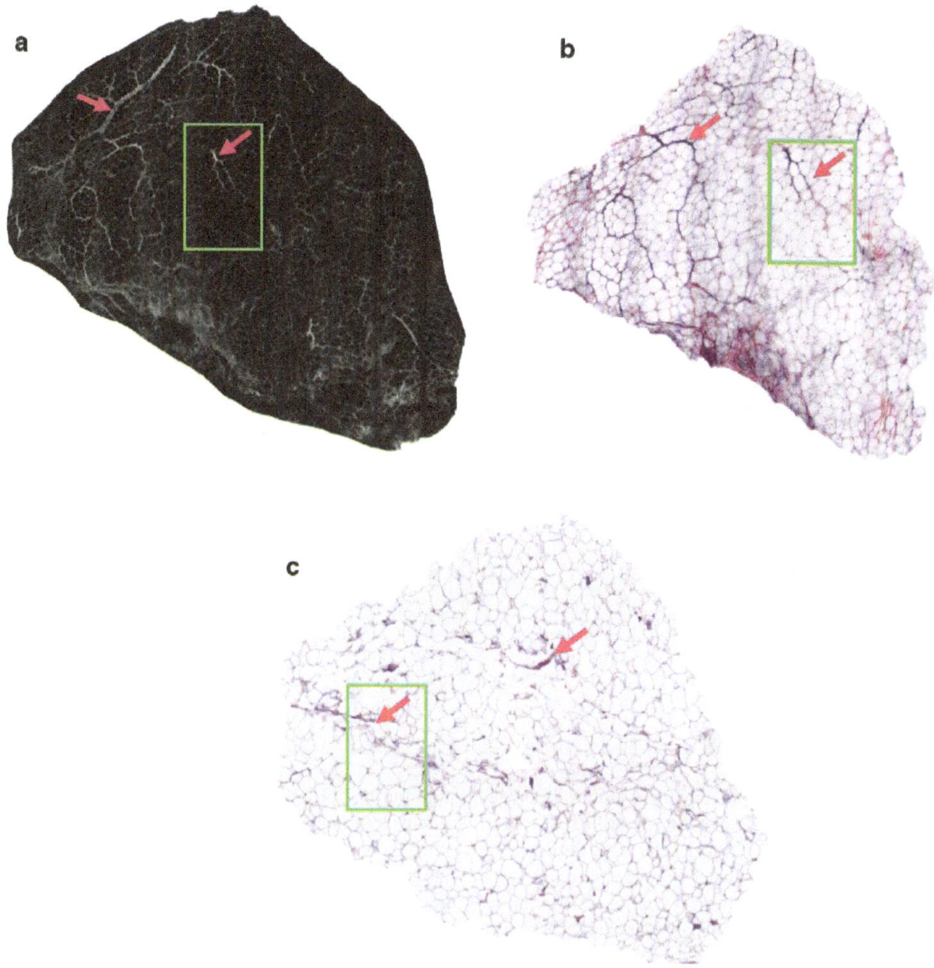

Fig. 6.2 Low magnification mosaic images of a lipoma excised from upper forearm of a 47-year-old female. **a** FCM (grayscale) image shows sheet of dark polygonal cells (green boxed area) with intervening thick bright ramifying fibrous strands (pink arrows). **b** On the corresponding DHE image these sheets (green boxed area) appear pale pinkish (grayish on FCM) and the fibrous strands (pink arrows) appear purple (bright on FCM). **c** Corresponding conventional H&E-stained image shows an excellent correlation with FCM image (**a**) and DHE image (**b**) for sheet of mature lipocytes and intervening fibrous septate (pink arrows). H&E magnification = 2x

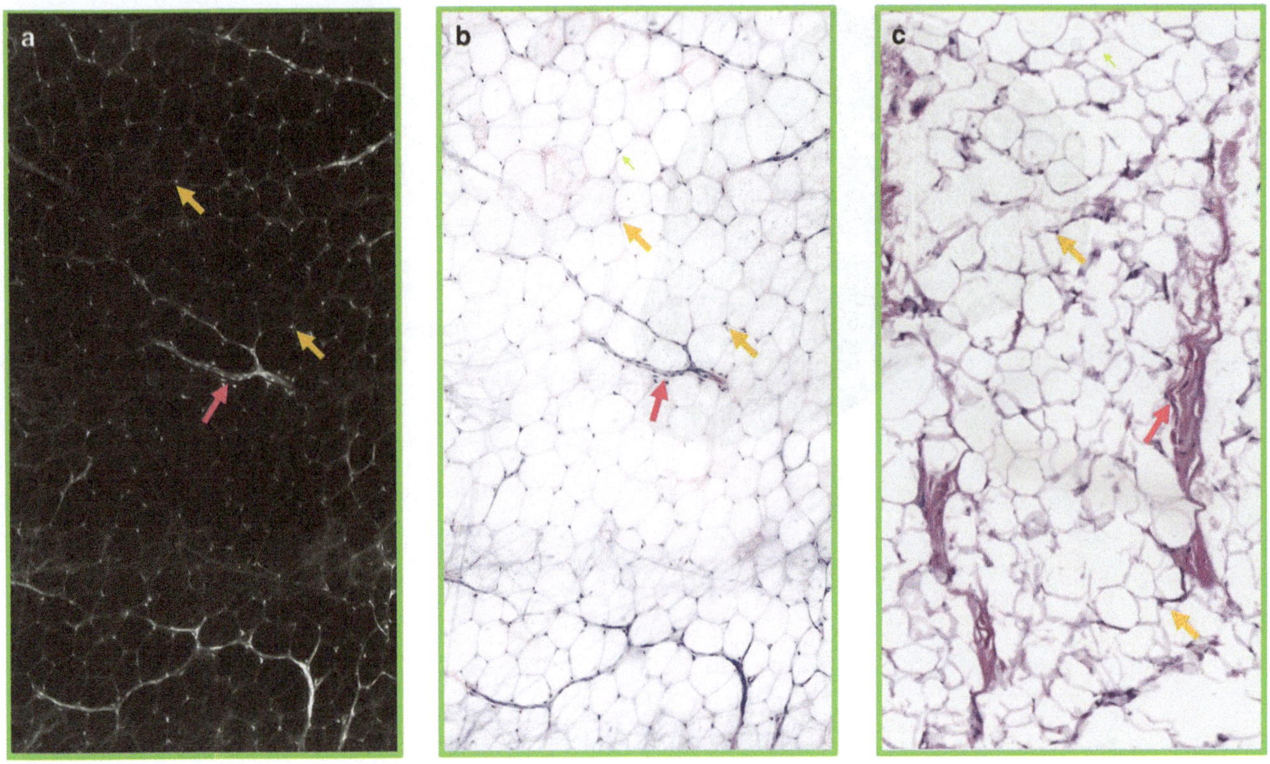

Fig. 6.3 Higher magnification sub-mosaic images of mature lipocytes in the lipoma obtained by zooming in the bright green boxed area in Fig. 6.2. **a** FCM image shows dark polygonal cells with thin grayish borders and small peripheral bright nuclei (orange arrows) with intervening bright fibrous bands (pink arrows). **b** On the corresponding DHE image, these lipocytes appear pale pinkish (dark on FCM) with thin purple borders and a small dark purple (bright on FCM) peripheral nucleus (orange arrows). The fibrous septate (pink arrows) appears purple color. **c** Corresponding conventional H&E-stained image shows an excellent correlation with FCM image (**a**) and DHE image (**b**) for mature lipocytes (orange arrows) and fibrous septate (pink arrows). H&E magnification = 10x

6.4 Epidermal Inclusion Cyst

Epidermal inclusion cysts (EIC), also called epidermoid cysts, are cystic structures located in the dermis. Clinically they present as a subcutaneous nodule with an apparent punctum on the skin surface. They are the most common cutaneous cystic lesions. EIC is also falsely called ‘sebaceous cyst’ even though there is no associated sebaceous structure.

EIC may appear sporadically or as a result of a preceding trauma causing epidermal implantation into the dermis. They can concomitantly be seen with acneiform lesion [12] and can be related to nevi [13]. Rarely, multiple EIC in abnormal locations (especially head & neck) can herald an underlying inherited condition such as Gardner syndrome [14]. Spontaneous resolution or recurrence after treatment is usually seen. They are removed when there is an inflammation or infection of the sac, causing discomfort to the patient.

On histopathology, EIC is seen as dermal sac-like lesions lined by stratified squamous epithelium. They arise from the follicular infundibulum and often exhibit a prominent granular layer with keratohyaline granules. The cyst lumen is filled with laminated pink keratinous material. Rupture of the cyst may initiate an intense inflammatory reaction composed of mostly neutrophils and foreign body giant cells in the surrounding dermis.

Table 6.5 summarizes the basic clinical–histopathological features of an epidermal inclusion cyst. Table 6.6 summarizes morphological features of epidermal inclusion cyst on FCM, DHE, and their comparison with histopathology (Figs. 6.4 and 6.5).

Table 6.5 Basics of an epidermal inclusion cyst

Epidemiology and patient demographics: Common in population between 2nd and 4th decades of life with a male predominance, M/F ratio: 2/1 [29]
Etiology: Mostly spontaneous occurrence, other include trauma, and rarely hereditary conditions (e.g., Gardner syndrome) [14, 30]
Clinical presentation:Usually located on the extremities, uncommon locations, and multiple lesions should raise suspicion for underlying diseases
Dermatoscopy: Dermal sac-like lesions lined by stratified squamous epithelium arising from follicular infundibulum with a prominent granular layer with keratohyalin granules and has laminated pink keratinous material in the center [12]

Table 6.6 Morphological features of epidermal inclusion cyst on FCM, DHE, and comparison with conventional histopathology

Fluorescence confocal microscopy (FCM) mode (grayscale images) (Figs. 6.4a and6.5a)	Digital H&E (combined FCM and RCM modes) (Figs. 6.4b and 6.5b)	Conventional H&E-stained images (Figs. 6.4c and 6.5c)
Cyst: A dark (nonfluorescent) cystic structure originating from the overlying bright (hyperfluorescent) epidermis seen in the dermis. The cyst has grayish (hypofluorescent) material within it and is lined by a bright (hyperfluorescent) epidermal layer. Note granular cell layer with keratohyalin granules cannot be identified in this mode	**Cyst:** A whitish cystic structure originating from purplish epidermis seen in the dermis. The cyst is lined by dark purple stratified epithelium and has pinkish purple material in the center. Note keratohyalin granules of the granular layer are seen as intense pink-clustered granules	**Cyst**: A dermal cystic (dark on FCM; whitish on DHE) lesion containing laminated keratin (grayish on FCM; pinkish purple on DHE) surrounded by stratified epithelium (bright on FCM; purple on DHE) with a prominent granular cell layer and keratohyalin granules (intense pink on DHE)

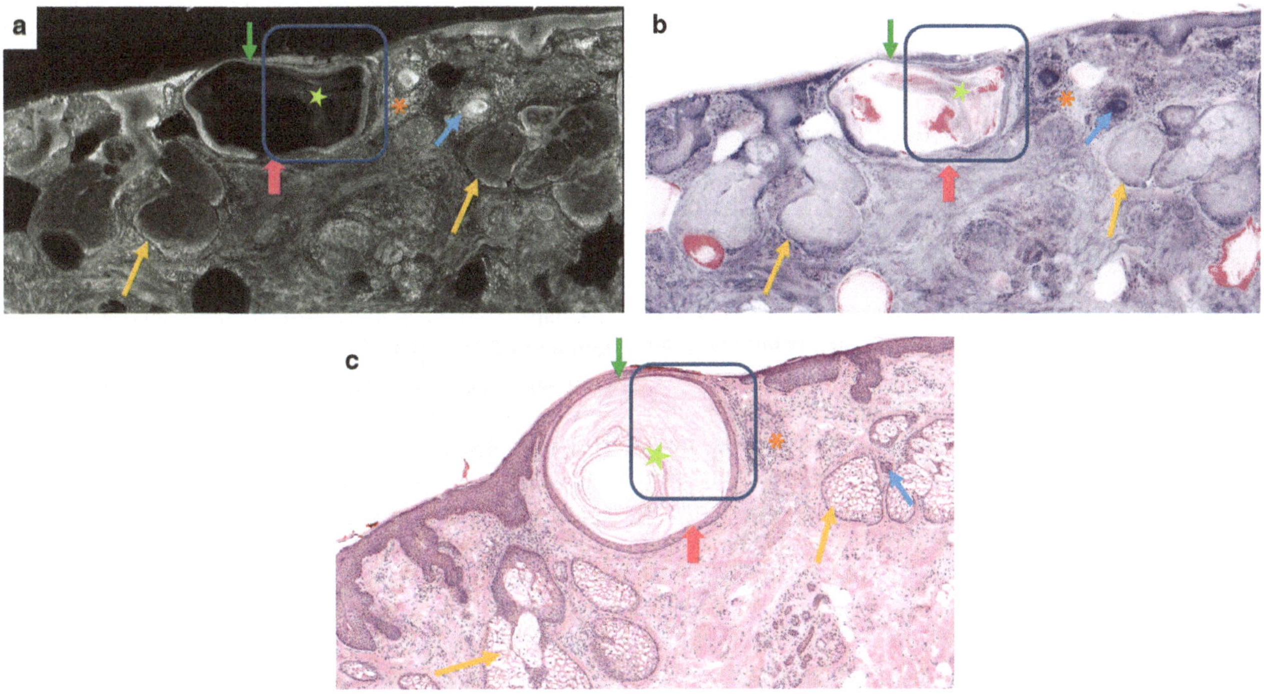

Fig 6.4 Low magnification mosaic images of an epidermal inclusion cyst (EIC) seen as an incidental finding in a negative Mohs margin section. **a** FCM (grayscale) image shows a dark cyst (pink arrow) with grayish acellular material (green star) within. The cyst is seen originating from the overlying bright epidermis (dark green arrow) and lined by the same epidermal layer (blue boxed area). In the underlying grayish dermis (yellow arrow), small bright inflammatory cells (orange asterisk), bright hair follicle (blue arrow), and grayish white lobules of sebaceous glands (orange arrows) are seen. **b** On the corresponding DHE image this cystic structure (pink arrow) is filled with pinkish purple (grayish on FCM) keratin material and lined by purple-colored (bright on FCM) epidermis (green arrow; blue boxed area). Within the purplish pink (grayish on FCM) dermis, scattered dark purple (bright on FCM) inflammatory cells (orange asterisk), hair follicle (blue arrow), and purplish white (grayish white on FCM) sebaceous glands (orange arrows) are seen. **c** Corresponding conventional H&E-stained image shows an excellent correlation with FCM image (**a**) and DHE image (**b**) for the EIC (pink arrow) with central whorled keratin (green star) and its epidermal lining (dark green arrow; blue boxed area), inflammatory cells (orange star), hair follicle (blue arrow), and sebaceous glands (orange arrows). H&E magnification = 4x

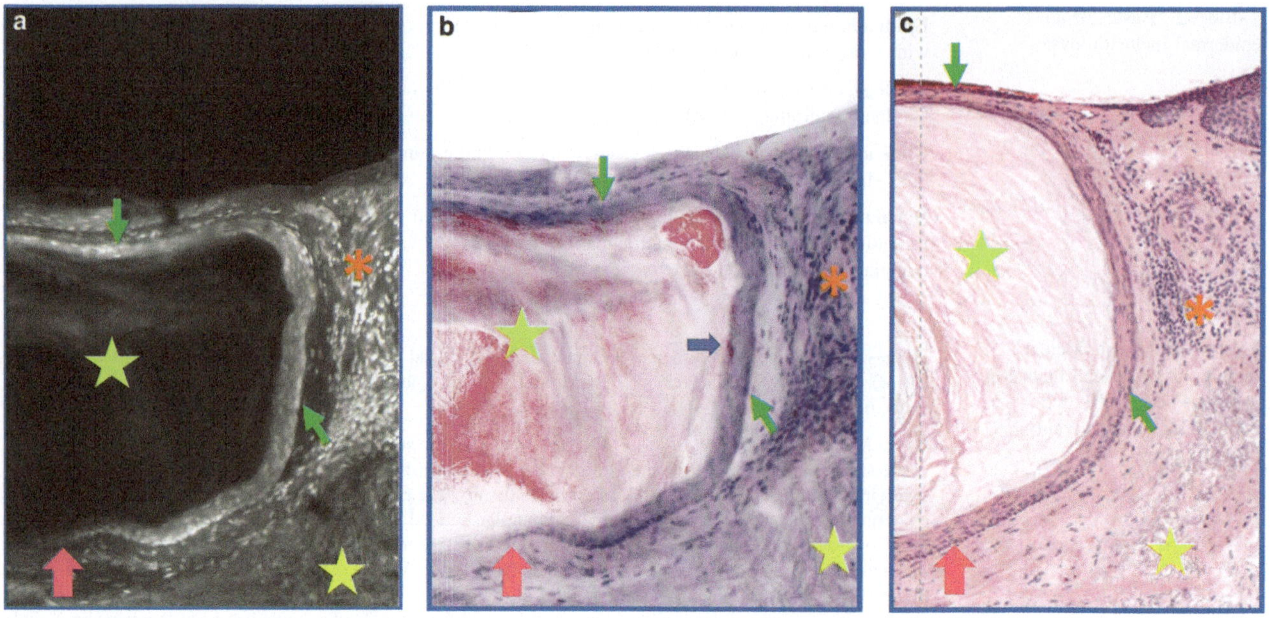

Fig 6.5 Higher magnification sub-mosaic images showing lining of the epidermal inclusion cyst obtained by zooming in the blue boxed area in Fig. 6.4. **a** FCM image shows bright stratified squamous epithelial lining (green arrows) of the dark EIC (pink arrow) and central grayish acellular material (green star). Surrounding the cyst, in grayish dermis (yellow star), clusters of bright inflammatory cells (orange asterisk) are seen. **b** On the corresponding DHE image, cyst (pink arrow) lining (green arrow) appears as purple (bright on FCM) and central acellular material (green star) appears with pinkish purple (grayish on FCM) acellular material (green star) and lined by purple (bright on FCM). *Note* in the DHE image the keratohyalin granules can be appreciated as intense pink granules (blue arrow) which are not visible in the FCM image. There is a cluster of purple color (bright on FCM) inflammatory cells (orange asterisk) adjacent to the cyst in the purplish dermis (yellow star). **c** Corresponding conventional H&E-stained image shows an excellent correlation with FCM image (**a**) and DHE image (**b**) for the features of EIC (pink arrow) with whorled central keratin (green star) and its epidermal lining (green arrow), and for surrounding inflammatory cells (orange asterisk) in the dermis (yellow star). H&E magnification = 20x

6.5 Fibroepithelial Polyp

Fibroepithelial polyps (FEP), also known as 'acrochordon', 'skin tag', 'soft fibroma' or 'cutaneous papilloma', are common benign skin tumors. They usually manifest as multiple, small 1–2 mm [15] pedunculated lesions with a narrow stack, mostly located on the intertriginous areas. Underlying metabolic syndrome, insulin resistance, diabetes, lipid abnormalities, pregnancy, or friction can play a role in their etiology [16–19]. Multiple FEP lesions can be associated with genetic disorder such as Birt-Hogg-Dube syndrome [20]. They are also seen in patients with Crohn's disease in the perianal area [21].

FEP becomes symptomatic due to friction and trauma, and/or infarction due to torsion of the lesion. Diagnosis of FEP is clinical; however, when they become giant, they should be differentiated from neurofibromas, dermal nevi, and polypoid seborrheic keratosis (EVCM features are given in Chap. 5). Another clinical differential diagnosis for FEP is fibroepithelioma of Pinkus, a rare variant of basal cell carcinoma, which clinically presents as skin-colored, pink sessile, or pedunculated lesions with a smooth surface. Histologically, fibroepithelioma of Pinkus consists of anastomosing strands-cords of basaloid cells connected to the epidermis, while FEP appears as a well-circumscribed polypoid lesion lined by normal epithelium and has a fibrous core with dense collagenous stroma.

Table 6.7 summarizes the basic clinical–histopathological features of a fibroepithelial polyp. Table 6.8 summarizes morphological features of fibroepithelial polyp on FCM, DHE, and their comparison with histopathology (Fig. 6.6).

Table 6.7 Basics of a fibroepithelial polyp

Epidemiology and patient demographics: Present in more than 50% of the population and increases with age
Etiology: Diabetes, metabolic syndrome, friction, pregnancy, obesity are risk factors; however, they can occur sporadically as well
Clinical presentation: Multiple, 1-2 mm pedunculated lesions with a narrow stack located on the intertriginous areas (axillary, inguinal folds, neck, genitalia, etc.)
Histopathology: A well-circumscribed polypoid lesion lined by normal epithelium and has a fibrous core with dense collagenous stroma

Table 6.8 Morphological features of fibroepithelial polyp on FCM, DHE, and comparison with conventional histopathology

Fluorescence confocal microscopy (FCM) mode (green scale images) (Fig. 6.6a)	Digital H&E (combined FCM and RCM modes) (Fig. 6.6b)	Conventional H&E-stained images. (Fig. 6.6c)
Polypoid lesion with a central dark (nonfluorescent) core composed of thin elongated to round, bright (hyperfluorescent) fibroblast nuclei and surrounded by normal bright (hyperfluorescent) epidermal layer	Polypoid lesion with a central light purple core composed of thin elongated to round, dark purple nuclei of fibroblasts, and surrounded by dark purple epidermal layer	Polypoid lesion with a fibrous core of dense collagenous stroma (dark on FCM; light purple on DHE) surrounded by epidermis (bright on FCM; dark purple on DHE)

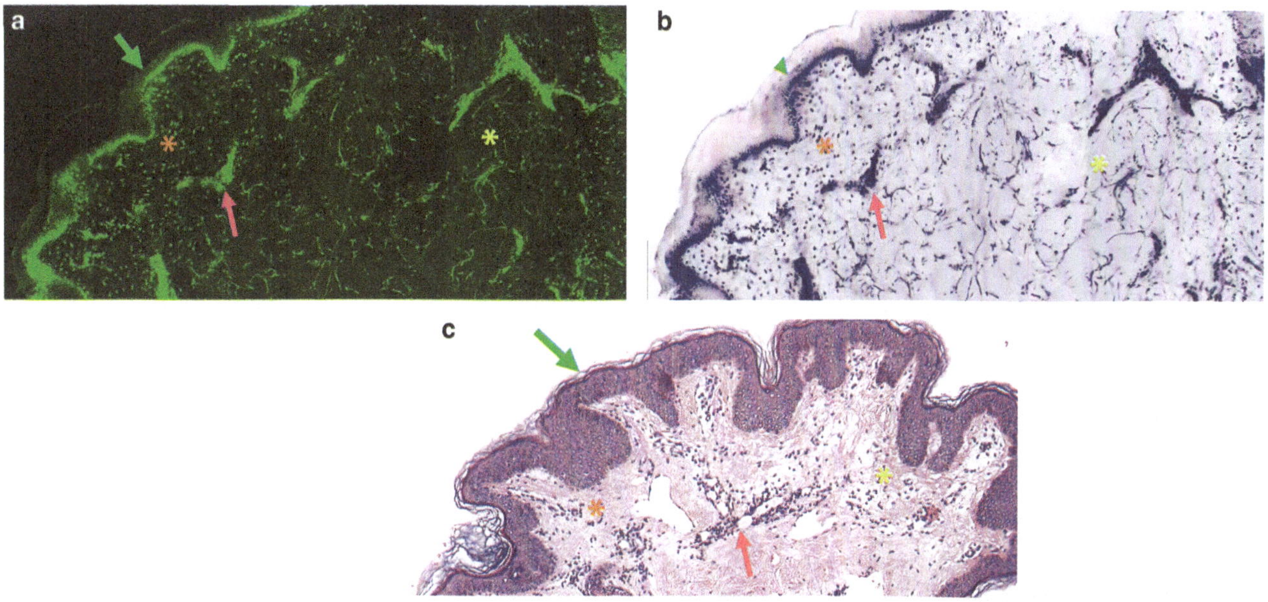

Fig 6.6 Medium magnification mosaic images of a fibrous epithelial polyp (FEP) from axilla of a 73-year-old male. **a** FCM (green scale) image shows a well-circumscribed polypoidal lesion lined by a normal bright epidermis (green arrow). The center of the lesion has a darkish loose stroma (yellow asterisk) with some bright fibrous strands (pink arrow) and bright scattered small inflammatory cells (orange asterisk). **b** On the corresponding DHE image the epidermal (green arrow) lining of this polypoid lesion appears purple (bright on FCM) and the central loose stroma (yellow asterisk) appears pale purplish (pale green on FCM) with few dark purple (bright on FCM) fibrous strands (pink arrow) and dark purple scattered small inflammatory cells (orange asterisk). **c** Corresponding conventional H&E-stained image shows an excellent correlation with FCM image (**a**) and DHE image (**b**) for the well-circumscribed polypoid lesion of FEP with a normal epidermal lining (green arrow) and a central core of loose stroma (yellow asterisk), fibrous strands (pink arrow), and sparse inflammatory cell infiltrate (orange asterisk). H&E magnification = 4x

6.6 Neurofibroma

Neurofibromas (NF) are benign hamartomas of the skin composed of Schwann cells, fibroblasts, mast cells, and nerve axons. NF are relatively common in adult population and they usually occur sporadically. The presence of plexiform NF or multiple NF with additional lesions (multiple café au-lait macules, axillary freckles, optic glioma, astrocytoma, neural crest tumor, etc.) herald the diagnosis of neurofibromatosis type-1, which is a genetic disorder caused by tumor suppressor gene mutation of NF-1 gene [22].

NF clinically present as soft, pink-colored papules less than 2 cm in diameter. When direct pressure is applied to the lesion it gets buried within the skin, so called 'button-hole sign'. Clinical differential diagnoses include neuroma, palisaded encapsulated neuroma (PEN), granular cell tumor, lipoma, EIC, fibroma, dermal nevi, and FEP [23].

Histologically, NFs are poorly circumscribed, unencapsulated nodular lesions that are seen in the dermis and subcutaneous tissue. There are different subtypes of NF including classic, cellular, myxoid, hyalinized, epithelioid, plexiform, diffuse, pigmented, granular cell, Pacinian, lipomatous neurofibroma, dendritic cell neurofibroma with pseudo-rosettes, and hybrid tumors [24, 25]. The most common histologic variant is the classic form, which is seen in the superficial dermis and creates a dome-shaped, polypoid lesion surrounded by a normal epidermis. The tumor is composed of haphazardly arranged spindle cells with wavy nuclei in a loose, pale stroma, and mast cells are numerously seen within the tumor [23].

Table 6.9 summarizes the basic clinical–histopathological features of a neurofibroma. Table 6.10 summarizes morphological features of neurofibroma on FCM, DHE, and their comparison with histopathology (Fig. 6.7).

Table 6.9 Basics of a neurofibroma

Epidemiology and patient demographics: Benign hamartomas, common in general population, no gender predominance
Etiology: Sporadic, multiple lesions of plexiform variant is seen in NF type 1
Clinical presentation: Solitary, skin-colored soft papules showing a 'button-hole' sign located on head & neck, trunk, or extremities
Histopathology: Poorly circumscribed unencapsulated dermal proliferation of neural and mesenchymal cells (Schwann cell, fibroblast, nerve axons, mast cells). Spindle-shaped cells with wavy nuclei with a loose mucinous stroma

Table 6.10 Morphological features of neurofibroma on FCM, DHE, and comparison with conventional histopathology

Fluorescence confocal microscopy (FCM) mode (green scale images) (Fig. 6.7a)	Digital H&E (combined FCM and RCM modes) (Fig.6.7b)	Conventional H&E-stained images (Fig.6.7c)
A nodular dermal lesion composed of haphazardly arranged bright (hyperfluorescent) spindle cells with wavy nuclei within a loose pale green (hypofluorescent) stroma	A dark purple nodular lesion in the dermis composed of haphazardly arranged spindle cells with dark purple wavy nuclei within a light purple loose stroma	A basophilic nodular lesion in the dermis consisting of spindle-shaped cells with wavy nuclei with a pale cytoplasm and wavy nuclei interlacing by thickened collagen bundles

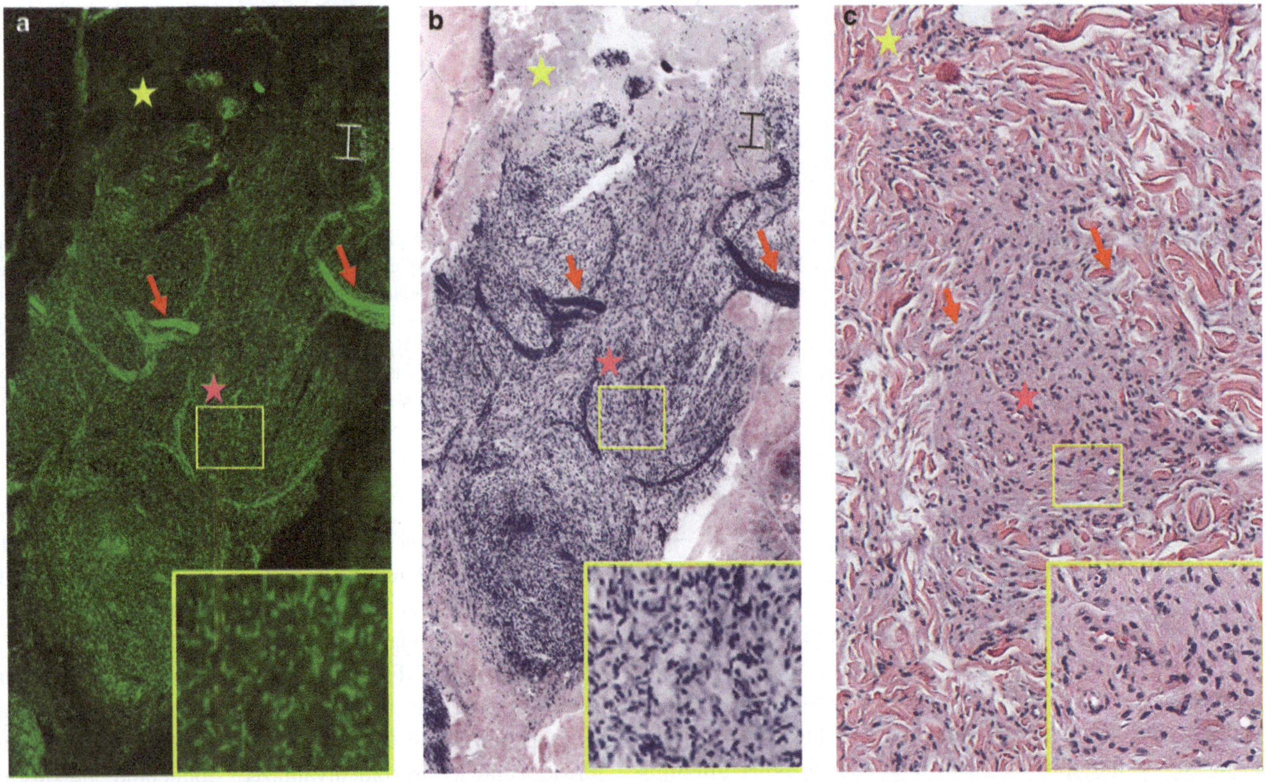

Fig 6.7 Low magnification sub mosaic images of a neurofibroma (NF) from upper back of a 43-year-old male. **a** FCM (green scale) image shows localized proliferation of thin, bright wavy spindle cells (pink star and yellow inset) within a dark (hypofluorescent) loose stroma and intermixed with thick bright fibrous bands (pink arrows). There is a darkish green skeletal muscle bundle (yellow star) at the upper left corner of the lesion (yellow star). **b** On the corresponding DHE image, these wavy spindle cell nuclei (pink star and yellow inset) appear dark purple (bright on FCM) within a purplish pink (dark on FCM) loose stroma and the fibrous bands (pink arrows) appears dark purple (bright on FCM). At the periphery of the lesion the bundle of skeletal muscle appears light purple (yellow star). **c** Corresponding conventional H&E-stained image of the neurofibroma shows an excellent correlation with FCM image (**a**) and DHE image (**b**) by showing proliferation of spindle cells with wavy nuclei (pink star and yellow inset) within a loose stroma and admixed with thick fibrous bundles (pink arrow). H&E magnification =10x

References

1. Farabi B, Rao BK. Reflectance confocal microscopy findings of vascular lesions and others. In: Rao BK, Jain M, Haroon A, Farabi B, editors. A beginner's guide to reflectance confocal microscopy. New York: Path Presenter Publications; 2020. (pp. Section D-Chapter 10).
2. Wells WA, et al. In vivo and ex vivo microscopy: moving toward the integration of optical imaging technologies into pathology practice. Arch Pathol Lab Med. 2019;143(3):288–98.
3. Doyle LA, Fletcher CD. Metastasizing "benign" cutaneous fibrous histiocytoma: a clinicopathologic analysis of 16 cases. Am J Surg Pathol. 2013;37(4):484–95.
4. Myers DJ, Fillman EP. Dermatofibroma, in StatPearls. Treasure Island, FL: StatPearls Publishing; 2021.
5. Zaballos P, et al. Dermoscopy of dermatofibromas: a prospective morphological study of 412 cases. Arch Dermatol. 2008;144 (1):75–83.
6. Alves JV, et al. Variants of dermatofibroma–a histopathological study. An Bras Dermatol. 2014;89(3):472–7.
7. Nicolo Rotolo MC. Stefano La Rosa, andrea imperatori, a rare case of incidental tracheal lipoma. Arch Bronconeumol. 2018;54 (12):630–2.
8. McTighe S, Chernev I. Intramuscular lipoma: a review of the literature. Orthop Rev (Pavia). 2014;6(4):5618.
9. Djuric-Stefanovic A, et al. Gastroduodenal lipomatosis in familial multiple lipomatosis. Med Princ Pract. 2017;26(2):189–91.
10. Granados DMM, de Bonadia LC, Bertuzzo CS, Steiner CE. Clinical and molecular investigation of familial multiple lipomatosis: variants in the HMGA2 gene. Clin Cosmet Investig Dermatol. 2020;13:1–10.
11. Rahmani G, McCarthy P, Bergin D. The diagnostic accuracy of ultrasonography for soft tissue lipomas: a systematic review. Acta Radiol Open. 2017;6(6):2058460117716704.
12. Zito PM, Scharf R, Epidermoid Cyst. In StatPearls. 2021, StatPearls Publishing, Copyright © 2021, StatPearls Publishing LLC.: Treasure Island (FL).
13. Cohen PR, R.P. Rapini, Nevus with cyst. A report of 93 cases. Am J Dermatopathol. 1993;15(3):229–34.
14. Hoang VT, et al. Overview of epidermoid cyst. Eur J Radiol Open. 2019;6:291–301.
15. Banik R, Lubach D. Skin tags: localization and frequencies according to sex and age. Dermatologica. 1987;174(4):180–3.
16. Ragunatha S, et al. Cutaneous disorders in 500 diabetic patients attending diabetic clinic. Indian J Dermatol. 2011;56(2):160–4.
17. Senel E et al. Acrochordons as a cutaneous sign of impaired carbohydrate metabolism, hyperlipidemia, liver enzyme abnormalities and hypertension: a case-control study. J Eur Acad Dermatol Venereol, 2011.
18. Boza JC, et al. Skin manifestations of obesity: a comparative study. J Eur Acad Dermatol Venereol. 2012;26(10):1220–3.
19. Winton GB, Lewis CW. Dermatoses of pregnancy. J Am Acad Dermatol. 1982;6(6):977–98.
20. Toro JR, et al. BHD mutations, clinical and molecular genetic investigations of Birt-Hogg-Dubé syndrome: a new series of 50 families and a review of published reports. J Med Genet. 2008;45 (6):321–31.
21. Bonheur JL, et al. Anal skin tags in inflammatory bowel disease: new observations and a clinical review. Inflamm Bowel Dis. 2008;14(9):1236–9.
22. Beth G Goldstein. A.O.G. Overview of Benign Skin Lesions. Apr 08, 2021 [cited 2021 June 1st]; https://www.uptodate.com/contents/overview-of-benign-lesions-of-the-skin#H1101420005.
23. Ferringer T. Neural Tumors. 3rd ed. Dermatology, ed. J. Bolognia, vol. 1. Elsevier, 2018.
24. Megahed M. Histopathological variants of neurofibroma. A study of 114 lesions. Am J Dermatopathol. 1994;16(5):486–95.
25. Messersmith L, Krauland K. Neurofibroma. in StatPearls. Treasure Island, FL, StatPearls Publishing; 2021.
26. Aust MC, et al. Posttraumatic lipoma: fact or fiction? Skinmed. 2007;6(6):266–70.
27. Charifa A, Azmat CE, Badri T. Lipoma pathology, in statpearls. Treasure Island, FL: StatPearls Publishing; 2021.
28. Johnson CN, et al. Lipomatous soft-tissue tumors. J Am Acad Orthop Surg. 2018;26(22):779–88.
29. Weir CB, St.Hilaire NJ. Epidermal inclusion cyst, in StatPearls. Treasure Island, FL: StatPearls Publishing; 2021.
30. Gardner EJ. Follow-up study of a family group exhibiting dominant inheritance for a syndrome including intestinal polyps, osteomas, fibromas and epidermal cysts. Am J Hum Genet. 1962;14(4):376–90.

Part IV
Malignant Keratinocyte Neoplasms

Morphological Features of Basal Cell Carcinoma on Ex Vivo Confocal Imaging and Histopathologic Correlation

7

Mercedes Sendín-Martín, Ucalene Harris, Matthew Moronta, Melissa Pulitzer, Erica Lee, Anthony Rossi, Chih-Shan Jason Chen, Kishwer Nehal, and Manu Jain

Abstract

Basal cell carcinoma is the most common skin cancer worldwide. Mohs micrographic surgery (MMS) is a specialized surgical procedure to treat and achieve complete clearance of BCCs. However, frozen sectioning performed during MMS is a time-consuming and a labor-intensive process. Ex vivo confocal microscopy (EVCM) is an attractive alternative to frozen sectioning during MMS as it can rapidly image fresh (unprocessed and un-sectioned) tissue, enabling tumor margin assessment in real-time. EVCM has shown high sensitivity and specificity for detecting BCCs. In this chapter, we have described the morphological features of common subtypes of BCC as visualized in various imaging modes on the EVCM (fluorescence confocal microscopy and digital H&E mode) and compared them with their corresponding.

Basal cell carcinoma (BCC) is the most common skin cancer worldwide [1]. Annually, two million new BCC cases are estimated in the USA [2], and the incidence rates are expected to rise due to aging population [3]. Exposure to UV radiation from sunlight is the primary etiologic agent for BCC, particularly among individuals with lighter skin prototypes (I to III) (Table 7.1).

Although BCCs are indolent tumors and rarely metastasis, it can be locally aggressive in nature [1] with a high destructive potential, affecting the surrounding structures including bone and vital organs (eyelids, earlobes, etc.).

Clinically, BCCs can have varied appearance based on their histopathological subtypes. Nodular BCCs (nBCC) are the most common subtypes accounting for 80% of the lesions [4]. They are primarily located on the face. Clinically, they present as shiny papules or nodules, which may have "rolled" borders and telangiectatic vessels. Superficial BCCs (sBCC) are mostly located on upper trunk and appears as slightly scaly plaques, pink to flesh-colored. Morpheaform or sclerosing BCCs are typically located in the mid-facial sites and appears as scar-like lesions with ill-distinct borders. Ulceration is a usual feature for all BCCs subtypes, and pigmentation can also be seen.

Dermoscopy is useful for the diagnosis of BCC. Several dermoscopic features of BCC have been described: large blue-grey ovoid nests, multiple blue-grey dots and globules, leaf-like areas, spoke wheel areas, concentric structures, arborizing vessels, short fine telangiectasias, shiny white and white-red structures, ulceration, multiple small erosions, and shiny white-red structureless areas [5].

On histopathology, there are various subtypes of BCCs. The most common subtypes including nBCC, sBCC, micronodular BCC, infiltrative BCC (iBCC), morpheaform or sclerosing BCCs, and infundibulocystic BCCs. nBCCs and sBCCs consist of varied sized nests of basaloid cells with peripheral palisading and clefting located in the dermis and attached to the basal layer of epidermis, respectively. Around the tumor nests, mucinous stroma can be seen. In iBCC, small irregular clusters of basaloid cells are seen in the dermis. Micronodular BCC is characterized by small

M. Sendín-Martín
Dermatology Department, Hospital Universitario Virgen del Rocío, Sevilla, Spain

U. Harris · M. Moronta · E. Lee · A. Rossi · C.-S.J. Chen · K. Nehal · M. Jain (✉)
Dermatology Service, Department of Medicine, Memorial Sloan Kettering Cancer Center (MSKCC), 530 E. 74th Street, New York, NY, USA
e-mail: jainm@mskcc.org

M. Pulitzer
Department of Pathology, Memorial Sloan Kettering Cancer Center (MSKCC), New York, NY, USA

M. Jain et al. (eds.), *Cutaneous Atlas of Ex Vivo Confocal Microscopy*,
https://doi.org/10.1007/978-3-030-89316-3_7

Table 7.1 Basics of basal cell carcinoma (BCC)

• **Epidemiology and patients demographics**: BCC is the most common skin cancer, with rising incidence [13]. Predominantly seen in older adults with fair skin phototype (I to III), characteristically in sun-exposed body sites.
• **Etiology**: UV exposure, childhood and adolescent sun exposure, radiation, genetics (PTCH, p53, BAX gene mutations).
• **Clinical Presentation**: Nodular BCCs are pink or flesh-colored papules with pearly quality, and often with telangiectatic vessel. Ulceration is frequent. Superficial BCCs typically present as slightly scaly plaques, light red to pink in color. Morpheaform or sclerosing BCCs usually present as atrophic plaques with ill-defined borders.
• **Dermatoscopy**: Large blue-grey ovoid nests, multiple blue-grey dots and globules, leaf-like areas, spoke wheel areas, concentric structures, arborizing vessels, short fine telangiectasias, shiny white and white-red structures, ulceration, multiple small erosions, shiny white-red structureless areas [5].
• **Histolopathology**: Several subtypes have been described. The most common subtypes are nodular (50–80%) and superficial (10–30%). Otherless common BCC subtypes (<10%) include infiltrative, morpheaform, infundibulocystic, fibroepithelial, or basosquamous patterns. Most of the BCCs exhibit more than one histopathologic pattern [3].

basaloid nests; while, sclerosing BCC have thin strands of basaloid cells embedded within a dense sclerotic stroma.

Mohs micrographic surgery (MMS) is a specialized surgical procedure designed to treat and achieve complete clearance of BCCs. MMS provides high cure rates [6] and allows surgeons to maximize tissue preservation while treating recurrent BCCs or BCCs located in the cosmetic sites, such as the face [7]. Frozen sections performed during MMS is a time consuming ($\sim$20 to 40 min /layer removal) and a labor intensive [8] process, which is currently a major limitation of this procedure.

Ex vivo confocal microscopy (EVCM) is an attractive alternative to frozen sectioning during Mohs surgery as it can rapidly image fresh (unprocessed and un-sectioned) tissue, enabling tumor margin assessment in real-time. EVCM has shown high sensitivity and specificity for detecting BCCs in fresh Mohs surgical excisions in several studies [9–12], paving the way to a future rapid bedside histopathologic guidance for the surgeons.

In this chapter, we have described the morphological features of common subtypes of BCC as visualized in various imaging modes on the EVCM device, including fluorescence confocal microscopy (FCM) mode and digital H&E (DHE) mode and compared them with their corresponding H&E-stained frozen sections (Table 7.2). For image acquisition, the latest generation of EVCM device (Vivascope 2500; Caliber ID, Rochester, NY, USA) was used to image fresh discarded BCC tissues from MMS (Figs. 7.1, 7.2, 7.3, 7.4, 7.5, 7.6, 7.7, 7.8, 7.9, 7.10, 7.11, 7.12, 7.13, 7.14 and 7.15).

Table 7.2 Morphological features of BCCs on FCM, and DHE modes and comparison with their corresponding histopathology

Fluorescence confocal microscopy (FCM) mode (grayscale image) [7, 11] (Figs. 7.1a, 7.2a, 7.3a, 7.4a, 7.5a, 7.6a, 7.7a, 7.8a, 7.9a, 7.10a, 7.11a, 7.12a, 7.13a, 7.14a and 7.15a)	Digital H&E (combined FCM and RCM modes) (Figs. 7.1b, 7.2b, 7.3b, 7.4b, 7.5b, 7.6b, 7.7b, 7.8b, 7.9b, 7.10b, 7.11b, 7.12b, 7.13b, 7.14b and 7.15b)	Conventional H&E-stained image (Figs. 7.1c, 7.2c, 7.3c, 7.4c, 7.5c, 7.6c, 7.7c, 7.8c, 7.9c, 7.10c, 7.11c, 7.12c, 7.13c, 7.14c and 7.15c)
Fluorescence: BCC nuclei visualized as bright white structures with 1000-fold enhanced contrast compared to the surrounding dermis	Fluorescent nuclear signal appears dark purple in DHE images	The bright signal on FCM and dark purple color on DHE corresponds to hematoxylin-stained hyperchromatic nuclei of BCC
Tumor demarcation: Tumor shape can be clearly drawn to separate bright cell nuclei in the tumor nodule from rest of the grayish dermis	**Tumor demarcation**: Dark purple tumor nodules appear well-demarcated in the background of pinkish collagen	**Tumor demarcation**: Tumor nodules of basaloid cells (bright on FCM; purple on DHE) can be readily identified in the dermis (grayish on FCM; pinkish on DHE)
Nuclear crowding: Higher nuclear density within the tumor nodule makes neoplastic islands appear bright	**Nuclear crowding**: Nuclear crowding appears as dark purple color nuclei within the BCC nodules	**Nuclear crowding**: Crowded nuclei of BCC nodules are hematoxylin-stained (bright on FCM; dark purple on DHE)
Peripheral palisading: Parallel arrangement of bright cells at the periphery of the tumor nodule	**Peripheral palisading**: Parallelly arranged dark purple nuclei at the periphery of the BCC nodules	**Peripheral palisading**: Hematoxylin-stained basaloid nuclei (bright on FCM; purple on DHE) at the periphery of BCC nodules
Clefting: Fluorescence-free (nonfluorescent) dark half-moon attached to the bright tumor mass	**Clefting**: Appears as white area adjacent to the dark purple BCC nodules	**Clefting**: Retraction space between tumor and the surrounding stroma created by washing off the mucin during tissue processing (dark on FCM; white on DHE)
Nuclear pleomorphism: Aberrant hyperfluorescent nuclei visualized as bright white structures	**Nuclear pleomorphism**: Varied size and shaped dark purple nuclei within the BCC nodules	**Nuclear pleomorphism**: Hematoxylin-stained tumor nuclei (bright on FCM dark purple on DHE) of varied in size and shape within the BCC nodules
Enlarged N/C ratio: BCC nests are visualized as enlarged bright (hyperfluorescent) masses with surrounding scant grayish (hypofluorescent) cytoplasm	**Enlarged N/C ratio**: Tumor cells have enlarged dark purple nuclei with scant pale pink cytoplasm	**Enlarged N/C ratio**: Higher nuclear/cytoplasmic ratio of BCC tumor cells (bright on FCM; dark purple on DHE) as compared to the normal keratinocytes
Tumoral Stroma: Within the tumor stroma inflammatory infiltrate may be visualized as bright (hyperfluorescent) dots in a dark (nonfluorescent/ hypofluorescent) background. This pattern is known as *"starry sky"* stroma	**Tumoral Stroma**: Inflammatory cells appear as purple small scattered or clustered dots within a pale pink collagen and can be easily distinguished from the dark purple and enlarged nuclei of BCC	**Tumoral Stroma**: Hematoxylin-stained lymphocytes (bright on FCM; purple on DHE) seen in the stroma surrounding the BCC nodules

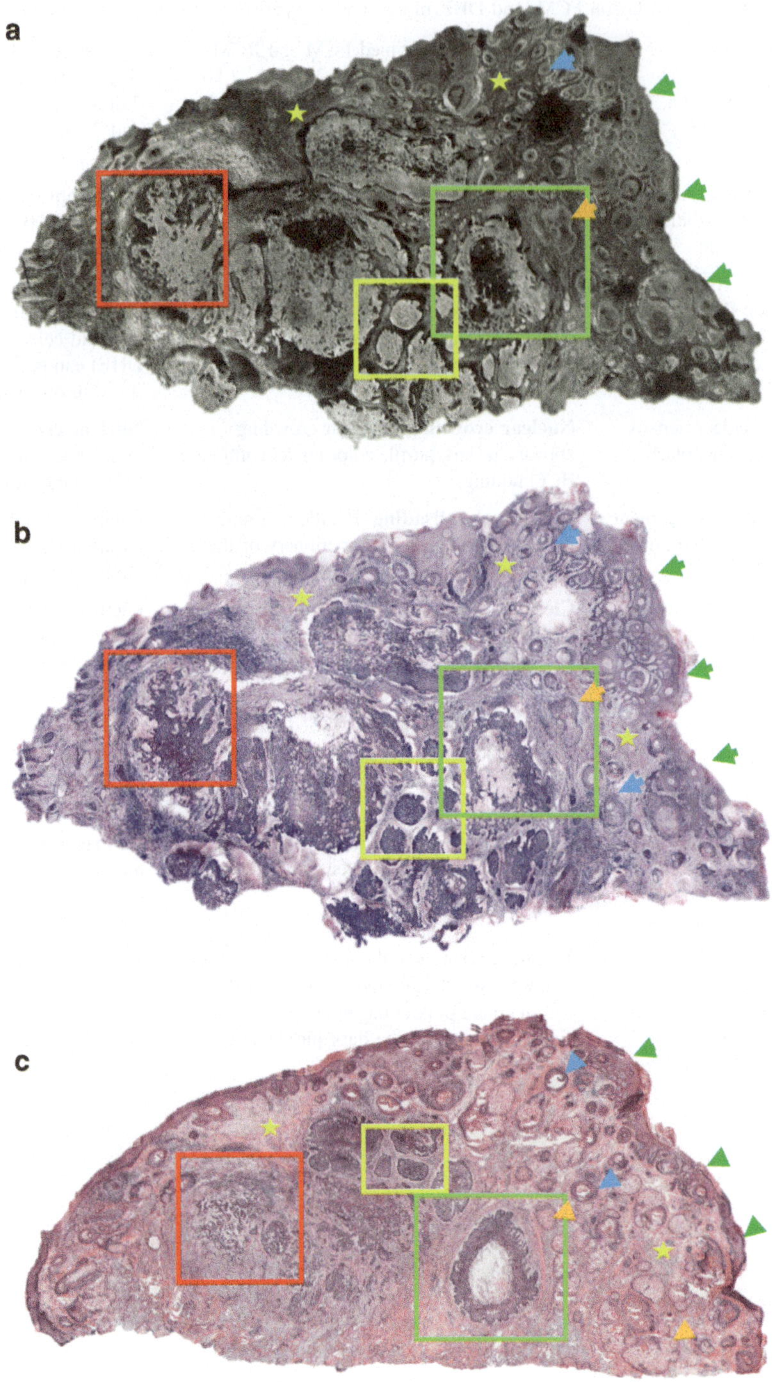

Fig. 7.1 Low magnification mosaic images of a nodular and infiltrative BCC from left nasal ala of an 84-year-old male. **a** FCM image shows well demarcated bright tumor nodules (boxed areas) within a grayish white dermis (yellow stars). These nodules can be clearly differentiated from the surrounding grayish white normal skin structures of hair follicles (blue arrows), sebaceous glands (orange arrow), and epidermis (green arrows). **b** On the corresponding digital H&E (DHE) image, BCC tumor nodules (boxed areas) appear as dark purple (bright on FCM) aggregates of varied sizes within a purplish pink (grayish white on FCM) dermis (yellow stars). These tumor nodules can be readily distinguished from the surrounding hair follicles (blue arrows), sebaceous glands (orange arrow), and epidermis (green arrows). **c** Corresponding conventional H&E stained image shows an excellent correlation with FCM (**a**) and DHE (**b**) images for the BCC tumor nodules (boxed areas), dermis (yellow stars), hair follicles (blue arrows), sebaceous glands (orange arrow), and epidermis (green arrows). H&E magnification = 2X

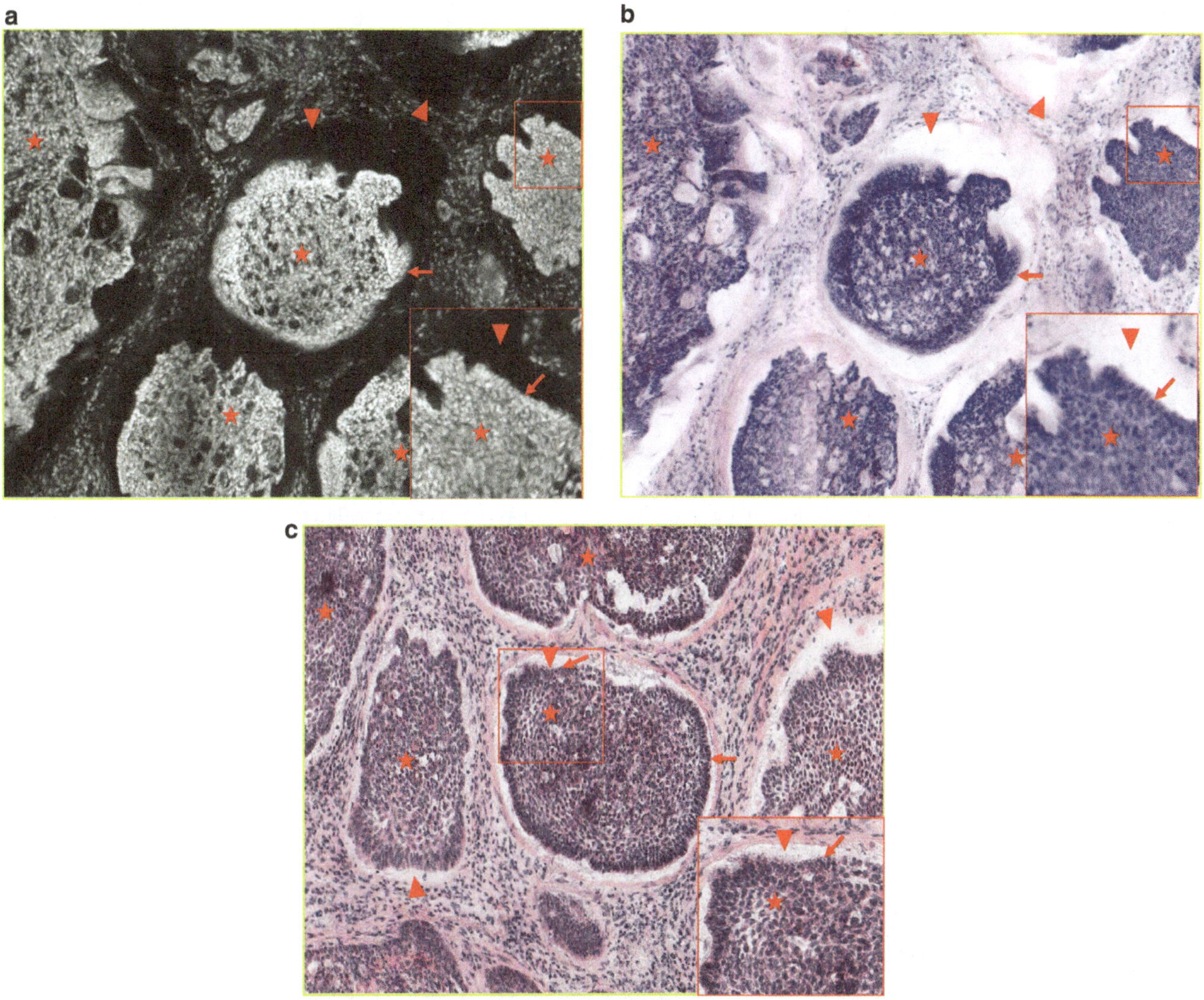

Fig. 7.2 Medium and high magnification submosaic images of nodular BCC showing cellular features obtained by digitally zooming in the yellow boxed area of mosaics in Fig. 7.1. **a** FCM image shows bright (hyperfluorescent) tumor nodules of BCC (red stars) composed of crowded nuclei with increased N:C ratio (inset). These bright nuclei form peripheral palisading (red arrow and inset) at the tumor periphery. Around each tumor nodule, clefting (red arrowheads and inset) appears as a dark space. **b** Corresponding digital H&E image shows these tumor islands (red stars) as dark purple (bright on FCM) well-defined aggregates. Tumor cells (inset) have nuclear pleomorphism and a high N:C ratio forming peripheral palisading (red arrow, red box). Clefting (red arrowheads) appears as a white space (dark on FCM) around individual tumor nodule. **c** Corresponding conventional H&E-stained image shows a good correlation with the FCM (**a**) and DHE (**b**) images for cellular features of the nodular BCC (red stars and inset) including palisading (red arrow and inset), and surrounding clefting (red arrowheads and inset). H&E magnifications = 10X, 20X (inset)

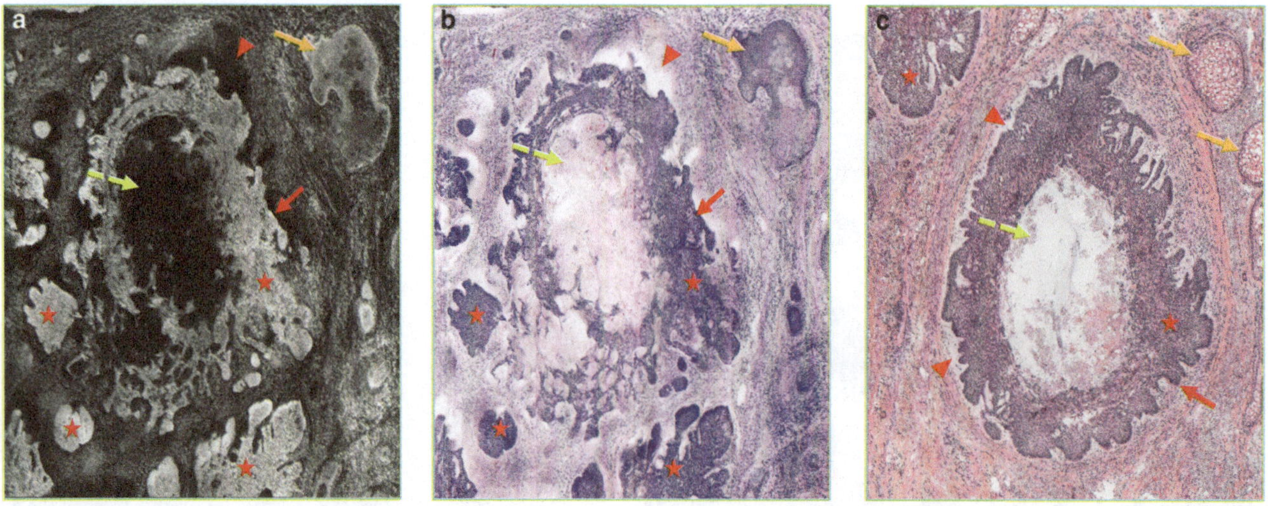

Fig. 7.3 Medium magnification submosaic images of a nodular BCC showing cystic degeneration obtained by digitally zooming in green boxed area of mosaics in Fig. 7.1. **a** FCM image shows bright well-demarcated tumor aggregates of nodular BCCs (red stars) with peripheral palisading (red arrow) and dark surrounding clefting (red arrowhead). The tumor nodule shows central dark area of cystic degeneration (yellow dashed arrow). These BCC tumor nodules appear distinct from the surrounding sebaceous gland (orange arrow) which appears less bright and lacks cellular features of BCC (enlarged N/C ratio, nuclear crowding, and palisading) and clefting. **b** On the corresponding digital H&E (DHE) image, BCC tumor nodules (red stars) appear as dark purple (bright on FCM) aggregates with palisading (red arrow) and the clefting (red arrowheads) appears as white space (dark on FCM). The cystic degeneration tumor necrosis (yellow dashed arrow) appears as pale pink (dark on FCM) acellular area. Sebaceous gland (orange arrow) appears pale purple in color (less bright on FCM). **c** Corresponding conventional H&E stained image shows an excellent correlation with FCM (**a**) and DHE (**b**) images for the BCC tumor nodules (red stars) with palisading (red arrow), clefting (red stars), and central necrosis (yellow dashed arrow), and for the sebaceous glands (orange arrows). H&E magnifications = 10X

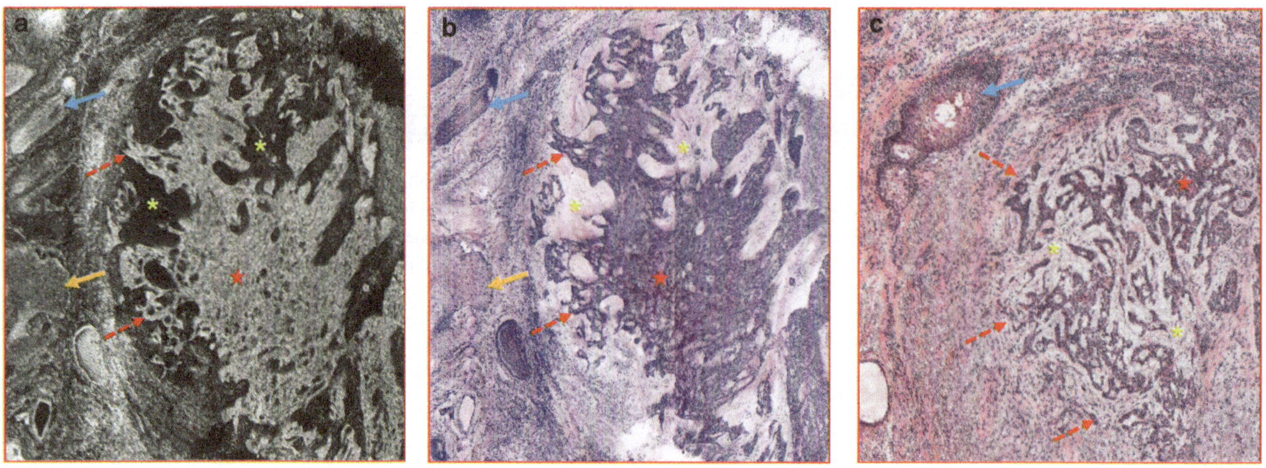

Fig. 7.4 Medium magnification submosaic images of a nodular infiltrative BCC with surrounding mucin obtained by digitally zooming in red boxed area of mosaics in Fig. 7.1. **a** FCM image shows a bright BCC tumor nodule (red star) with irregular cords at the periphery (red dashed arrows), which is surrounded by a dark area of mucin (yellow asterisks). The sebaceous gland (orange arrow) and hair follicle (blue arrow) have a sparse less bright smaller nuclei as compared to the nuclei of BCC. **b** On the corresponding digital HE (DHE) image, BCC nodule (red star) and cords (red dashed arrows) appears dark purple (bright on FCM). The surrounding mucin (yellow asterisks) is visualized as pale pink (dark on FCM) homogenous area (red asterisks). Sebaceous gland (orange arrow) and hair follicle (blue arrow) can be clearly visualized. **c** Corresponding conventional H&E stained image shows an excellent correlation with the FCM (**a**) and DHE (**b**) images for the features of BCC (red star) with mostly infiltrative cords (red dashed arrows), surrounding mucin (yellow asterisks) and a pilosebaceous unit (blue arrow). Note, nodular pattern of BCC is not evident on H&E due to a relatively deeper section. H&E magnification = 10X

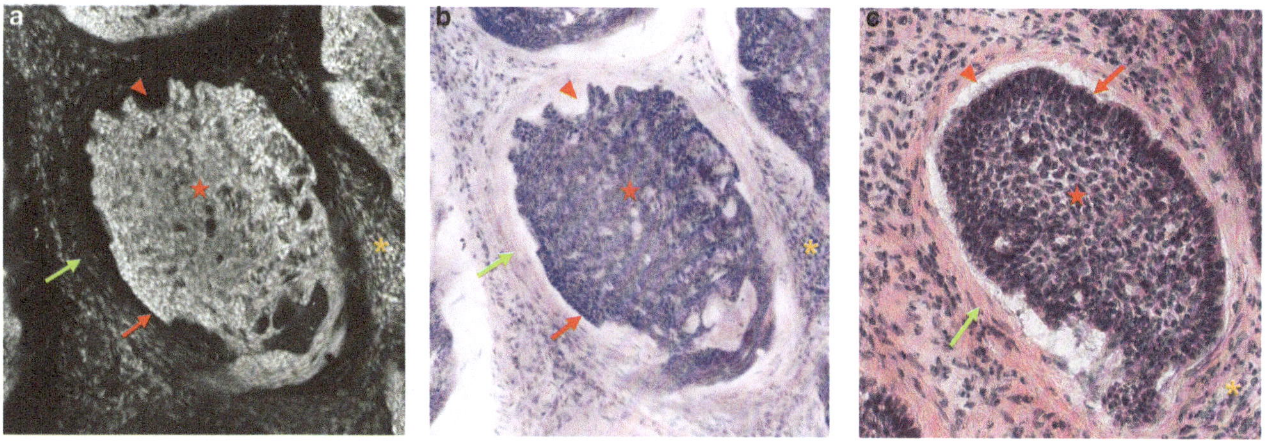

Fig. 7.5 High magnification submosaic images of another nodular BCC showing surrounding collagenous stroma. **a** FCM image shows a bright well-demarcated BCC tumor nodule (red star) with cellular features of BCC, including peripheral palisading (red arrow). Around the nodule, one can appreciate organized bundles of grayish collagen fibers (fluorescent green arrow) and a foci of inflammatory cells (orange asterisk). **b** On the corresponding digital H&E (DHE) image, the nodule (red star) appears well-demarcated and dark purple (bright on FCM) in color with cellular features of BCC and palisading. The organized collagen fibers (fluorescent green arrow) appears pink (grayish on FCM) and inflammatory foci (orange asterisk) appears purple in color (bright on FCM). **c** Corresponding conventional H&E-stained image shows an excellent correlation with the FCM (**a**) and DHE (**b**) images for the BCC tumor nodule (red star) with palisading (red arrow), and surrounding collagen bundles (fluorescent green arrow) and inflammation (orange asterisk). H&E magnification = 40X

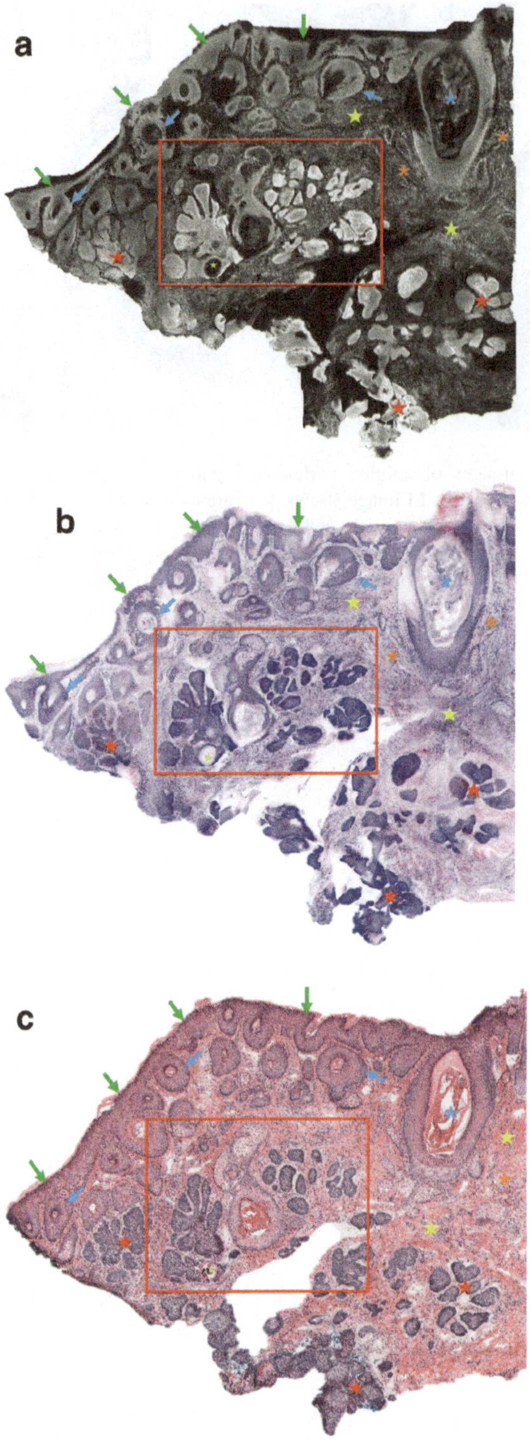

Fig. 7.6 Low magnification submosaic images of a nodular BCC showing focal keratinization obtained from left nasal ala of a 34-year-old male. **a** FCM image shows well-demarcated bright and crowded aggregates of BCC tumor nodules (red stars and red boxed area) that can be readily distinguished from the less bright surrounding hair follicles (blue arrows), and epidermis (green arrows). Within one of the tumor nodule, a grayish dark cyst-like space (fluorescent green star) can be seen. One of the hair follicle infundibulum is dilated and is filled with grayish material (blue asterisk). The surrounding grayish dermis (yellow stars) has extensive inflammatory infiltrate of small bright cells (orange asterisks). **b** On the corresponding DHE image, aggregates of nodular BCC (nBCC; red boxed area and red stars) appears dark purple in color (bright on FCM) BCC, while the hair follicles (blue arrows) and epidermis (green arrows) appears paler purple (less bright on FCM). The dark cyst-like space (fluorescent green star) within the tumor nodule appears as lamellated pale purple keratin (grayish dark on FCM). The keratin within the dilated hair infundibulum (blue asterisk) also appears purplish in color (blue asterisk). Inflammatory infiltrate (orange asterisks) appears as small purple (bright on FCM) dots in a pinkish (grayish on FCM) dermis (yellow stars). **c** Corresponding conventional H&E stained image shows correlative features with the FCM (**a**) and DHE (**b**) images for the BCC tumor nodules (red stars and boxed area) with focal keratinization (fluorescent green star), hair follicles (blue asterisk), dilated hair follicle infundibulum (blue star), epidermis (green arrows), and Inflammatory infiltrate (orange asterisk) in the dermis (yellow stars). H&E magnifications = 4X

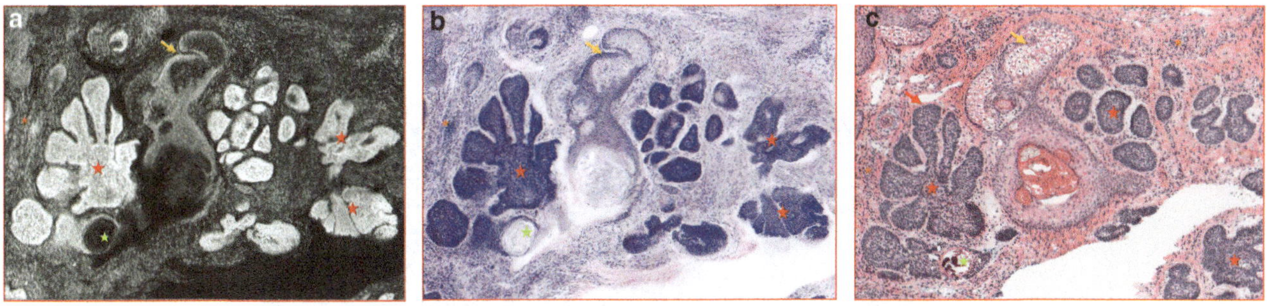

Fig. 7.7 High magnification submosaic images of the nodular BCC with focal keratinization obtained by digitally zooming in the red boxed area of mosaics in Fig. 7.6. **a** FCM image shows well-demarcated bright and crowded aggregates of BCC tumor nodules (red stars) with grayish dark cyst-like space (fluorescent green star) in the center of one of the nodules. Small bright inflammatory cells (orange asterisk) are seen surrounding the tumor nodules. A pilosebaceous gland (orange arrow) can be easily distinguished from the adjacent BCC tumor nodules. **b** On the corresponding digital H&E (DHE) image, the tumor aggregates of BCC (red stars) appears dark purple (bright on FCM) and the central keratinization (fluorescent green star) appears as lamellated pale purple (grayish dark on FCM) material. The inflammatory cells clusters (orange asterisk) appears purple in color (bright on FCM) and the pilosebaceous gland (orange arrow) as pale purple. **c** Corresponding conventional H&E-stained image, correlates well with the FCM (**a**) and DHE (**b**) images for BCC tumor nodules (red stars) with keratinization (fluorescent green star), inflammatory cells (orange asterisk), and pilosebaceous gland (orange arrow). H&E magnifications = 10X

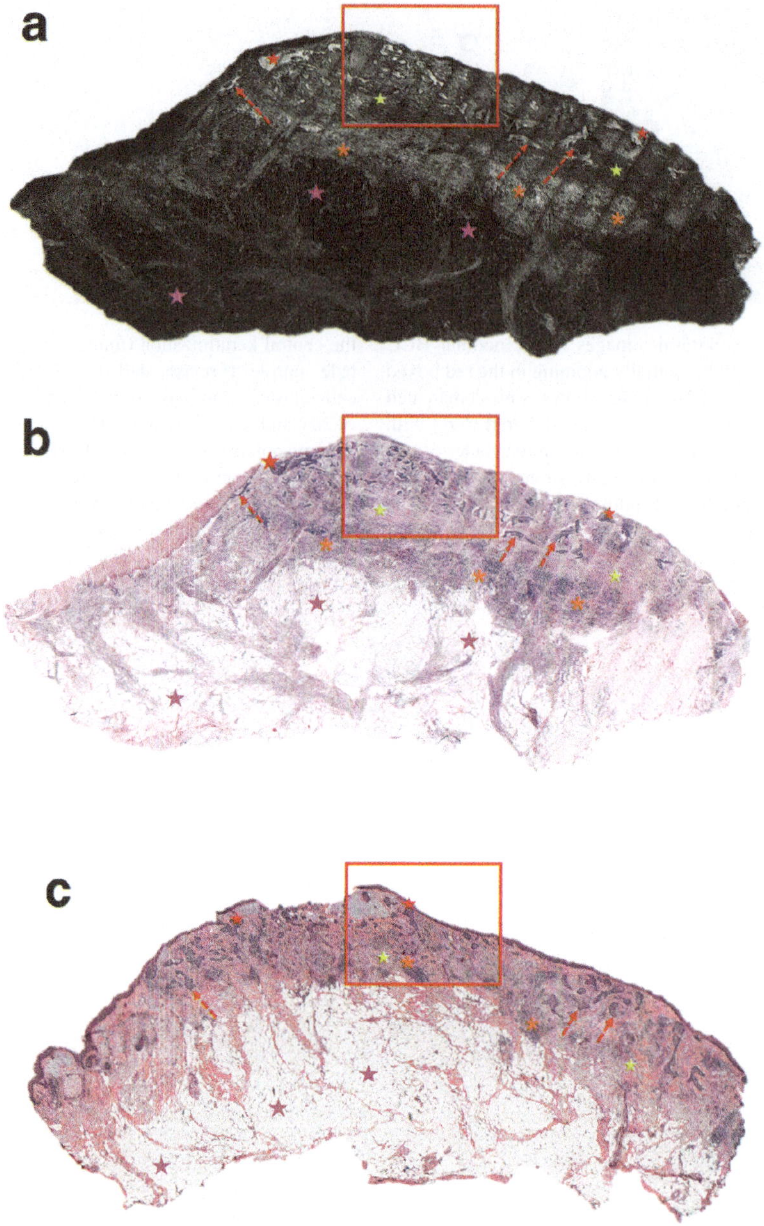

Fig. 7.8 Low magnification mosaic images of a nodular and infiltrative BCC (nBCC and iBCC) obtained from right cheek of a 95-year-old male. **a** FCM image shows larger bright tumor aggregates (red boxed area, and red stars) of nBCC and bright thin irregular strands (red dashed arrows) of iBCC in the grayish dermis (yellow stars). Even at this low magnification, the thin bright strands of iBCC can be readily identified within the "starry sky" appearance of inflammatory cells (orange asterisks). Subcutis appears as dark lobules (pink stars) and is free from BCC infiltration. **b** On the corresponding DHE image, both the large aggregates (red stars) of nBCC and thin strands of iBCC (red dashed arrows) appears as dark purple in color (bright on FCM) within a purplish pink (grayish on FCM) dermis (yellow stars). The inflammatory cells (orange asterisks) appears as clusters of small purple color (bright on FCM) dots, which can be readily distinguished from the thin iBCC strands (red dashed arrows). Subcutis (pink stars) appears as white lobulated (dark on FCM) structures of adipocytes. **c** Corresponding conventional H&E-stained image demonstrates an excellent correspondence with the FCM (**a**) and DHE (**b**) images for the nodular (red star) and infiltrative components of BCC (red boxed area and red dashed arrows), and for the surrounding dermis (yellow stars) with inflammation (orange stars), and the subcutis (pink stars). H&E magnifications = 2X

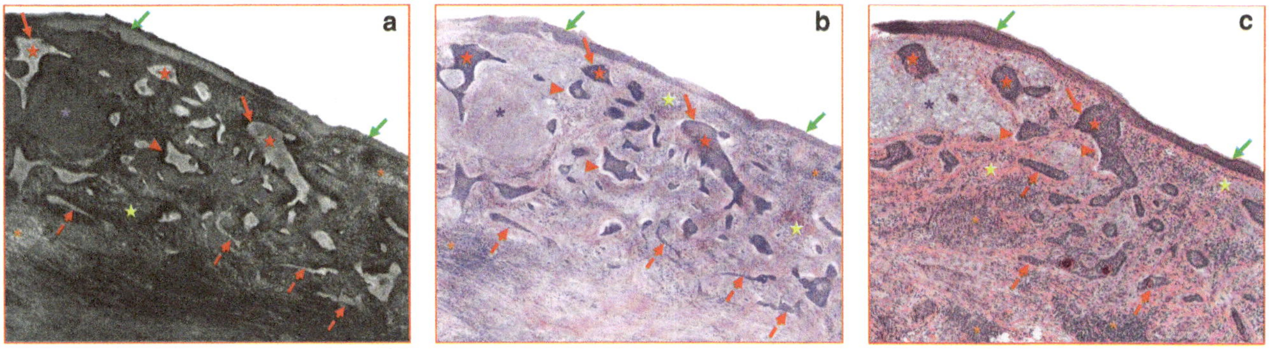

Fig. 7.9 Medium magnification submosaic images of the nodular and infiltrative BCC showing cellular features obtained by digitally zooming in the red boxed area from the mosaics in Fig. 7.8. **a** FCM image shows both the nodular (red stars) and infiltrate BCC (red dashed arrows) as bright tumor aggregates within a relatively dark grayish dermis (yellow star). Nodular BCC (red stars) is surrounded by dark areas of clefting (red arrowheads). Within the tumor aggregates, crowded and bright nuclei forms peripheral palisading (red arrows). Palisading and clefting are less distinct around the thin strands of iBCCs; however, these tumor strands can be readily distinguished from surrounding small bright inflammatory cell aggregates (orange asterisks) due to their angulated and irregular shape. In the superficial dermis, a grayish crumpled area of solar elastosis (purple asterisk) can be seen. Epidermal nuclei (green arrows) appear less bright and less crowded compared to the BCCs. **b** On the corresponding DHE image, nBCC (red stars) and iBCC (red dashed arrows) appears as dark purple (bright on FCM) large aggregates (red stars) and thin irregular strands (red dashed arrows), respectively. Solar elastosis (purple asterisk) appear as crumpled pale purple (grayish on FCM) material in the dermis (yellow star). Inflammatory cells (orange asterisks) appears purple (bright on FCM). **c** Corresponding conventional H&E-stained image shows an excellent correlation with FCM (**a**) and DHE (**b**) images for the morphological features of nBCC (red stars) and iBCC (red dashed arrows), clefting (red arrowheads), palisading (red arrow), epidermis (green arrows), inflammatory cells (orange asterisks), and solar elastosis (purple asterisk). H&E magnifications = 10X

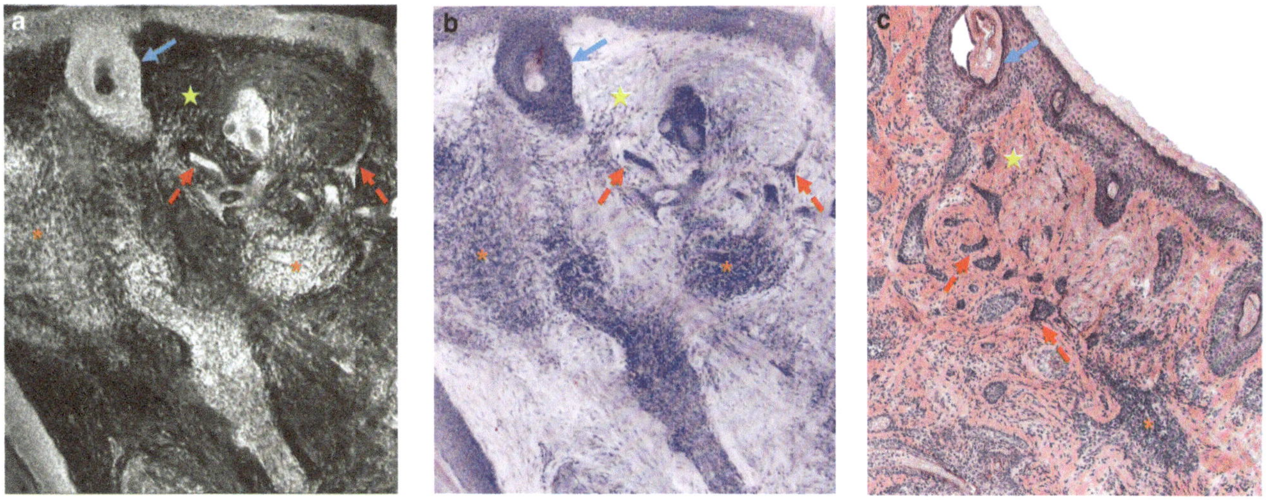

Fig. 7.10 Medium magnification submosaic images of another infiltrative BCC (iBCC) showing subtle thin tumor strands obtained from left supra nasal tip of a 74-year-old female. **a** FCM image shows thin bright strands of iBCC (red dashed arrows), which can be readily discerned in extensive bright inflammatory cell clusters (orange asterisks) within a grayish dark dermis (yellow star). A hair follicle (blue arrow) can also be identified. **b** Digital HE image shows these thin strands (red dashed arrows) as dark purple in color (bright on FCM) in the surrounding pinkish purple (grayish on FCM) dermis (yellow star). These cohesive thin strands of iBCC can be readily contrasted from the surrounding dense purple color (bright on FCM) inflammatory infiltrate (orange asterisks). The hair follicle (blue arrow) appears pale purple in color. **c** Corresponding conventional H&E-stained image shows a close correspondence with the FCM (**a**) and DHE (**b**) images for iBCC strands (red dashed arrows), inflammatory cells (orange asterisk) in dermis (yellow star), and a hair follicle (blue arrow). H&E magnifications = 10X

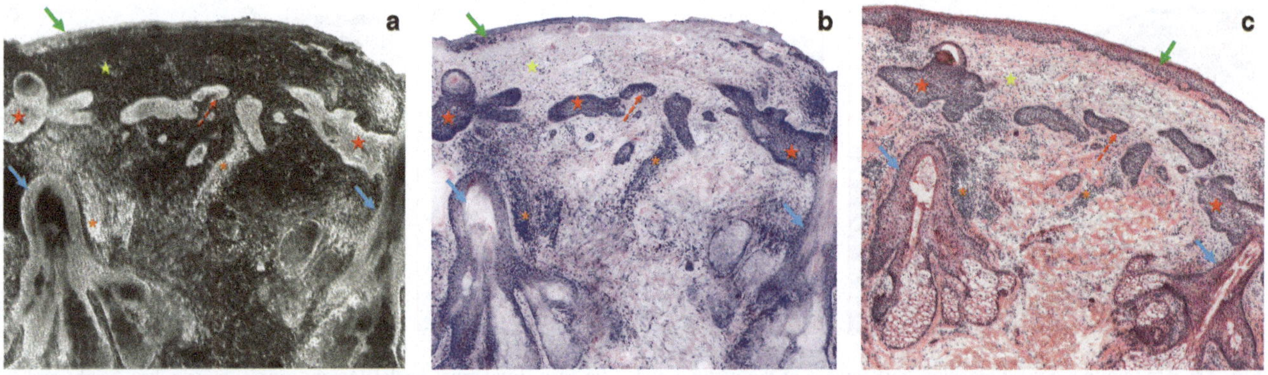

Fig. 7.11 Medium magnification mosaic images of another nodular and infiltrative BCC (nBCC and iBCC) showing perifollicular origin obtained from left scalp of a 69-year-old male. **a** FCM image shows bright tumor aggregates of nBCC (red stars) and infiltrative BCCs (red dashed arrow). One of the nodules, on the right side of the image, is seen originating from the lining of a pilosebaceous gland (blue arrow). Overlying epidermis (green arrow) is bright with sparse nuclei in comparison to the crowded nuclei of BCC aggregates (red stars and red dashed arrow). In the grayish dark dermis (yellow star), bright clusters of inflammatory cells (orange asterisks) are identified. **b** On the corresponding digital (H&E) image, the tumor nodules (red stars) and infiltrative BCCs strands (red dashed arrow) appears dark purple in color (bright on FCM). Perifollicular (blue arrows) involvement by one of the larger dark purple nodule (red star) is evident. Epidermis (green arrow) has small purple sparse nuclei, while clusters of small purple (bright on FCM) inflammatory cells (orange asterisks) stands out in purplish pink (yellow star). **c** Corresponding conventional H&E stained tissue section shows close resemblance to the DHE (**a**) image for superficial and infiltrative BCC components (red stars) within the dermis (yellow asterisk), inflammatory cells (orange asterisks), pilosebaceous glands (orange arrows) and hair follicles (blue arrows) and overlying epidermis (green arrow). H&E magnifications = 10X

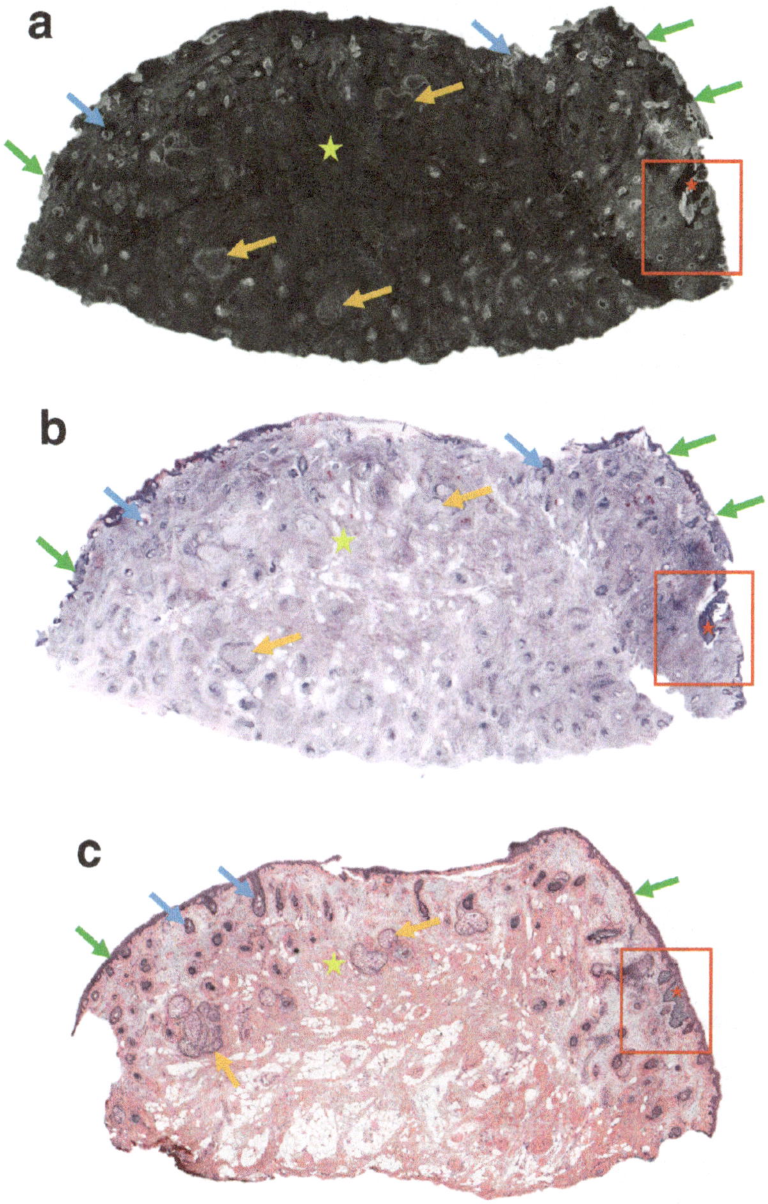

Fig. 7.12 Low magnification mosaic images of a superficial BCC obtained from right cheek of a 60-year-old male. **a** FCM image shows a bright well-demarcated sBCC tumor aggregate (red star and red boxed area) originating from the overlying normal epidermis (green arrows). Even at this scanning magnification tumor aggregate appears distinct from the relatively less bright epidermis and other normal structures such as sebaceous glands (orange arrows) and hair follicles (blue arrows). Dermis appears grayish dark (yellow star). **b** On the corresponding DHE, sBCC tumor nodule (red star, red boxed area) appears dark purple in color (bright on FCM). Sebaceous glands (orange arrows), hair follicles (blue arrows), and epidermal nuclei (green arrows) appear less intense purple in color (less bright on FCM) and can be readily distinguished from the dark purple aggregate of sBCC. Dermis appears pinkish purple (grayish dark on FCM). **c** Corresponding conventional H&E stained image shows an excellent correspondence to the sBCC tumor foci (red star and red boxed area), sebaceous gland (orange arrows), hair follicles (blue arrows), dermis (yellow star), and overlying epidermis (green arrows). H&E magnifications = 4X

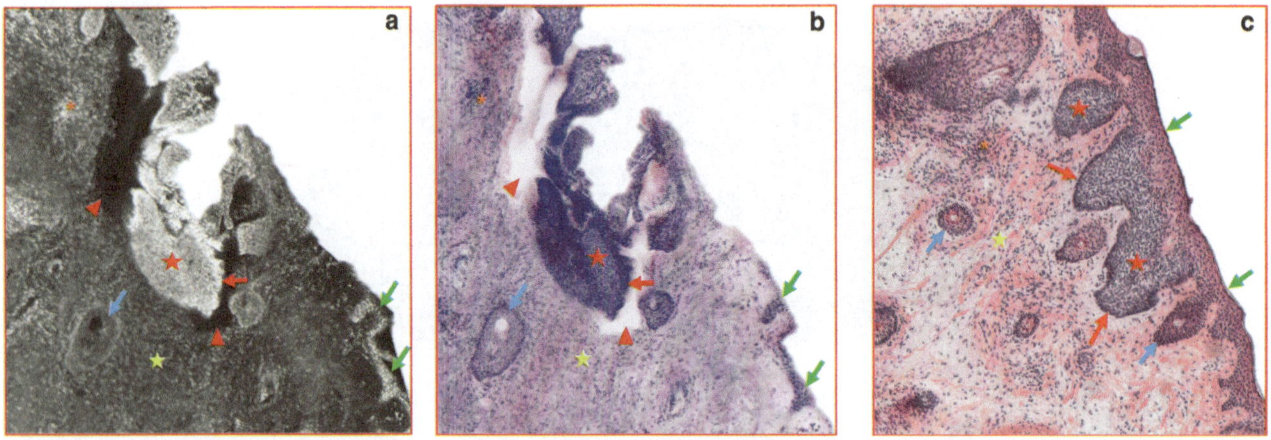

Fig. 7.13 High magnification submosaic images of the superficial BCC showing cellular features obtained by digitally zooming in the red boxed area from the mosaics in Fig. 7.12. **a** FCM image shows a bright tumor aggregate of superficial BCC (sBCC) (red star) with high N:C ratio, nuclear crowding, and peripheral palisading (red arrow). Surrounding this bright tumor aggregate is a dark area of clefting (red arrowheads). This bright tumor aggregate can be readily distinguished from the sparsely nucleated and less bright epidermis (green arrows), and hair follicles (blue arrow). There is a sprinkle of bright small cells inflammatory cells (orange asterisk) throughout the grayish dermis (yellow stars). **b** On the corresponding DHE image, aggregate of sBCC (red star) appears dark purple (bright on FCM) and well-defined. The cellular features of increased N:C ratio, overcrowded nuclei, and peripheral palisading (blue arrow) are distinct. This tumor aggregate appears distinct from the normal epidermis (green arrows) and hair follicle (blue arrow) which have sparse and lighter purple (less bright on FCM) nuclei. Hugging the lower edge of this tumor nodule is a whitish space (dark on FCM) that corresponds to clefting (red arrowheads). Inflammatory cells (orange asterisk) appears as small purple color (bright on FCM) dots in the dermis (yellow star). **c** Corresponding conventional H&E stained image shows correlative cellular features of sBCC (red stars) including peripheral palisading (red arrow), normal epidermis (green arrows), hair follicles (blue arrow), and dermal (yellow star) inflammatory cells (orange asterisk). H&E magnifications = 20X

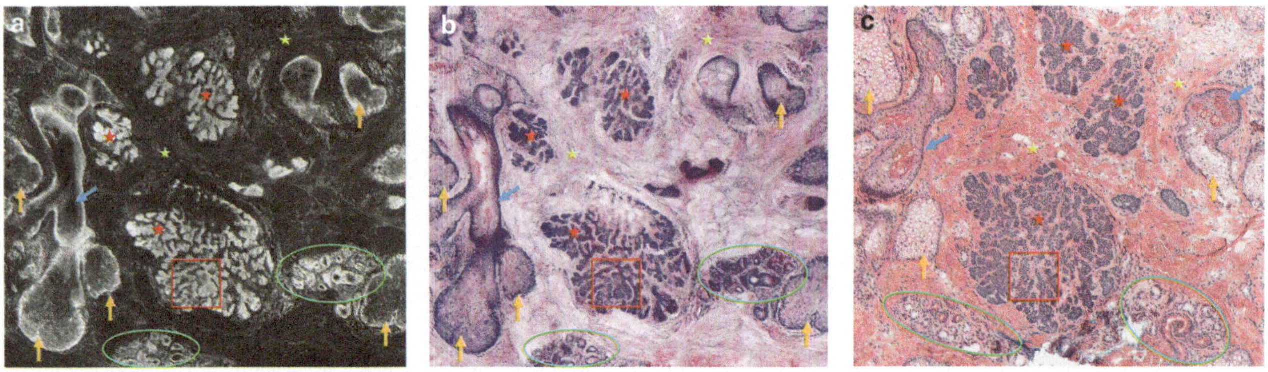

Fig. 7.14 Medium magnification submosaic images of an infundibulocystic BCC obtained from right lower canthus of a 59-year-old male. **a** FCM image shows a multiple tumor foci with anastomosing cords and strands (red stars) within a grayish dark dermis (yellow stars). These tumor cords and strands can be clearly distinguished from the surrounding hair follicles (blue arrow), sebaceous glands (orange arrows), and eccrine ducts (fluorescent green circles). **b** On the corresponding DHE image, clusters of tumor cords and strands (red stars) appears dark purples (bright on FCM) within a pinkish (grayish on FCM) dermis (yellow stars). The hair follicle (blue arrow), and sebaceous glands (orange arrows) appears paler purple with sparse nuclei, while the eccrine ducts (fluorescent green circles) appears dark purple but with less nuclear crowding than BCC. **c** Corresponding conventional H&E-stained image shows an excellent correlation with the FCM (**a**) and DHE (**b**) images for the tumor (red stars), hair follicle (blue arrow), sebaceous glands (orange arrows), eccrine ducts (fluorescent green circles) and dermis (yellow stars). H&E magnifications = 10X

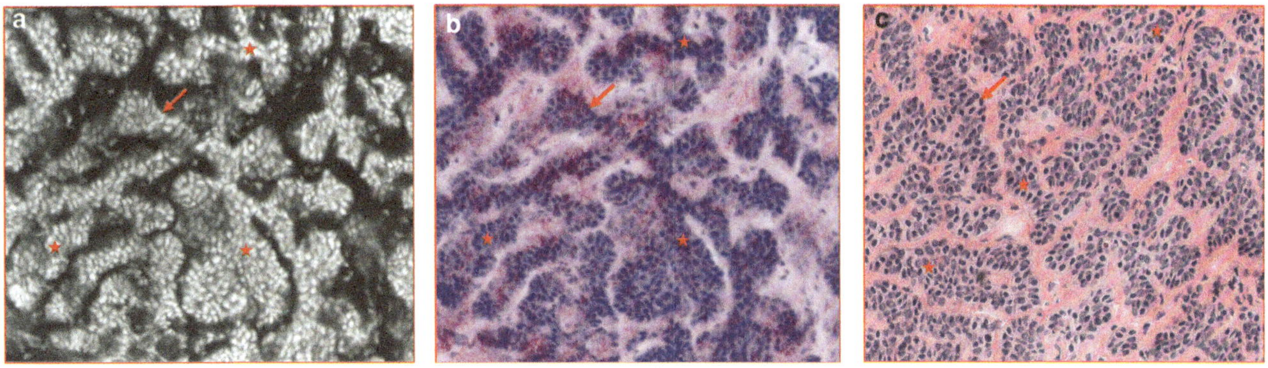

Fig. 7.15 High magnification submosaic images of the infundibulocystic BCC showing cellular features obtained by digitally zooming in the red boxed area in Fig. 7.14. **a** FCM image shows bright anastomosing cords and strands of BCC (red stars) composed of nuclei with increased N:C ratio and peripheral palisading (red arrows). **b** On the corresponding digital HE (DHE) image, these cords and strands of BCC (red stars) appears dark purple composed of nuclei with increased N:C ratio and peripheral palisading (red arrows). **c** Corresponding conventional H&E-stained image shows an excellent correlation with the FCM (**a**) and DHE (**b**) images for the cellular details of BCC (red arrowheads). H&E magnifications = 40X

References

1. Nehal KS, Bichakjian CK. Update on keratinocyte carcinomas. N Engl J Med. 2018;379:363–74. https://doi.org/10.1056/NEJMra1708701.
2. Asgari MM, Moffet HH, Ray GT, Quesenberry CP. Trends in basal cell carcinoma incidence and identification of high-risk subgroups, 1998–2012. JAMA Dermatol. 2015;151:976–81. https://doi.org/10.1001/jamadermatol.2015.1188.
3. Cameron MC, Lee E, Hibler BP, Barker CA, Mori S, Cordova M, et al. Basal cell carcinoma: epidemiology; pathophysiology; clinical and histological subtypes; and disease associations. J Am Acad Dermatol. 2019;80:303–17. https://doi.org/10.1016/j.jaad.2018.03.060.
4. Scrivener Y, Grosshans E, Cribier B. Variations of basal cell carcinomas according to gender, age, location and histopathological subtype. Br J Dermatol. 2002;147:41–7. https://doi.org/10.1046/j.1365-2133.2002.04804.x.
5. Reiter O, Mimouni I, Dusza S, Halpern AC, Leshem YA, Marghoob AA. Dermoscopic features of basal cell carcinoma and its subtypes: a systematic review. J Am Acad Dermatol. 2020. https://doi.org/10.1016/j.jaad.2019.11.008.
6. Van Loo E, Mosterd K, Krekels GAM, Roozeboom MH, Ostertag JU, Dirksen CD, et al. Surgical excision versus Mohs' micrographic surgery for basal cell carcinoma of the face: a randomised clinical trial with 10 year follow-up. Eur J Cancer. 2014;50:3011–20. https://doi.org/10.1016/j.ejca.2014.08.018.
7. Jain M, Rajadhyaksha M, Nehal K. Implementation of fluorescence confocal mosaicking microscopy by "early adopter" Mohs surgeons and dermatologists: recent progress. J Biomed Opt. 2017;22:24002. https://doi.org/10.1117/1.JBO.22.2.024002.
8. Rajadhyaksha M, Menaker G, Flotte T, Dwyer PJ, Âlez SG. Confocal examination of nonmelanoma cancers in thick skin excisions to potentially guide Mohs micrographic surgery without frozen histopathology. J Invest Dermatol. 2001;117:1137–43. https://doi.org/10.1046/j.0022-202x.2001.01524.x.
9. Karen JK, Gareau DS, Dusza SW, Tudisco M, Rajadhyaksha M, Nehal KS. Detection of basal cell carcinomas in Mohs excisions with fluorescence confocal mosaicing microscopy. Br J Dermatol. 2009;160:1242–50. https://doi.org/10.1111/j.1365-2133.2009.09141.x.
10. Gareau DS. Feasibility of digitally stained multimodal confocal mosaics to simulate histopathology. J Biomed Opt. 2009;14:034050. https://doi.org/10.1117/1.3149853.
11. Bennàssar A, Carrera C, Puig S, Vilalta A, Malvehy J. Fast evaluation of 69 basal cell carcinomas with ex vivo fluorescence confocal microscopy. JAMA Dermatol. 2013;149:839. https://doi.org/10.1001/jamadermatol.2013.459.
12. Mu EW, Lewin JM, Stevenson ML, Meehan SA, Carucci JA, Gareau DS. Use of digitally stained multimodal confocal mosaic images to screen for nonmelanoma skin cancer. JAMA Dermatol. 2016;152:1335–41. https://doi.org/10.1001/jamadermatol.2016.2997.
13. Rubin AI, Chen EH, Ratner D. Basal-cell carcinoma. N Engl J Med. 2005;353:2262–9. https://doi.org/10.1056/NEJMra044151.

Squamous Cell Carcinoma Features on Ex Vivo Confocal Imaging and Histopathologic Correlation

8

Javiera Pérez-Anker, Raquel Albero-González, and Josep Malvehy

8.1 Introduction

The availability of an instantaneous histological diagnosis through the use of ex vivo confocal microscopy (evCM) has completely changed pathology analysis not only of the skin but also in general pathology [1–4].

Squamous cell carcinoma (SCC) is a malignancy caused by keratinocytes from squamous tissues, both of the skin, mucosa (for example the cervix), and semi-mucosal membranes, and also of different organs such as the lungs, pancreas, bladder, prostate, and thyroid glands, among others [5, 6].

Cutaneous SCC is the second most frequent skin cancer. At onset, it is only locally aggressive, but in approximately 5% of cases shows a metastatic progression (especially in immunosuppression conditions). Therefore, an early and accurate histopathological diagnosis is essential for the correct management of the disease [7, 8]. It has different clinical presentations according to anatomical location, the histologic subtype, and the degree of differentiation.

Until recently, evCM criteria for SCC were reported based on now superseded technology and using a Fluorescence laser [9, 10].

In this chapter, we will describe the most important diagnostic criteria for cutaneous SCC with evCM in digital Hematoxylin & Eosin (H&E) and compare them with conventional histopathology, as reported in the literature using fusion confocal microscopy (fluorescence and reflectance lasers at the same time) for basal cell carcinoma and SCC [1–4, 11].

All the images shown in this chapter have been acquired using the VivaScope® 2500 M-Gen4; MAVIG, Munich, Germany. The stain protocol applied in each lesion was that previously described [11] with the combination of acetic acid and acridine orange, and this is further explained in Chap. 16 of this atlas. The specimens were flattened and processed according to the standard procedure reported in the literature [12, 13]. Moreover, all the images will be presented in digital H&E, in order to compare the histologic diagnostic features, as has previously been reported in the literature for normal skin, basal cell carcinoma (BCC), SCC, and dermatofibrosarcoma protuberans [14–16]. More details of the advantages of this image visualization mode are described in Chaps. 3 and 16.

A summary of all the salient comparisons between both techniques is shown in Table 8.1.

J. Pérez-Anker (✉) · J. Malvehy
Dermatology Department, Hospital Clínic de Barcelona, Barcelona, España
e-mail: perez12@clinic.cat

R. Albero-González
Department of Pathology, Hospital Clínic de Barcelona, Barcelona, España

8.2 Actinic Keratosis

The histological characteristics of these "pre-malignant lesions" are, for some researchers, quaternization and alteration in some areas of the epidermis. From a cytologic point of view (nuclear and cytoplasm features), the same alterations are found in an infiltrating SCC [17], but they are not present in all the epidermis layers. In fact, they are normally present at the basal layer. There is a focal loss of the granular layer and the presence of parakeratosis. The proliferation of the atypical keratinocytes at the basal layer is evident not only in the epidermis, but frequently, also in the follicles and sebaceous glands [2].

Actinic keratosis (AK) is an intraepithelial neoplastic lesion considered premalignant, as it could potentially progress to infiltrating SCC. Characteristically, the main histological criteria are the presence of parakeratosis over the atypical epidermis, which often spares follicular infundibulum and acrosyringium, and atypia of the keratinocytes, in a solar elastosis context. Keratinocytic atypia starts at the basal cell layer but can extend to the spinous layer causing loss of polarity, mitotic activity, and varying loss of the granular layer

M. Jain et al. (eds.), *Cutaneous Atlas of Ex Vivo Confocal Microscopy*,
https://doi.org/10.1007/978-3-030-89316-3_8

Table 8.1 Ex vivo confocal microscopy SCC features. Double stain protocol [2, 11]

Feature	Digital H&E	H&E
Intercellular bridges	Observed at high magnification (due to cellular disaggregation)	Apparent (due to cellular disaggregation)
Keratinization	Varying degrees of keratinization	Varying degrees of keratinization
Dyskeratosis	Premature keratinization visible at medium–high magnification	Premature keratinization visible at low-medium magnification
Parakeratosis	Identifiable at low magnification	Identifiable at low magnification
Pleomorphism	Varying degrees of differentiation (most tumors classified as well to moderately differentiated)	Varying degrees of differentiation (most tumors classified as well to moderately differentiated)
Desmoplasia	Feature not clearly evident	Fibrous reactive stroma can be very prominent
Borders	Pushing or infiltrative	Pushing or infiltrative
Loss of granular layer	Identifiable at lower magnification	Identifiable at lower magnification
Mitotic figures	Easy to detect at higher magnification	Easy to detect at higher magnification
Stromal inflammation	Moderate to dense lymphoplasmocytic stroma infiltrate (characteristic)	Moderate to dense lymphoplasmocytic stromal infiltrate (characteristic)
Dendritic cells	Identifiable at lower magnification as pink projections	Only observed with special staining

and does not involve the full thickness of the epidermis, as when this occurs, the lesion is categorized as an in-situ SCC. It can be seen not only in the epidermis but also in the follicles and sebaceous glands. From a cytologic point of view, the nuclei are hyperchromatic, pleomorphic, and irregular in size. The morphological changes between atypical and non-atypical cells are clearly evident (Figs. 8.1 and 8.2) [2].

8.3 In Situ Squamous Cell Carcinoma

In in situ SCC, the normal maturation of the different epidermal layers is lost, and there is full-thickness epidermal involvement by atypical keratinocytes. The neoplastic lesion is confined to the epidermis and superficial adnexa and do not infiltrate through the underlying dermis. Epidermal thickness is variable, usually with hyperkeratosis, parakeratosis, acanthosis, and loss of the granular layer. Because of the alteration in the normal keratinization, atypical keratinocytes typically show eosinophilic cytoplasm, and crowded, enlarged, and pleomorphic nuclei which are typically heterochromatic. Dyskeratotic cells are present, and keratin pearls can be observed according to the differentiation of the tumor. Necrosis and mitotic figures can also be seen, characteristically extending to the epidermal surface (Fig. 8.3) [2].

All these characteristics are equally observed in conventional H&E and digital H&E of evCM. However, despite the morphological characteristics being the same, the colors vary according to the stain protocol applied before scanning in evCM and also to different laser intensities. The cytoplasm details which are often observed are basophilic although they may also be eosinophilic. Parakeratosis and dyskeratotic cells can also be seen either basophilic or also eosinophilic (Fig. 8.4) [2].

Regarding Bowen´s disease, taking as reference the 4th Edition of the WHO Classification of Skin Tumors, histologically the terms "Bowen´s disease" and "In situ SCC" are now used as synonyms to describe in situ SCC either in sun-damaged or sun-protected skin.

In Bowen´s disease, there is also full thickness involvement of atypical keratinocytes in the totality of the epidermis. The rete ridges become broad, and the epidermis is usually acanthotic. Dyskeratotic cells are less commonly observed while hyperkeratosis and parakeratosis may also be seen. Atypical keratinocytes can be vacuolated [2].

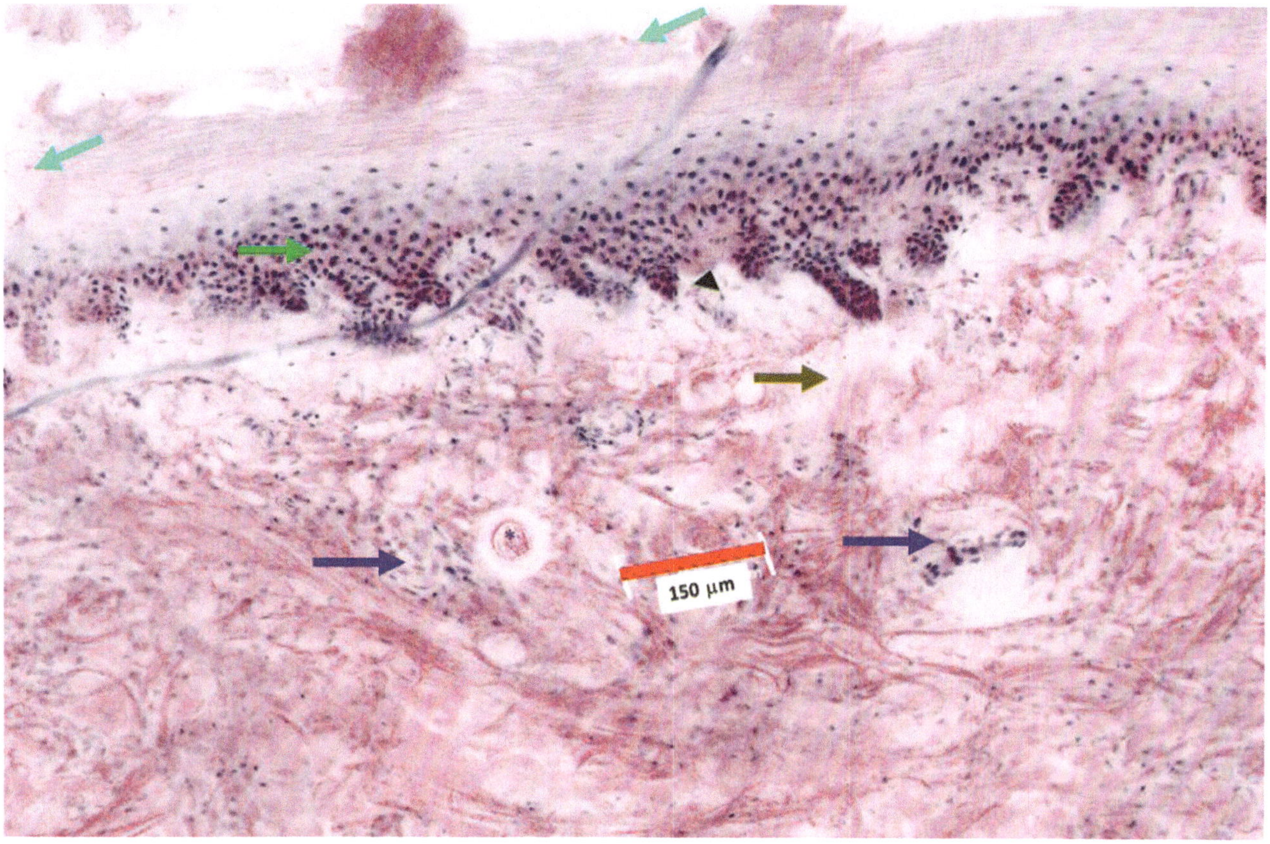

Fig. 8.1. Actinic Keratosis. EVCM. Vessels (purple arrow), solar elastosis (brown arrow), epidermis (dark green arrow), parakeratosis (turquoise arrows), cellular atypia (black triangle)

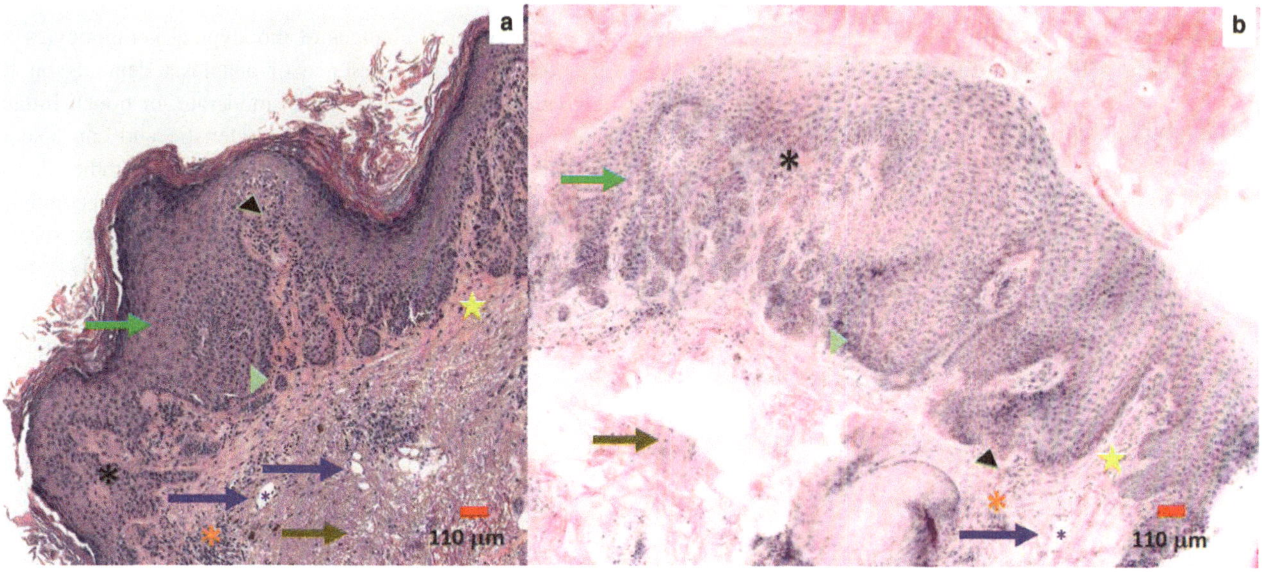

Fig. 8.2. Solar lentigo and AK. Granular layer less evident in EVCM. **a** Traditional H&E. **b** EVCM. Epidermis (dark green arrows), DEJ (light green triangles), dermis (yellow stars), inflammatory cells (dark orange asterisks), solar elastosis (brown arrows), vessels (purple arrows), lumen (purple asterisks), dyskeratotic cells (black asterisks), cellular atypia (black triangles)

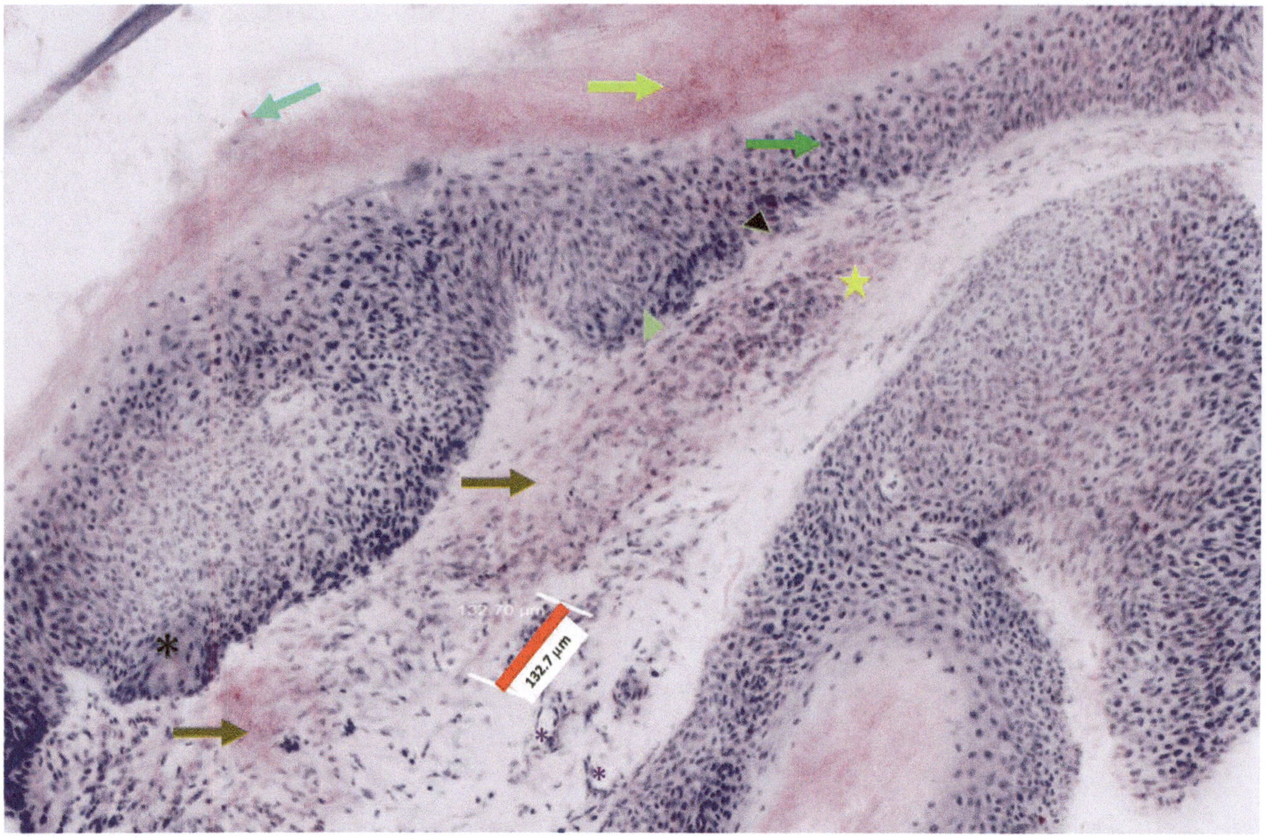

Fig. 8.3 In situ SCC. Bowen´s disease. EVCM. Epidermis (dark green arrow), DEJ (light green triangle), dermis (yellow star), hyperkeratosis (light green arrow), vascular lumen (black arrows), solar elastosis (brown arrows), parakeratosis (turquoise arrows), cellular atypia (black triangle), dyskeratotic cells (black asterisk)

8.4 Infiltrating Squamous Cell Carcinoma

In infiltrating SCC, atypical cells arise from the epidermis and progress, leading to an invasion of the dermis and its deepest components. Some infiltrating SCCs may not have a clearly observable in situ SCC component. Tumor cells are arranged in nests, strands, and sheets. Typically, intercellular bridges can be observed due to cellular disaggregation. All the morphological changes of the atypical keratinocytes are also observed. The presence of anaplasia depends on the degree of differentiation (good, moderate, or poor). Inflammation around the tumor is characteristic and can also be present around blood vessels, lymphatics, and other deeper components which could be a useful clue to recognizing fragments of tumor, especially during Mohs surgery (Table 8.1; Figs. 8.5, 8.6, 8.7, 8.8, 8.9 and 8.10) [2].

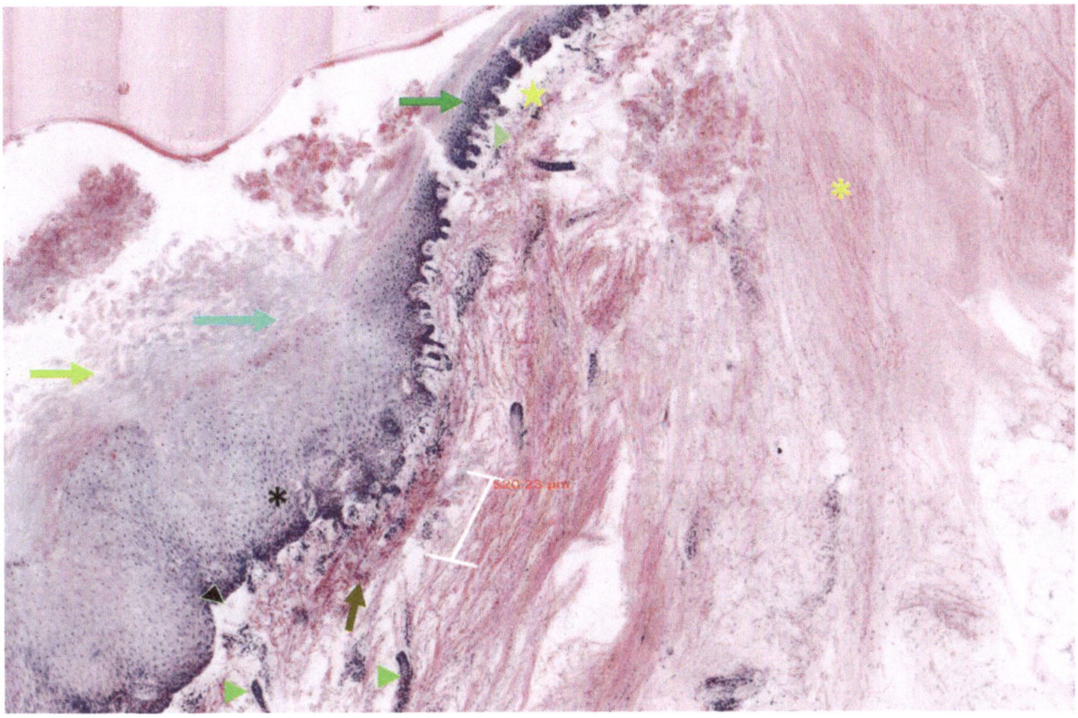

Fig. 8.4. In situ SCC with AK at the periphery. EVCM. Epidermis (dark green arrow), DEJ (light green triangle), papillary dermis (yellow star), reticular dermis (yellow asterisk), hyperkeratosis (light green arrow), parakeratosis (turquoise arrow), eccrine and glands ducts (fluorescent green triangles), solar elastosis (brown arrow), cellular atypia (black triangle), dyskeratotic cells (black asterisk)

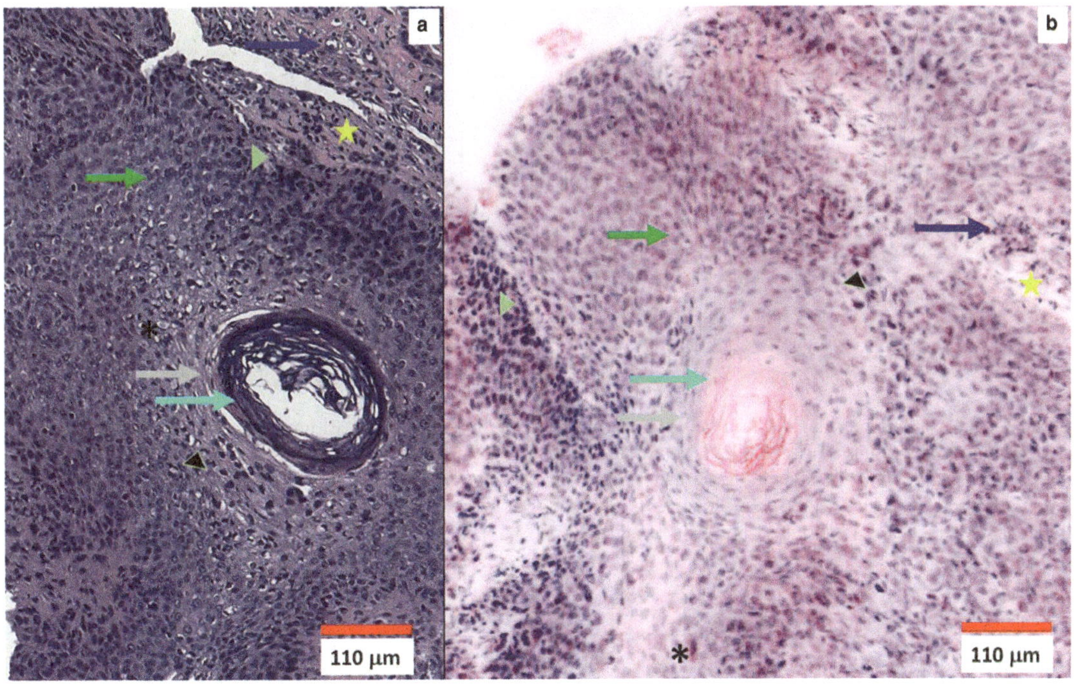

Fig. 8.5. Keratin pearl. **a** Traditional H&E. **b** EVCM. Full microscopic features can be seen in both images. Epidermis (dark green arrows), DEJ (light green triangles), dermis (yellow stars), vessels (purple arrows), keratin pearl (light grey arrows), parakeratosis (turquoise arrows), cellular atypia (black triangles), dyskeratotic cells (black asterisks)

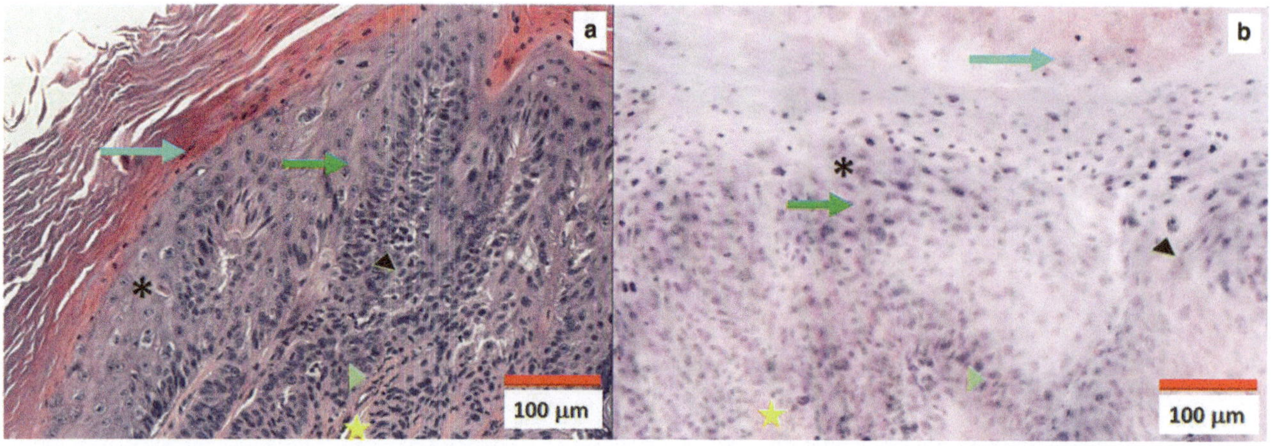

Fig. 8.6. Nuclear atypia in a well-differentiated squamous cell carcinoma. **a** Traditional H&E. **b** EVCM. Nuclear atypia is better visualized in EVCM. Epidermis (dark green arrows), DEJ (light green triangles), dermis (yellow stars), parakeratosis (turquoise arrows), cellular atypia (black triangles), dyskeratotic cells (black asterisks)

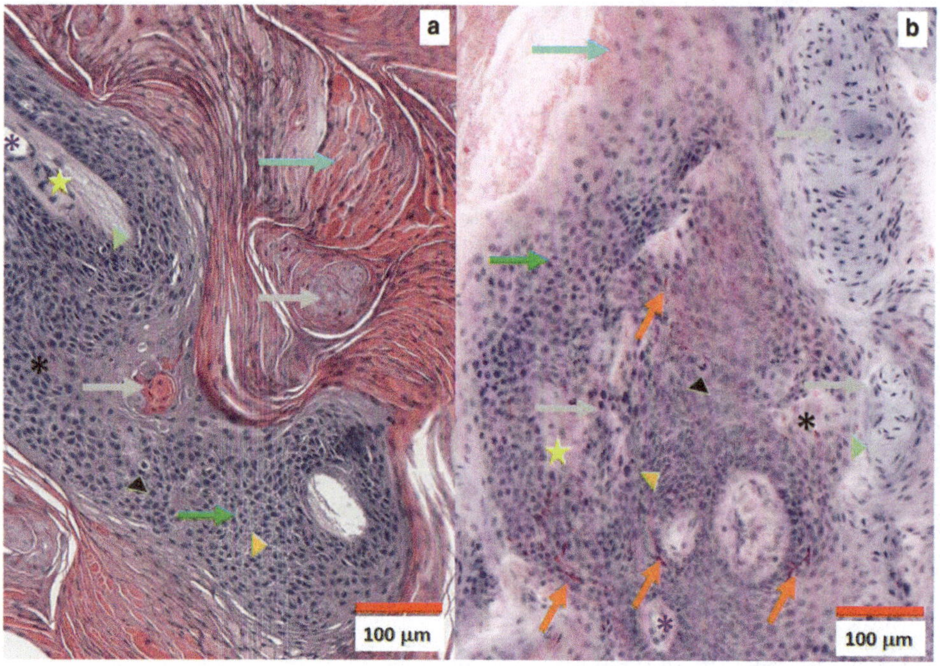

Fig. 8.7. Cellular details and dendritic cells in a well- differentiated squamous cell carcinoma. **a** Traditional H&E. **b** EVCM. Architectural disarrangement in both images with alteration in the normal maturation. Epidermis (dark green arrows), DEJ (light green triangles), papillary dermis (yellow stars), parakeratosis (turquoise arrows), keratin pearls (light grey arrows). Lumen vessel (purple asterisks), cellular atypia (black triangles), dyskeratotic cells (black asterisks), intercellular bridges (yellow triangles). Dendritic cells only observed in evCM (orange arrows)

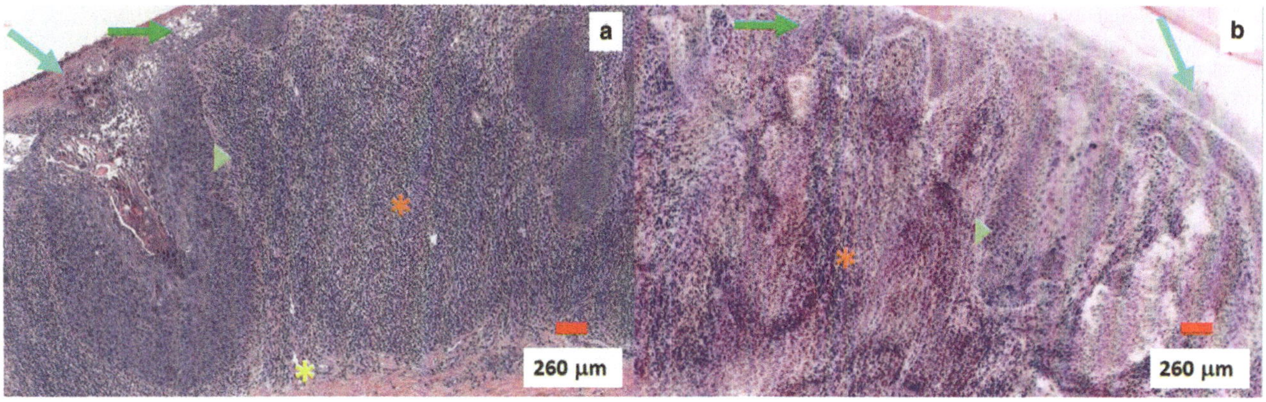

Fig. 8.8. Dense inflammation hiding an endophytic SCC. **a** Traditional H&E. **b** EVCM. Nuclear details are better visualized in EVCM at this magnification. Epidermis (dark green arrows), DEJ (light green triangles), papillary dermis (yellow asterisks), parakeratosis (turquoise arrows), and inflammatory cells (dark orange asterisks)

Fig. 8.9. Keratoacanthoma. **a** Traditional H&E. **b** EVCM. Epidermis (dark green arrows), DEJ (light green triangles), dermis (yellow asterisks), inflammatory cells (dark orange asterisks), hyperkeratosis (light green arrows)

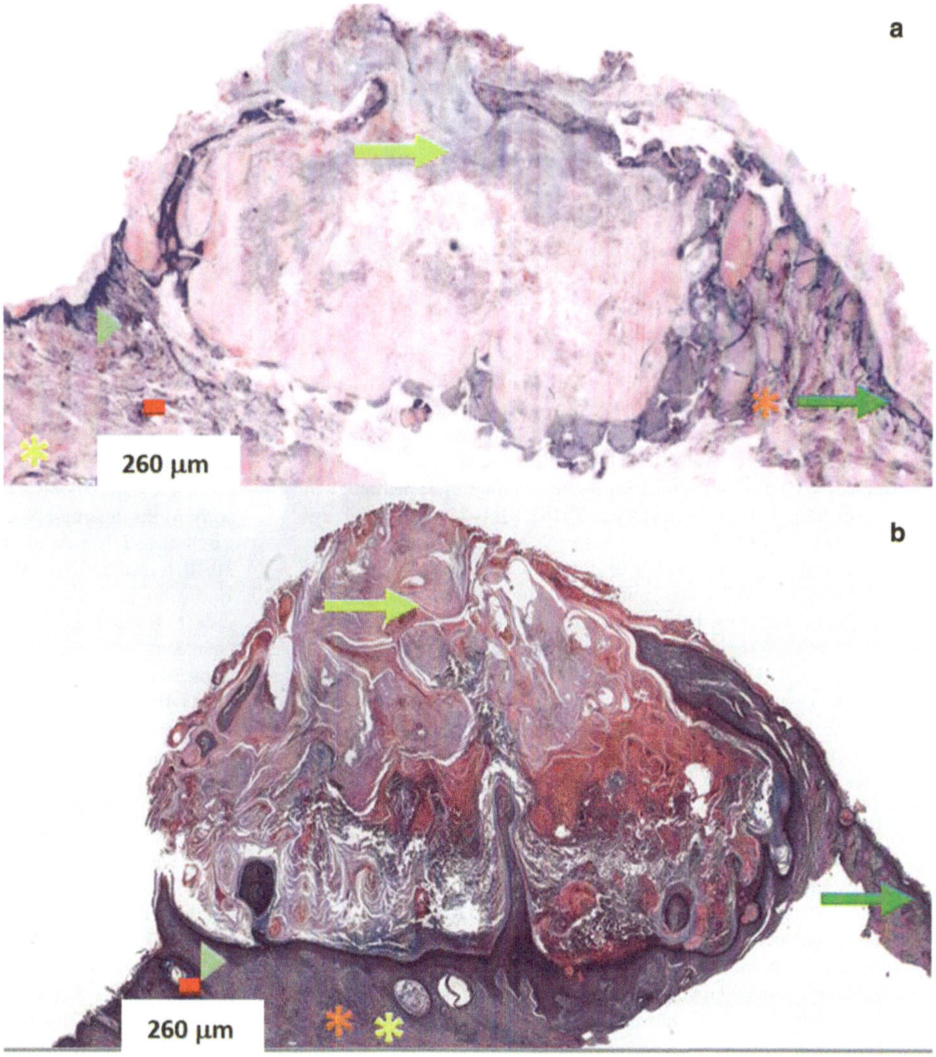

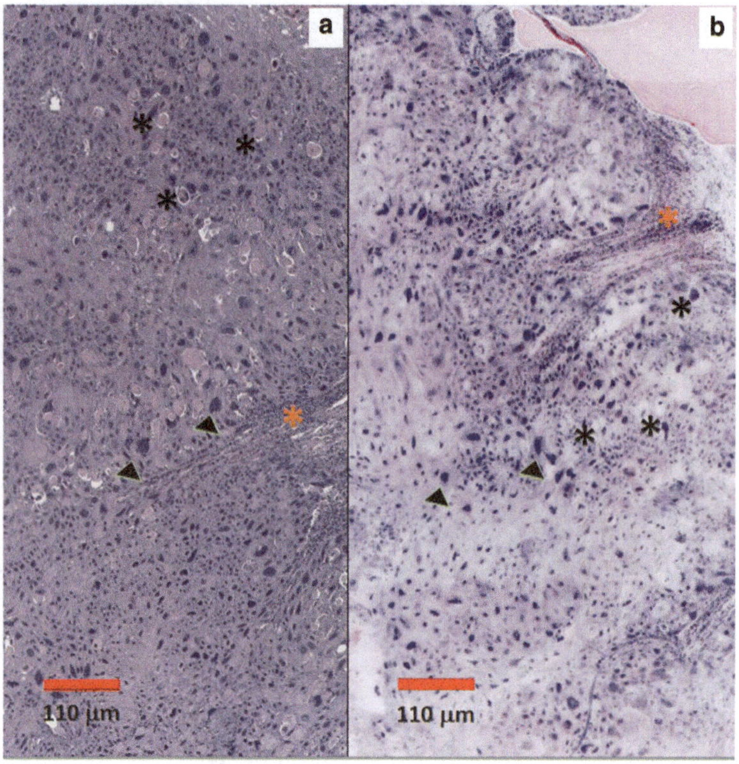

Fig. 8.10. Acantholytic squamous cell carcinoma. Cytological details. **a** Traditional H&E. **b** EVCM. Moderate cellular pleomorphism is observed in both techniques with an enlargement of the nuclei. Inflammatory cells (dark orange), acantholytic dyskeratotic cells (black asterisks), cellular atypia (black triangles)

References

1. Pérez-Anker J, Malvehy J, Moreno-Ramírez D. Ex vivo confocal microscopy using fusion mode and digital staining: changing paradigms in histological diagnosis. Actas Dermo-Sifiliograficas 2020;111. https://doi.org/10.1016/j.ad.2019.05.005.
2. Malvehy J, Pérez-Anker J, Toll A, Pigem R, García A, Alós L, et al. Ex vivo confocal microscopy: revolution in fast pathology in dermatology. Br J Dermatol. 2020;183:1011–25. https://doi.org/10.1111/bjd.190173.
3. Villarreal JZ, Pérez-Anker J, Puig S, Pellacani G, Solé M, Malvehy J, et al. Ex vivo confocal microscopy performs real-time assessment of renal biopsy in non-neoplastic diseases. J Nephrol. 2020. https://doi.org/10.1007/s40620-020-00844-8.
4. Panarello D, Compérat E, Seyde O, Colau A, Terrone C, Guillonneau B. Atlas of Ex Vivo prostate tissue and cancer images using confocal laser endomicroscopy: a project for intraoperative positive surgical margin detection during radical prostatectomy. Eur Urol Focus. 2019:1–18. https://doi.org/10.1016/j.euf.2019.01.004.
5. Sánchez-Danés A, Blanpain C. Deciphering the cells of origin of squamous cell carcinomas. Nat Rev Cancer. 2018;18:549–61. https://doi.org/10.1038/s41568-018-0024-5.
6. Dotto P (Department of Biochemistry U of L, Rustgi, Anil K. (Division of Gastroenterology, Departments of Medicine and Genetics, Abramson Cancer Center U of PPS of M. Squamous cell cancers: a unified perspective on biology and genetics. Cancer Cell. 2016;29:622–37. https://doi.org/10.1016/j.ccell.2016.04.004. Squamous.
7. Tejera-Vaquerizo A, Cañueto J, Llombart B, Martorell-Calatayud A, Sanmartín O. Predictive value of sentinel lymph node biopsy in cutaneous squamous cell carcinoma based on the AJCC-8 and brigham and women's hospital staging criteria. Dermatologic Surgery : Official Publication for American Society for Dermatologic Surgery [et Al]. 2020;46:857–62. https://doi.org/10.1097/DSS.0000000000002170.
8. Tejera-Vaquerizo A, Descalzo-Gallego MA, Otero-Rivas MM, Posada-García C, Rodríguez-Pazos L, Pastushenko I, et al. Cancer incidence and mortality in spain: a systematic review and meta-analysis. Actas Dermo-Sifiliograficas. 2016;107:318–28. https://doi.org/10.1016/j.ad.2015.12.008.
9. Longo C, Ragazzi M, Gardini S, Piana S, Moscarella E, Lallas A, et al. Ex vivo fluorescence confocal microscopy in conjunction with Mohs micrographic surgery for cutaneous squamous cell carcinoma. J Am Acad Dermatol. 2015;73:321–2. https://doi.org/10.1016/j.jaad.2015.04.027.
10. Hartmann D, Krammer S, Bachmann MR, Mathemeier L, Ruzicka T, Bagci IS, et al. Ex-vivo confocal microscopy features of cutaneous squamous cell carcinoma. J Biophotonics 2018;11:1–7. https://doi.org/10.1002/jbio.201700318.
11. Anker JP, Ribero S, Yélamos O, García A, Alos L, Alejo B, et al. Basal cell carcinoma characterisation using fusion ex vivo confocal microscopy: a promising change in conventional skin histopathology. Br J Dermatol. 2020;182:468–76. https://doi.org/10.1111/bjd.18239.
12. Pérez-Anker J, Puig S, Malvehy J. A fast and effective option for tissue flattening: Optimizing time and efficacy in ex vivo confocal microscopy. Journal of the American Academy of Dermatology 2020;82. https://doi.org/10.1016/j.jaad.2019.06.041.
13. Pérez-Anker J, Toll A, Puig S, Malvehy J. Six steps to reach optimal scanning in ex vivo confocal microscopy. J Am Acad Dermatol. 2021 Jan 19:S0190–9622(21)00193–6. https://doi.org/10.1016/j.jaad.2021.01.044. Epub ahead of print. PMID: 33476729.
14. Toloza Salech C, Pérez-Anker J, La Rotta Higuera E, Toll A, García A, Alós LL, et al. Tinción digital en diagnóstico histológico de tumors cutáneos con microscopía confocal ex vivo de fusión. Poster presented at: 47th Congreso nacional de Dermatología y Venerología (AEDV); June 5–8, 2019; Barcelona, Spain.

15. Pérez-Anker J, Toll A, Puig S, García A, Alós Ll, Malvehy J. Real time visualization of mitosis and nuclear pleomorphism in fresh tissue using ex vivo confocal microscopy. Poster presented at: 28th Congress of the European Academy of Dermatology and Venerology (EADV); October 9–13, 2019; Madrid, Spain.
16. Toloza Salech C, Malvehy J, Puig S, García A, Alós Ll, Toll A, et al. A proof of concept: digital H & E in the histological diagnosis of cutaneous tumors with ex vivo fusion confocal microscopy. In: Poster presented at 28th congress of the European Academy of Dermatology and Venerology (EADV); October 9–13, 2019; Madrid, Spain.
17. Ackerman AB, Mones JM, Academy A, York N. Solar (actinic) keratosis is squamous cell carcinoma. Br J Dermatol. 2006;155:9–22. https://doi.org/10.1111/j.1365-2133.2005.07121.x.

Part V
Melanocytic Lesions

9 Acquired Nevi: Junctional, Compound, and Dermal

Daniela Hartmann

9.1 Basics of Acquired Nevi

Ex vivo confocal microscopy (EVCM) enables bedside histology and offers the surgeon a direct intraoperative tissue examination. This innovative, ultra-rapid diagnostic tool may be extended beyond non-melanoma skin cancer, especially basal cell carcinoma, to other indications including melanocytic lesions. Current knowledge and experience on the use of EVCM in melanocytic lesions are limited but shows promising horizons. Examples of junctional, compound, and dermal nevi with classic EVCM features are presented in this chapter (Table 9.1; Figs. 9.1, 9.2, 9.3 and 9.4).

9.1.1 Basics of Acquired Nevi

- Definition: Benign, congenital or acquired, melanocytic lesions are mostly brown to black colored, rarely they appear as reddish or skin-colored. They may be located only in the epidermis (junctional melanocytic nevi) or both in epidermis and dermis (compound type of melanocytic nevi) or exclusively in the dermis (dermal melanocytic nevi) [1].
- Patient demographics: Light-skinned population (Caucasian) presents a greater number of melanocytic nevi than dark-pigmented races of African-Americans and Asians. On the contrary, melanocytic nevi on the palms and soles, as well as in the nail bed are more common in African-Americans and Asians than in light-skinned races [1–3].
- Histopathology: Melanocytic nevi show junctional and/or dermal proliferation of melanocytes. There are three types of nevocytes primarily based on their shape, size, and their location (superficial to deeper layers): *Type A cells (epithelioid)* are large round to oval with abundant cytoplasm, located mostly at the junction or in the papillary dermis and forms cohesive nests; (2) *Type B cells (lymphocytoid)* cells are smaller with decreased cytoplasmic content, located deeper than type A cells in the dermis, and forms nests, cords, and strands or distributed diffusely; (3) *Type C cells (Schwannian)* are elongated or spindle-shaped, located deeper than Type B cell in the dermis. In nevi, these cells are seen transiting from type A through type C, a phenomenon called "maturation". Nevocytes can also be seen surrounding the skin appendages but do not destroy them [1, 4].
- Nevi variants: There are several nevi types that differs in clinical and histological appearance. Spitz nevus is a benign neoplasm of epithelioid and/or spindle melanocytes that usually appears in childhood. Spitz nevi may often have clinical, dermoscopic, confocal, and histological features that may sometimes make it difficult to distinguish them from melanomas [5]. Report on confocal features of Spitz nevus using EVCM have been previously presented [6]. However, a larger number of Spitz nevi are necessary to be studied using this device to better define their features.

D. Hartmann (✉)
Department of Dermatology and Allergy, University Hospital, LMU Munich, Frauenlobstrasse 9-11, 80337 Munich, Germany
e-mail: Daniela.Hartmann@med.uni-muenchen.de

M. Jain et al. (eds.), *Cutaneous Atlas of Ex Vivo Confocal Microscopy*,
https://doi.org/10.1007/978-3-030-89316-3_9

Table 9.1 Acquired melanocytic nevi features on FCM, DHE, and corresponding conventional H&E

Normal melanocytic nevi features on FCM, DHE, and corresponding conventional H&E		
Fluorescence confocal microscopy (FCM) green scale mode [6]	Digital H&E (combined FCM and RCM modes)	Conventional H&E-stained image
Fluorescence in the stratum corneum: White (green*) glossy and brightly shining, blurred patch with a horizontally striped pattern	The bright white (green*) striped patches on FCM appear dark purple on DHE	Inclusions of lost pigment in stratum corneum
Fluorescence in the epidermis: Nuclei of the tumor cells visualized as bright white (green*) pearl-necklace-like dots and groups of dots in the basal epidermal layer with strongly enhanced contrast compared to the surrounding dermis (Fig. 9.1)	The bright white (green*) dots/group of dots on FCM appear dark purple (high fluorescence signal) or dark to light pink (low fluorescence signal; dominant reflectance signal) on DHE	Small nests and occasionally single cell proliferations of melanocytes present in the basal epidermal layer
Fluorescence in the dermo-epidermal junction: White (green*) brightly shining evenly shaped and evenly spread 'tongues' of epidermis reaching into the papillary dermis (Fig. 9.1)	The bright white (green*) dots/group of dots on FCM appear dark purple (high fluorescence signal) or dark to light pink (low fluorescence signal; dominant reflectance signal) on DHE	Rete ridges with melanocytic proliferations
Fluorescence in the dermis: Homogeneous nests and strands of white (green*) evenly and brightly shining dots gradually decreasing in size downwards within the dermis. The nest shape can be clearly drawn to separate bright fluorescent cell nuclei from the rest of the tissue. (Figs. 9.2, 9.3 and 9.4)	Usually dark purple (strong fluorescence signal) well- demarcated nests and strands of purple dots gradually decreasing in size downwards within the dermis set in a background of pale-pink collagen (reflectance signal)	Melanocytic nests in the dermis with signs of maturation
Other fluorescent structures in the dermis: Slightly less shining (compared to melanocytes), light grey (dull green*), large, round or oval structures with granular-like cytoplasm (recognition of the granular character might be easier in reflectance mode)	Dark to light purple (depending on melanin content and intensity of the fluorescence signal) large, round or oval structures with granular-like cytoplasm	Melanophages
Other fluorescent structures in the dermis: Tiny monomorphic white (green*) brightly shining dots (smaller in comparison to the above-described structures) (Fig. 9.3)	Tiny monomorphic dark purple dots	Inflammatory cells

Abbreviations CLSM = confocal laser scanning microscopy, FCM = fluorescent confocal microscopy, RCM = reflectance confocal mode, DHE = digital haematoxylin-eosin-like staining, H&E = hematoxylin-eosin staining, * color of the fluorescent signal depending on the generation of the ex vivo CLSM device

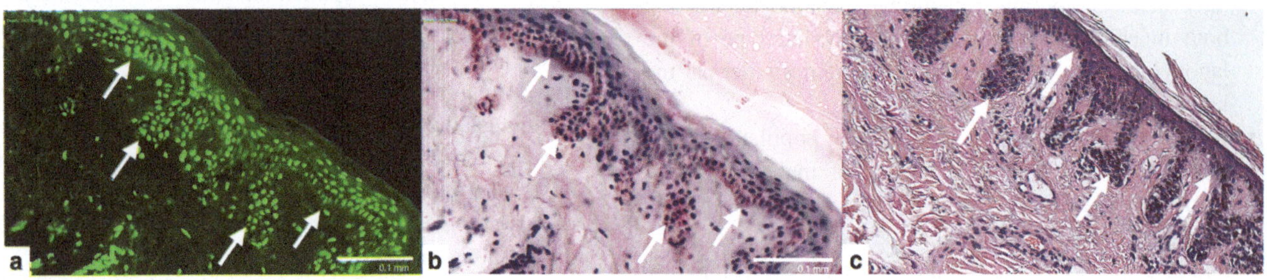

Fig. 9.1 Junctional melanocytic nevus displaying melanocytic proliferations, mostly gathered in small nests, in the basal layers of the epidermis (arrow) with no signs of transepidermal migration examined in two modes of the ex vivo confocal laser scanning microscopy (**a** fluorescence, **b** DHE) and comparison with their corresponding H&E-stained image (**c**). DHE; digital hematoxylin–eosin-like staining. H&E magnification = 20x

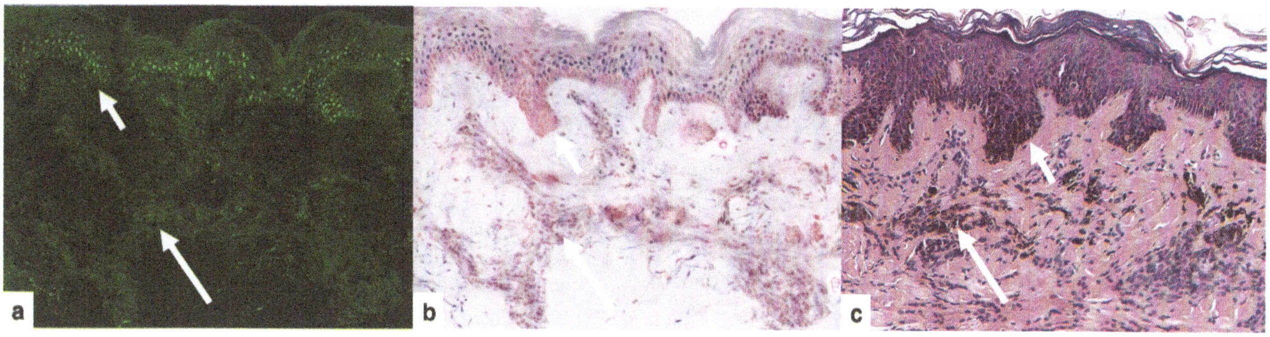

Fig. 9.2 Detailed (higher magnification) view of a compound melanocytic nevus examined in two modes of the ex vivo confocal laser scanning microscopy (**a** fluorescence, **b** DHE) and on corresponding H&E-stained image (**c**) presenting melanocytic proliferations in the basal layers of the epidermis (short arrow), as well as in the dermis (long arrow) gathered in melanocytic nests and cords showing signs of maturation in the dermis. DHE; digital hematoxylin–eosin-like staining. H&E magnification = 20x

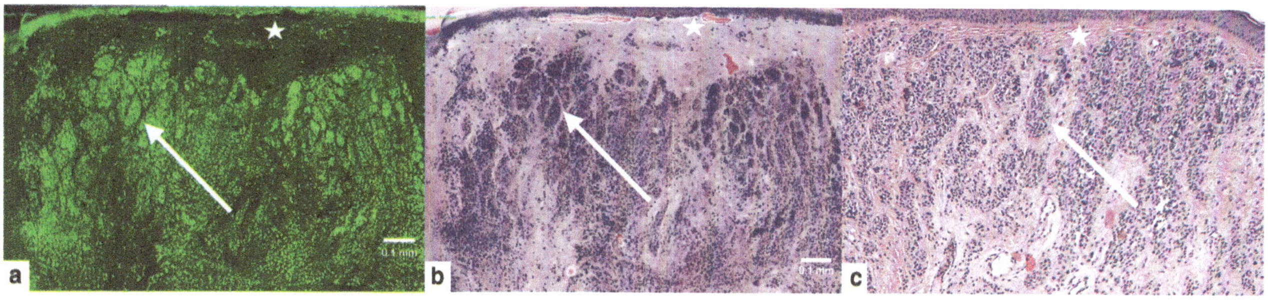

Fig. 9.3 Dermal melanocytic nevus examined in two modes of the ex vivo confocal laser scanning microscopy (**a** fluorescence, **b** DHE) and on corresponding H&E-stained image (**c**) presenting melanocytic proliferations in the dermis (long arrow) clustered as nests and cords showing signs of maturation in the dermis. No melanocytic proliferation is present in the epidermis. A melanocytes-free zone in the papillary dermis is clearly visible (star). DHE; digital hematoxylin–eosin-like staining. H&E magnification = 10x

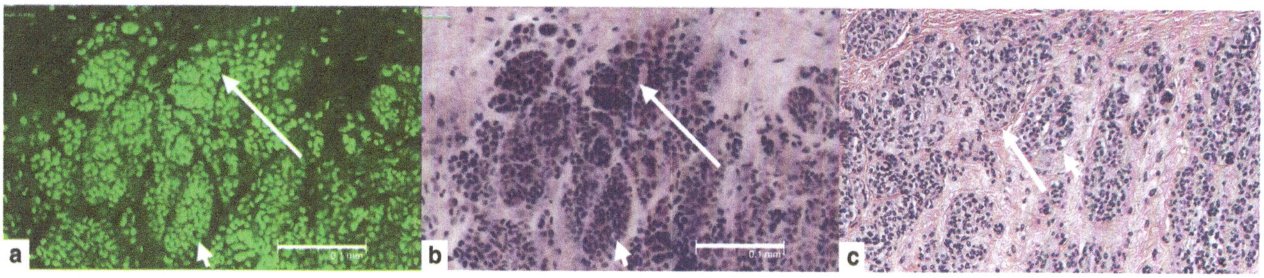

Fig. 9.4 Detailed (higher magnification) view of a dermal melanocytic nevus examined in two modes of the ex vivo confocal laser scanning microscopy (**a** fluorescence, **b** DHE) and on corresponding H&E-stained image (**c**) focusing on signs of maturation in the dermis. In the upper parts of the dermis large round melanocytes (long arrow) are clustered as nests. These melanocytes become smaller in size in the lower parts of the dermis (short arrow) and are organized in nests and cords. DHE = digital haematoxylin–eosin-like staining. H&E magnification = 20x

References

1. Hauschild A, Egberts F, Garbe C, Bauer J, Grabbe S, Hamm H, et al. expert group "Melanocytic nevi" Melanocytic nevi. J Dtsch Dermatol Ges. 2011;9(9):723–34.
2. English DR, Armstrong BK. Melanocytic nevi in children I anatomic sites and demographic and host factors. Am J Epidemiol. 1994;139(4):390–401.
3. MacKie RM, English J, Aitchison TC, Fitzsimons CP, Wilson P. The number and distribution of benign pigmented moles (melanocytic naevi) in a healthy British population. Br J Dermatol. 1985;113 (2):167–74.
4. Lund HZ, Stobbe, GD. The natural history of the pigmented nevus; factors of age and anatomic location. Am J Pathol. 1949;25 (6):1117–55, incl 4 pl.
5. Casso EM, Grin-Jorgensen CM, Grant-Kels JM. Spitz nevi. J Am Acad Dermatol. 1992; 27(6 Pt 1), 901–913.
6. Hartmann D, Ruini C, Mathemeier L, Bachmann MR, Dietrich A, Ruzicka T, et al. Identification of ex-vivo confocal laser scanning microscopic features of melanocytic lesions and their histological correlates. J Biophotonics. 2017;10(1):128–42.

Dysplastic Nevi

10

Daniela Hartmann

10.1 Basics of Dysplastic Nevi

Distinguishing normal and dysplastic nevi, both clinically and histologically, may sometimes be challenging and even more so in the ex vivo confocal laser scanning microscopy (ex vivo CLSM). The innovative bedside examination of freshly excised tissue has previously been applied also on melanocytic lesions [1–3]. We present current knowledge and images of dysplastic melanocytic nevi studied in ex vivo CLSM (Table 10.1, Figs. 10.1 and 10.2).

Table 10.1 Dysplastic melanocytic nevi features on FCM, DHE and corresponding conventional H&E

All normal melanocytic nevi features described in Chap. 9 may occur. Additional features often found in dysplastic nevi are listed below		
Fluorescence confocal microscopy (FCM) mode (1)	Digital H&E (combined FCM and RCM modes)	H&E correlation
Fluorescence in the epidermis: Nuclei of the tumor cells visualized as bright white (green*) dots of variable brightness and sizes (Figs. 10.1, and 10.2)	The bright white (green*) dots on FCM appear dark purple (high fluorescence signal) or dark to light pink (low fluorescence signal; dominant reflectance signal) on DHE. The variable cell size may be better visible in the DHE image than in the FCM	Atypical melanocytes
Fluorescence in the dermo-epidermal junction: White (green*) brightly shining 'tongues' of the epidermis interconnected horizontally in the dermo-epidermal junction and/or papillary dermis (Fig. 10.2)	The bright white (green*) interconnected 'tongues' of the epidermis on FCM appear most usually dark purple (high fluorescence signal) or dark to light pink (low fluorescence signal; dominant reflectance signal) on DHE	Horizontal confluence of the rete ridges (bridge-building) with melanocytic nests
Other features in the dermo-epidermal junction: Dark grey to black homogeneous band located directly subepidermal	Light purple to pink homogeneous band located directly subepidermal	Concentric fibrosis
Other fluorescent structures in the dermis and subcutis: White (green*) brightly shining tiny dots in the papillary, reticular dermis and/or subcutis, surrounding the tumor, often ascending, and infiltrating the epidermis (Fig. 10.1)	Dark purple tiny dots in the papillary, reticular dermis and/or subcutis, surrounding the tumor, often ascending, and infiltrating the epidermis	Inflammatory infiltrate surrounding the tumor and often masking the dermo-epidermal junction

CLSM = confocal laser scanning microscopy, FCM = fluorescent confocal microscopy, DHE = digital hematoxylin–eosin-like staining, H&E = hematoxylin–eosin staining, * color of the fluorescent signal depending on the generation of the ex vivo CLSM device

D. Hartmann (✉)
Department of Dermatology and Allergy, University Hospital, LMU Munich Frauenlobstrasse 9-11, 80337 Munich, Germany
e-mail: Daniela.Hartmann@med.uni-muenchen.de

M. Jain et al. (eds.), *Cutaneous Atlas of Ex Vivo Confocal Microscopy*,
https://doi.org/10.1007/978-3-030-89316-3_10

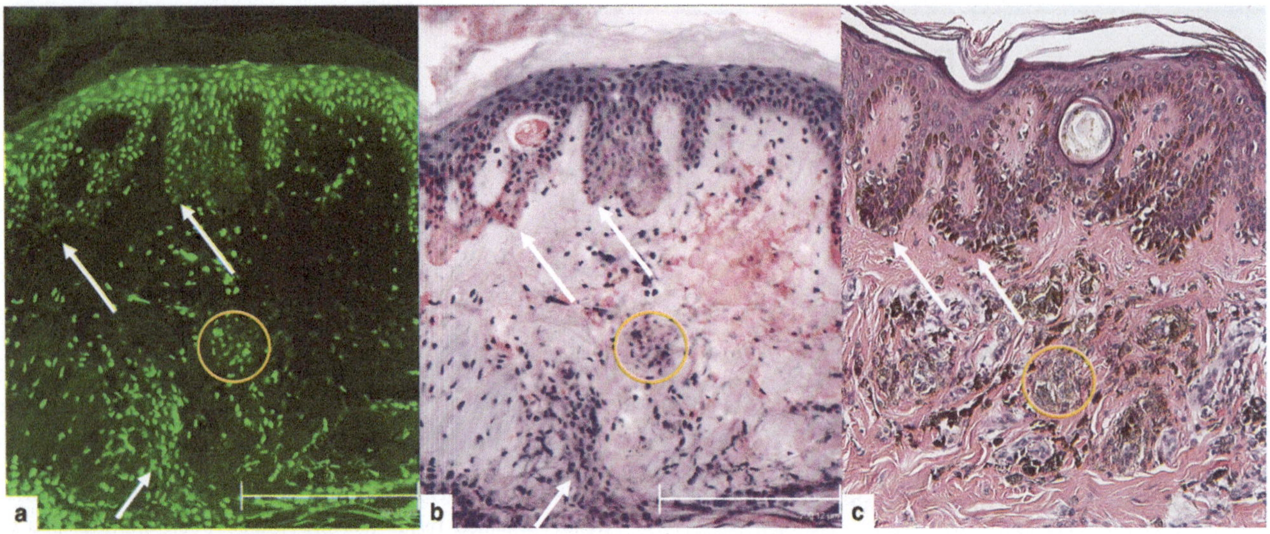

Fig. 10.1 Overview of a dysplastic compound melanocytic nevus displaying slightly chaotic architecture of the nevus consisting of melanocytic proliferations in the basal layers of the epidermis with varying levels of fluorescence and varying sizes (long arrow), as well as discrete inflammatory infiltrate (short, thick arrow) in the dermis. The images show dermal component of the compound nevus with normal appearing dermal nests (circled) of melanocytes showing typical signs of maturation and no signs of cellular or nuclear atypia, examined in two modes of the ex vivo confocal laser scanning microscopy (**a** fluorescence, **b** DHE) and comparison with their corresponding H&E-stained image (**c**). DHE = digital hematoxylin–eosin-like staining. H&E magnification = 20x

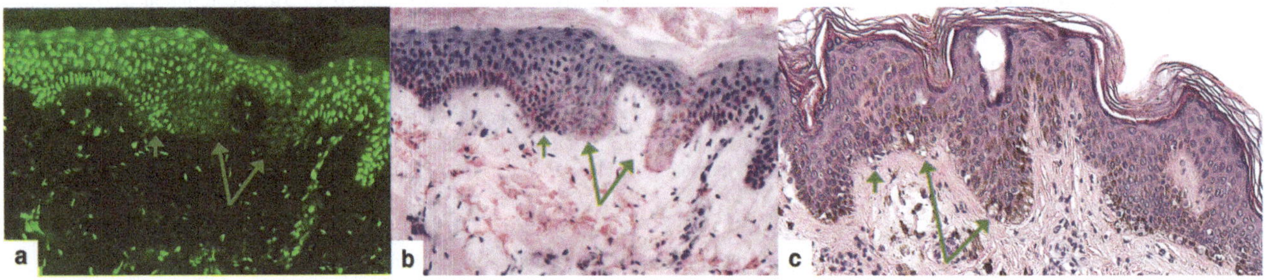

Fig. 10.2 Detailed view of a dysplastic compound melanocytic nevus presenting horizontal confluence of the rete ridges (short, thick arrow) and varying brightness of the fluorescence (long arrow) of the epidermal melanocytic proliferations, examined in two modes of the ex vivo confocal laser scanning microscopy (**a** fluorescence, **b** DHE) and comparison with their corresponding H&E-stained image (**c**). DHE = digital hematoxylin–eosin-like staining. H&E magnification = 20x

10.1.1 Basics of Dysplastic Nevi

- Definition: Clinically abnormal, congenital or acquired, melanocytic tumor with an increased probability of converting into melanoma. Various criteria in clinical examination, as well as dermoscopy, confocal microscopy, and histopathology are generally applied for the nevus analysis. If criteria are insufficient to diagnose a melanoma, the term dysplastic nevi is often being used [4–6].
- Epidemology: Incidence:100–150/100,000 inhabitants annually. The prevalence of dysplastic nevi is given as 1.8–4.6% of the population [6, 7].
- Histopathology: Dysplastic nevi of the junction type present irregularly spaced melanocytic nests, but also individually distributed (epithelioid) melanocytes in the

junction zone. More rarely, nests or single-cell-melanocytes are also found suprabasally. Melanocytic nests tend to have horizontal confluence (bridging). Occasional signs of cellular atypia may be seen. The compound dysplastic nevi present epidermal changes similar to the junction dysplastic nevi. Additionally, it has proliferation of melanocytes in the dermis, which show signs of maturation. Stromal reactions in the form of concentric or lamellar fibroplasia and inflammatory infiltrate may occur [8].

References

1. Hartmann D, Ruini C, Mathemeier L, Bachmann MR, Dietrich A, Ruzicka T, von Braunmühl T. Identification of ex-vivo confocal laser scanning microscopic features of melanocytic lesions and their histological correlates. J Biophotonics. 2017;10(1):128–42.
2. Hartmann D, Krammer S, Vural S, Bachmann MR, Ruini C, Sárdy M, Ruzicka T, Berking C, von Braunmühl T. Immunofluorescence and confocal microscopy for ex-vivo diagnosis of melanocytic and non-melanocytic skin tumors: a pilot study. J Biophotonics. 2018;11(3).
3. Hartmann D, Krammer S, Ruini C, Ruzicka T, von Braunmühl T. Correlation of histological and ex-vivo confocal tumor thickness in malignant melanoma. Lasers Med Sci. 2016;31(5):921–7.
4. Hauschild A, Egberts F, Garbe C, Bauer J, Grabbe S, Hamm H, Kerl H, Reusch M, Rompel R, Schlaeger M. Expert group "Melanocytic nevi" Melanocytic nevi. J Dtsch Dermatol Ges. 2011;9(9):723–34.
5. Farber MJ, Heilman ER, Friedman RJ. Dysplastic nevi. Dermatol Clin. 2012;30(3):389–404.
6. Roesch A, Landthaler M, Vogt T. The dysplastic nevus: separate entity, melanoma precursor or diagnostic dilemma? Hautarzt. 2003;54(9):871–83; quiz 884–5.
7. Naeyaert JM, Brochez L. CPDN. N Engl J Med. 2003;349(23):2233–40.
8. Roth ME, Grant-Kels JM, Ackerman AB, Elder DE, Friedman RJ, Heilman ER, Maize JC, Sagebiel RW. The histopathology of dysplastic nevi Continued controversy. Am J Dermatopathol. 1991;13(1):38–51.

11 Melanoma

Daniela Hartmann

Fast diagnostic methods in melanoma have been gaining importance in the last few decades due its malignant nature and its rising incidence [1]. Being able to safely distinguish normal or dysplastic nevi from melanoma intraoperatively and immediately decide further therapeutic steps would potentially decrease the number of surgical procedures [2], as well as associated risk of complications. Possible use of ex vivo confocal laser scanning microscopy (ex vivo CLSM) in melanoma diagnostics, including fast immunofluorescence [3] and intraoperative tumor thickness measurement [4], as well as up-to-date experience (Table 11.1) together with examples of melanoma images (Figs. 11.1, 11.2, 11.3 and 11.4) are presented and discussed.

11.1 Basics of Melanoma

- Definition: Malignant, invasive melanocytic skin tumor clinically presenting mostly as a deep brown to blue-blackish, brown-reddish, or even pigment-free (amelanotic melanoma) nodule or plaque with an early tendency to metastasize (strongly depending on tumor thickness). Different types of melanoma are distinguished on the basis of clinical and histological criteria (superficial spreading melanoma, nodular melanoma, lentigo maligna melanoma, acral lentiginous melanoma, and others) [1].
- Epidemiology: The incidence in a fair skinned population in Europe and North America is estimated to be 15/100,000 people annually [1].
- Histopathology: Melanomas have various histologic features including presence of atypical melanocytes singly or as nests with varying shape and size. These atypical cells can invade local structures causing their destruction. Atypical mitosis in the tumor cells, trans-epidermal melanocytic migration (TEM), asymmetry of the tumor, ill-defined tumor, horizontal confluence of the nests, lack of maturation of tumor cells in the deep dermis, vascular and/or perineural invasion, as well as spread of melanoma nests along the epithelial adnexal structures. Ulceration of the tumor, solar elastosis in the dermis, and strong peritumoral inflammatory infiltrate may occur [5].
- Immunofluorescence: First pilot studies on the combination of ex vivo CLSM and fluorescent-tagged antibodies have indicated new opportunities for fast and specific tissue examination [3, 6–11]. There is a particular interest in developing such techniques for the examination of melanocytic lesions to distinguish malignant from benign lesions. First report on the use of fluorescent-tagged S100-antibody and fluorescent-tagged Melan-A-antibody proved the possibility of such a concept and at the same time showed the complexity of such implementation within intraoperative setting [3]. Further studies on the use of immunofluorescence in the ex vivo CLSM are necessary.
- Confocal tumor thickness MEASUREMENT: Ex vivo CLSM enables intraoperative measurement of tumor thickness. First pilot study showed promising results and very good correlations to the tumor thickness on histopathology tissue Sect. (4). Larger studies on confocal tumor thickness measurement are needed to validate these results.

D. Hartmann (✉)
Department of Dermatology and Allergy, University Hospital, LMU Munich Frauenlobstrasse, 9-11, 80337 Munich, Germany
e-mail: Daniela.Hartmann@med.uni-muenchen.de

M. Jain et al. (eds.), *Cutaneous Atlas of Ex Vivo Confocal Microscopy*,
https://doi.org/10.1007/978-3-030-89316-3_11

Table 11.1 Melanoma features on FCM, DHE, and corresponding conventional H&E

In cases of melanomas, especially arising in nevi, all normal and/or dysplastic melanocytic nevi features described in Chaps. 9 and 10 may occur. Additional features often found in melanoma are listed below		
Fluorescence confocal microscopy mode (FCM) mode (green scale images) [2]	Digital H&E (combined FCM and RCM modes)	Conventional H&E-stained images
Fluorescence in the epidermis: Nuclei of the tumor cells visualized as bright white (green*) dots of variable shape, brightness and sizes ascending in the epidermis and/or spread throughout the entire epidermis (Figs. 11.1 and 11.2)	The bright white (green*) dots on FCM appear dark purple (high fluorescence signal) or dark to light pink (low fluorescence signal; dominant reflectance signal) on DHE. The tumor cells are ascending throughout or infiltrating the whole epidermis. The variable cell size may be better visible in DHE than in FCM	Atypical melanocytes with TEM and/or consumption of the epidermis by atypical melanocytes
Fluorescence in the dermo-epidermal junction and dermis: White (green*) 'tongues' of the epidermis of variable shape and brightness, interconnected horizontally at the dermo-epidermal junction and/or papillary dermis (Figs. 11.1 and 11.2)	Dark purple to dark pink 'tongues' of the epidermis of variable shape and brightness, interconnected horizontally at the dermo-epidermal junction and/or papillary dermis	Atypical melanocytic nests with horizontal confluence
Fluorescence in the epithelial adnexal structures (e.g. hair follicles): Bright white (green*) dots of variable shape, brightness and sizes spreading along the epithelium of the adnexal structures (Fig. 11.1)	Dark purple dots, surrounded by light purple to pink cytoplasm, of variable shape and sizes spreading along the epithelial adnexal structures	Spread of melanoma cells and/or nests along the epithelium of the adnexal structures
Fluorescent structures in the dermis and subcutis: Homogeneous, greyish (light to dark green*), slightly shining structureless mass (Fig. 11.1)	Homogeneous, light to dark purple structureless mass	Solar elastosis
Other fluorescent structures in the dermis and subcutis: Heterogeneous nests of varying shape, sizes and brightness consisting of atypical melanocytes in shades of grey (light to dark green*) (Fig. 11.3)	Heterogeneous purple to pink nests of varying shapes and sizes consisting of atypical melanocytes	Dermal nests of atypical melanocytes showing cellular and nuclear polymorphism without sign of maturation
Nuclear fluorescent structures: Bizarre shaped bright white (bright green*) chromatin structures in the nuclei (Fig. 11.4)	Bizarre shaped dark purple chromatin structures in the nuclei	Atypical mitosis in the melanoma cells

FCM = fluorescent confocal microscopy, DHE = digital hematoxylin–eosin-like staining, H&E = hematoxylin–eosin staining, TEM = trans-epidermal melanocytic migration, * color of the fluorescent signal depending on the generation of the ex vivo CLSM device

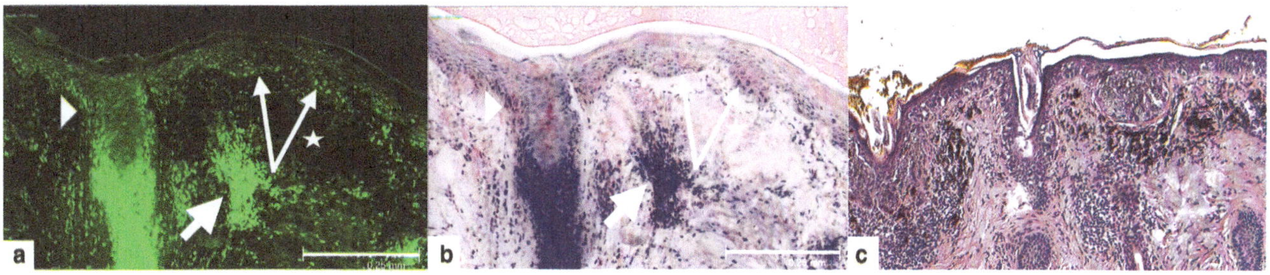

Fig. 11.1 Superficial spreading melanoma examined in two modes of the ex vivo confocal laser scanning microscopy (**a** fluorescence, **b** DHE), and comparison with their corresponding H&E-stained image (c) presenting nests of varying size and shape consisting of atypical melanocytes in the epidermis and dermo-epidermal junction (long arrow), as well as strong inflammatory infiltrate (short, thick arrow), and solar elastosis (star) in the dermis. Melanocytic spread along the follicle is highlighted with an arrowhead. DHE = digital hematoxylin–eosin-like staining. H&E magnification = 10x

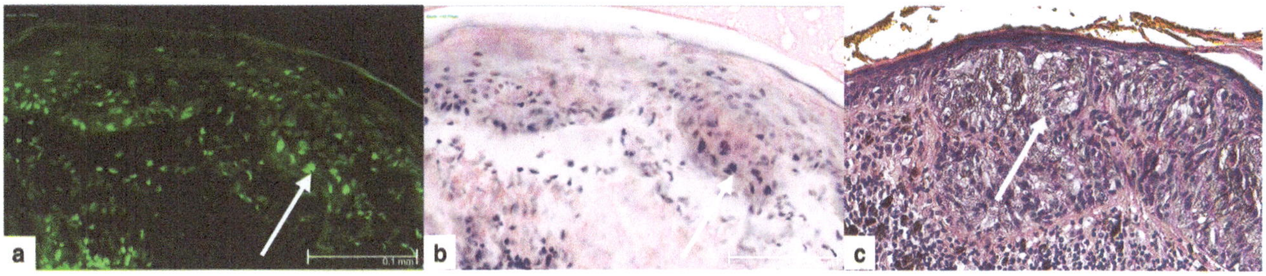

Fig. 11.2 Detailed view of a superficial spreading melanoma examined in two modes of the ex vivo confocal laser scanning microscopy (**a** fluorescence, **b** DHE) and comparison with their corresponding H&E-stained image (**c**), presenting atypical melanocytes clustered in nests as well as single cell proliferates in the epidermis and dermo-epidermal junction (long arrow) showing signs of transepidermal melanocytic migration DHE = digital hematoxylin–eosin-like staining. H&E magnification = 20x

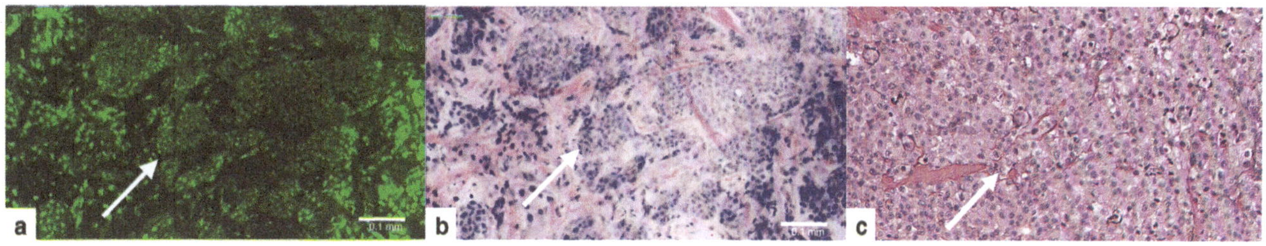

Fig. 11.3 Dermal part of a nodular melanoma examined in two modes of the ex vivo confocal laser scanning microscopy (**a** fluorescence, **b** DHE) and comparison with their corresponding H&E-stained image (**c**), presenting dermal nests of varying size and shape consisting of atypical melanocytes (arrow) showing varying levels of fluorescence signal, as well as cellular and nuclear pleomorphism. DHE = digital hematoxylin–eosin-like staining. H&E magnification = 20x

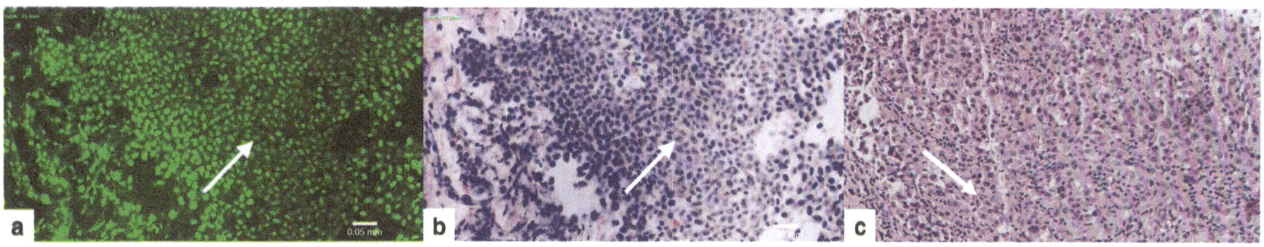

Fig. 11.4 Melanoma metastasis examined in two modes of the ex vivo confocal laser scanning microscopy (**a** fluorescence, **b** DHE) and comparison with their corresponding H&E-stained image (**c**), presenting masses of atypical melanocytes (arrow) with multiple atypical mitoses. DHE = digital hematoxylin–eosin-like staining. H&E magnification = 20x

References

1. Leitlinienprogramm Onkologie. Deutsche Krebsgesellschaft, Deutsche Krebshilfe, AWMF. S3-Leitlinie Prävention von Hautkrebs, Langversion 1.1. 2014. http://leitlinienprogramm-onkologie.de/Leitlinien.7.0.html. Accessed 03 Feb 2021.
2. Hartmann D, Ruini C, Mathemeier L, Bachmann MR, Dietrich A, Ruzicka T, von Braunmühl T. Identification of ex-vivo confocal laser scanning microscopic features of melanocytic lesions and their histological correlates. J Biophotonics. 2017;10(1):128–42.
3. Hartmann D, Krammer S, Vural S, Bachmann MR, Ruini C, Sárdy M, Ruzicka T, Berking C, von Braunmühl T. Immunofluorescence and confocal microscopy for ex-vivo diagnosis of melanocytic and non-melanocytic skin tumors: A pilot study. J Biophotonics. 2018 Mar;11(3).
4. Hartmann D, Krammer S, Ruini C, Ruzicka T, von Braunmühl T. Correlation of histological and ex-vivo confocal tumor thickness in malignant melanoma. Lasers Med Sci. 2016;31(5):921–7.
5. Wade TR, White CR Jr. The histology of malignant melanoma. Med Clin North Am. 1986;70(1):57–70.
6. Bağcı IS, Aoki R, Krammer S, Ruzicka T, Sárdy M, Hartmann D. Ex vivo confocal laser scanning microscopy: An innovative method for direct immunofluorescence of cutaneous vasculitis. J Biophotonics. 2019 Sep;12(9):e201800425.
7. Bağcı IS, Aoki R, Krammer S, Ruzicka T, Sárdy M, French LE, Hartmann D. Ex vivo confocal laser scanning microscopy for bullous pemphigoid diagnostics: new era in direct immunofluorescence? J Eur Acad Dermatol Venereol. 2019;33(11):2123–30.
8. Krammer S, Krammer C, Salzer S, Bağci IS, French LE, Hartmann D. Recurrence of Pemphigus Vulgaris Under Nivolumab Therapy. Front Med (Lausanne). 2019;12(6):262.
9. Bağcı IS, Aoki R, Krammer S, Vladimirova G, Ruzicka T, Sárdy M, French LE, Hartmann D. Immunofluorescence and histopathological assessment using ex vivo confocal laser scanning microscopy in lichen planus. J Biophotonics. 2020 Dec;13(12): e202000328.
10. Bağcı IS, Aoki R, Vladimirova G, Ergün E, Ruzicka T, Sárdy M, French LE, Hartmann D. New-generation diagnostics in inflammatory skin diseases: Immunofluorescence and histopathological assessment using ex vivo confocal laser scanning microscopy in cutaneous lupus erythematosus. Exp Dermatol. 2020 Dec 21.
11. Bağcı IS, Aoki R, Vladimirova G, Sárdy M, Ruzicka T, French LE, Hartmann D. Simultaneous immunofluorescence and histology in pemphigus vulgaris using ex vivo confocal laser scanning microscopy. J Biophotonics. 2021 Jan 24:e202000509.

Part VI

Inflammatory Lesions

Features of Lichen Planus and Psoriasis on Ex Vivo Confocal Imaging and Histopathologic Correlation

12

Laura Bertoni, Camilla Reggiani, Paola Azzoni, Luca Reggiani Bonetti, and Giovanni Pellacani

Abstract

Lichen Planus is an idiopathic inflammatory skin and mucous disease, typically characterized by pruritic, violaceous papules that favour the extremities and histologically featured by a dense, band-like lymphocytic infiltrate and keratinocyte apoptosis with destruction of the epidermal basal cell layer. Ex vivo confocal imaging shows dense band of bright small and round nuclei distributed in the dermis immediately below the epidermis. Psoriasis is an immune-mediated polygenic skin disorder in which the most common clinical setting includes sharply demarcated, scaly, erythematous plaques involving scalp, elbows, and knees. The histological findings include acanthosis with elongated rete ridges, hypogranulosis, hyper- and para-keratosis, dilated blood vessels and perivascular infiltrate oflymphocytes with neutrophils singly or within aggregates in the epidermis. Ex vivo fluorescent confocal imaging shows small, flat, bright nuclei distributed throughout the parakeratosic epidermis and subcorneal or intracorneal Munro microabscesses composed by irregular polylobate bright nuclei representing the neutrophils. In both Lichen Planus and Psoriasis on digital H&E mode all bright fluorescent structures appear dark and purple-blue, with a correspondence to hematoxylin stained hyperchromatic nuclei.

L. Bertoni (✉) · C. Reggiani · P. Azzoni
Department of Surgical, Medical, Dental and Morphological Sciences with Interest in Transplant, Oncological and Regenerative Medicine, University of Modena and Reggio Emilia, Modena, Italy
e-mail: laura.bertoni@unimore.it

L. R. Bonetti
Department of Medical and Surgical Sciences for Children and Adults, University of Modena and Reggio Emilia -AOU Policlinico of Modena, Modena, Italy

G. Pellacani
Dermatology Clinic, Department of Clinical Internal, Anesthesiological and Cardiovascular Sciences, SapienzaUniversity of Rome, Rome, Italy

12.1 Introduction

Ex vivo confocal microscopy (EVCM) is an innovative imaging tool that can be used intraoperatively to obtain real-time images of untreated excised tissue with almost histologic resolution. EVCM has been successfully applied to surgical skin specimens to analyze neoplastic and non-neoplastic lesions, as well as to assess surgical margins of melanoma and non-melanoma skin cancers [1–3]. This device, as already demonstrated in literature, can help to make a real time diagnosis of inflammatory skin diseases [4]. This technique has been validated on gray-scale (black and white) images using older generation tool employing FCM modealone and recently on digital (purple and pink) H&E images obtained using newer generation microscope that combines fluorescence (FCM) and reflectance (RCM) mode. All images (FCM and digital H&E) shown in this chapter have been acquired using the newer generation of FCM (Vivascope 2500, 4th Gen®, MAVIG Gmb). In this chapter, we will summarize morphological features and terms usedto describe psoriasis and lichen planus in both FCM and digital H&E (DHE) modes along with their histopathological-correlations.

12.2 Psoriasis

Psoriasis is an immune-mediated polygenic skin disorder which can be caused by various environmental triggers such as trauma, infections, and medications. The prevalence of psoriasis is likely to be 2% of the world's population [11]. Psoriasis can be a systemic disease in which up to 20–30% of patients have an association with psoriatic arthritis. There are many clinical types of psoriasis including chronic plaque, guttate, erythrodermic, pustular, flexural, and nail. The most common type is the chronic plaque psoriasis characterized by sharply demarcated, scaly, erythematous plaques, and the most common sites of involvement are scalp,

M. Jain et al. (eds.), *Cutaneous Atlas of Ex Vivo Confocal Microscopy*,
https://doi.org/10.1007/978-3-030-89316-3_12

elbows, and knees. The histopathological findings of psoriasis include acanthosis with elongated rete ridges, hypogranulosis, hypergranulosis and parakeratosis, dilated blood vessels and perivascular infiltrate of lymphocytes, and presence of neutrophils in the epidermis as single cells or as aggregates (Tables 12.1 and 12.2).

Table 12.1 Clinical (epidemiology, patient demographics, and genetics) and histopathological features of psoriasis

Basics of Psoriasis
• **Epidemiology**: 2% of the world population [12] • **Patient demographics**: Psoriasis can appear at any age, but there are two peaks of age onset: 20-30 and 50-60 years • **Genetics**: HLA-Cw6 (early onset patients). Other HLA alleles may be associated with different variants of psoriasis • **Histopathology**: Hyper- and parakeratosis, acanthosis of epidermis with elongated rete ridges, hypogranulosis, dilated vessels and perivascular infiltrate of lymphocytes and with neutrophils in epidermis as single cells or aggregates forming Munro microabscesses

Table 12.2 Psoriasis: morphological features on FCM, DHE and corresponding H&E

Morphological features of Psoriasis on FCM, DHE and corresponding conventional H&E		
Fluorescence Confocal Microscopy (FCM) mode (grayscale images) (Figs. 12.1a, 12.2a and 12.3a)	Digital H&E (combined FCM and RCM modes) (Figs. 12.1b, 12.2b and 12.3b)	Conventional H&E-stained images (Figs. 12.1c, 12.2c and 12.3c) [5–8]
Fluorescence: the nuclei appear bright (fluorescence) while other structures are grey (reflectance)	The fluorescent structure appears dark blue while reflectant structure are light pink	These intense fluorescent (on FCM) and purple pseudo-color (on DHE) structures correspond to hematoxylin-stained nuclei
Parakeratosis: sheets of keratin are evident with small, flat, bright nuclei distributed throughout the entire layer. Cell nuclei of the epidermis are similarly hyper-reflective with bright appearance	**Parakeratosis**: epidermal nuclei and parakeratotic nuclei within the keratin appear purple blue. Keratin layers are light violet	**Parakeratosis**: diffuse parakeratosis consisting of keratin with nuclei present throughout the entire layer of stratum corneum
Acanthosis and elongated crests resemble a comb; the nuclei of epidermal cells appear white and bright	**Acanthosis of epidermis with elongated rete ridges** appears blue dark nuclei of epidermal cells	**Acanthosis with elongated rete ridges** are particularly evident in H&E stained sections
Munro microabscesses: subcorneal or intracorneal pustules containing neutrophils that appears as irregular polylobated bright nuclei surrounded by black empty spaces	**Munro microabscesses**: intracorneal pustules that appear as clusters of darkish blue nuclei of the neutrophils surrounded by white empty spaces	**Munro microabscesses**: the intense fluorescent (on FCM) and purple pseudocolor (on DHE) structures correspond to hematoxylin-stained nuclei of neutrophils
Dermal inflammation mostly lymphocytes with small round hyper-reflective nuclei	**Dermal inflammation** lymphocytes appear as dark blue dot-like structures in the upper part of dermis	**Dermal inflammation**: inflammatory cells predominantly lymphocytes, present in the dermis
Prominent vessels: elongated vessels run tortuously from the dermis and touch the crests of the epidermis. Small bright nuclei of endothelial cells delimit the lumen of the vessels	**Prominent vessels**: dilated capillaries in papillary dermis appear as white tortuous vessels surrounded by flat dark blue nuclei	**Prominent vessels**: dilated capillaries in papillary dermis, which appear as tortuous near the lower limit of the epidermis. These vessels are lined by flat endothelial cells

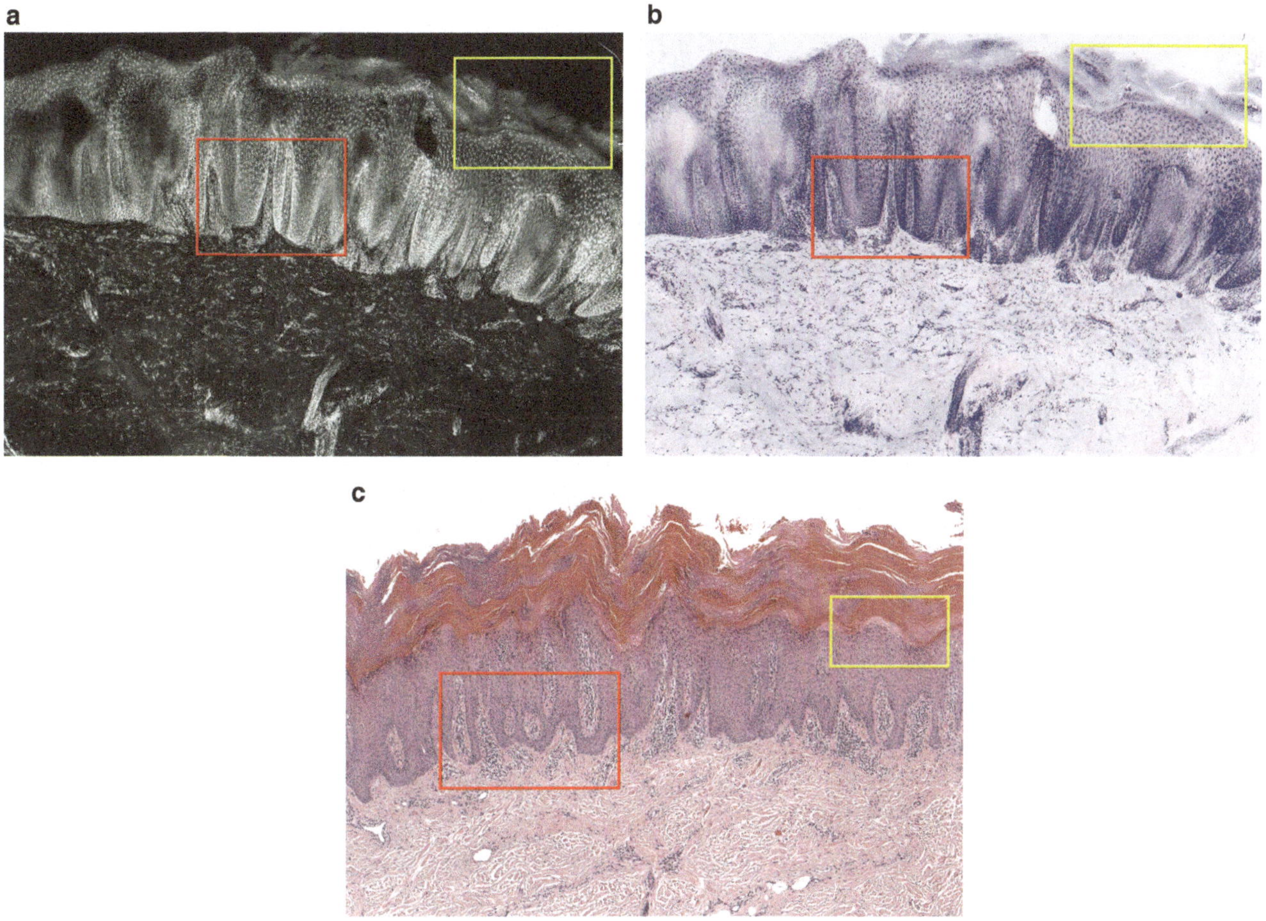

Fig. 12.1. Low magnification images of psoriasis: Features on FCM compared with the corresponding DHE and conventional H&E stained tissue of a psoriasis lesion from left hand of a 54 years old male. **a** FCM mosaic image shows the epidermis with acanthosis and elongated rete ridges resembling a "comb" (red box). The nuclei of the epidermal cells appear bright. On the surface, hyperkeratosis and parakeratosis are visible (yellow box). In the dermis, it is possible to observe the bright nuclei of the lymphocytes and dilated vessels in a dark background. **b** The corresponding DHE image shows acanthosis of the epidermis with elongated rete ridges, a "comb" (red box). The nuclei of the epidermal cells appear dark blue. On the surface, we can appreciate hyperkeratosis and parakeratosis (yellow box). In the dermis, prominent capillaries and chronic inflammatory infiltrate composed of lymphocytes are visible; the nuclei of lymphocytes are small and dark blue. **c** Corresponding image of the psoriasis in H&E stained Sect. (10X) shows acanthosis and downward lengthening of the rete ridges, resembling a comb (psoriasiform hyperplasia) (red box). The epidermis above the papillae is thinned especially at the tips of the dermal papillae, and there is a loss of stratum granulosum. On the surface, parakeratosis and hyperkeratosis is widespread (yellow box). In the dermis, dilated, prominent capillaries and a mixed chronic inflammatory infiltrate of lymphocytes, macrophages, and neutrophils are observed

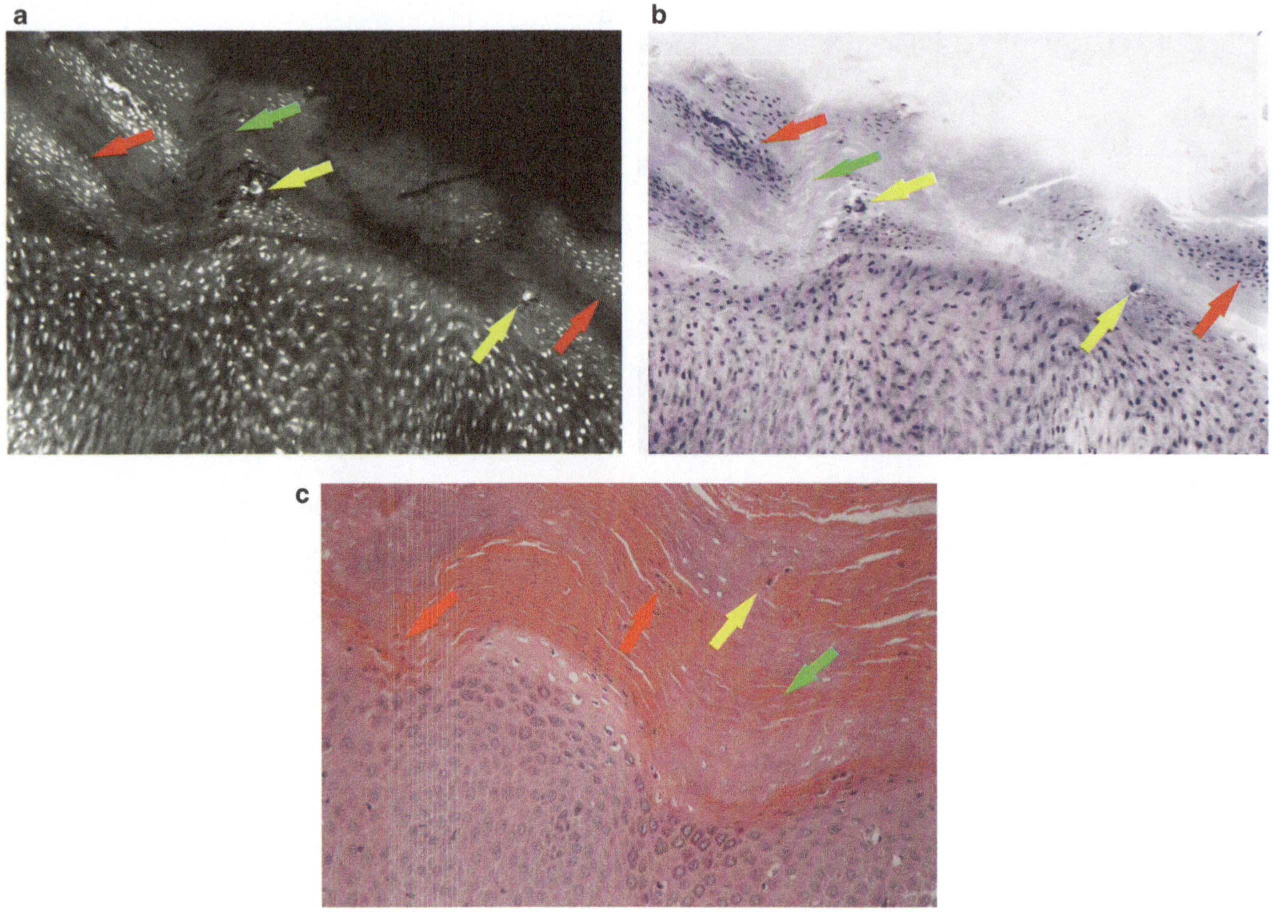

Fig. 12.2. High magnification images of psoriasis: Detailed features of psoriasis on FCM submosaic compared with corresponding DHE and conventional H&E-stained tissue section obtained from corresponding yellow boxed area in Fig. 12.1. **a** On FCM, the hyperkeratosis appears as a thickened stratum corneum (green arrow), and a diffuse parakeratosis is evident as small flat bright nuclei distributed throughout the layer (red arrows) in a background of gray sheet of keratin. Cell nuclei of the epidermis have a similar hyper-reflective bright appearance. Epithelial nuclei are within by black empty spaces in which thin corneal laminas appear distinct. Subcorneal or intracorneal pustules containing neutrophils (Munro microabscesses) are appreciable (yellow arrows). The nuclei of the neutrophils are irregular polylobated and are mixed with necrotic debris. **b** Digitally zoomed DHE image of the yellow square area shown in Fig. 12.1b. The epidermis shows acanthosis, hyperkeratosis (green arrow), and parakeratosis. Epidermal nuclei and parakeratotic nuclei (red arrows) within the keratin appear purple-blue. Intracorneal pustules containing neutrophils (Munro microabscesses) are evident (yellow arrows). The polylobed nuclei of the neutrophils mixed with cellular necrotic debris appear darkish blue. **c** In the corresponding H&E-stained Sect. (20X) diffuse parakeratosis (red arrows) is evident at a higher magnification consisting of nuclei distributed throughout the hyperkeratotic layer (green arrow). Yellow arrow indicates intra-corneal Munro microabscesses

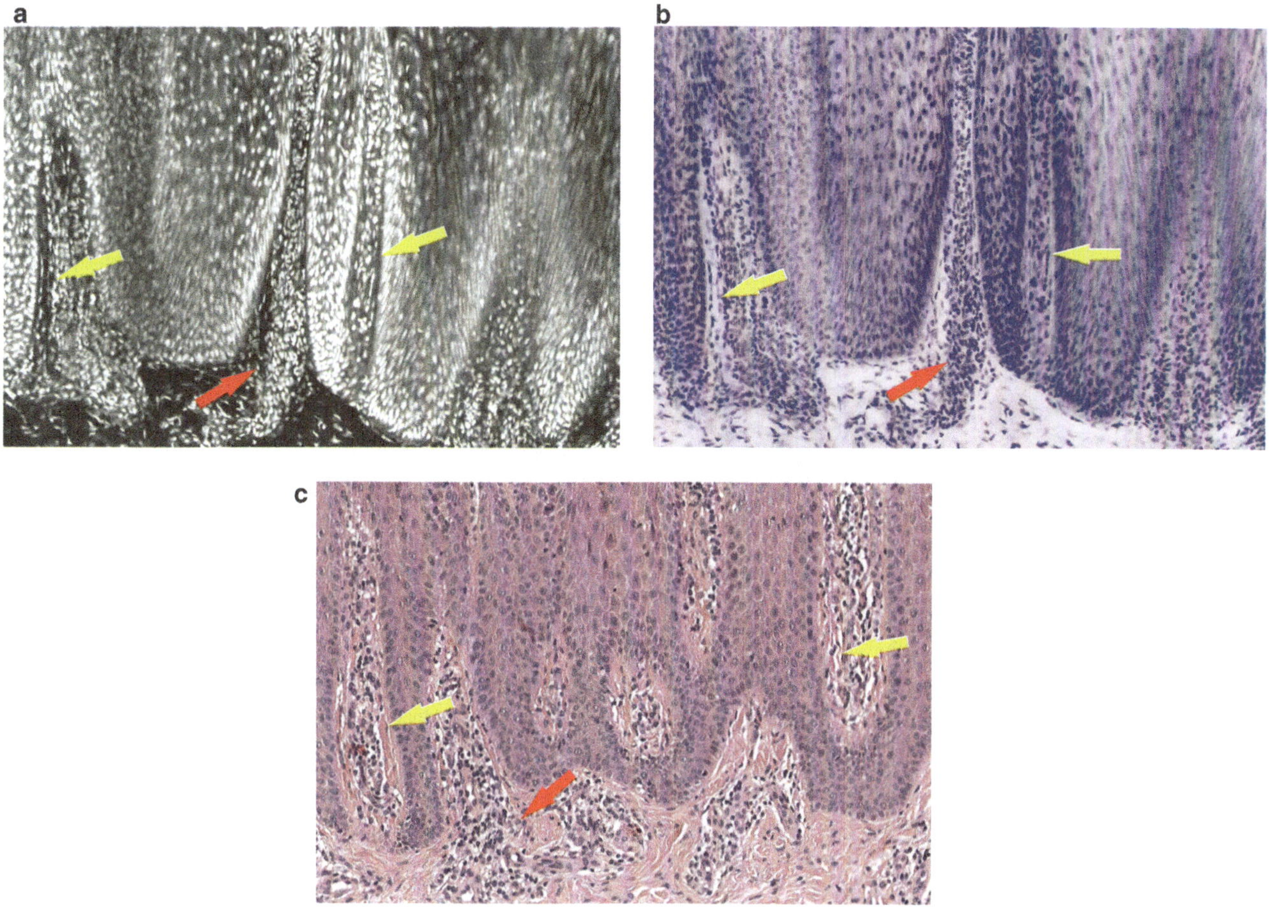

Fig. 12.3. High magnification images of psoriasis: Detailed features of psoriasis on FCM submosaic compared with the corresponding DHE and conventional H&E-stained image from the red boxed area in Fig. 12.1. **a** Digitally zoomed FCM submosaic image shows elongated vessels (yellow arrows) in the dermis running tortuously and touching the crests of the epidermis. The lumen is black, while the nuclei of endothelial cells appear flat and bright. Dermal inflammation is visible and is represented mainly by lymphocytes that appear as small and bright nuclei (red arrow). **b** In the DHE image, obtained by digitally zooming in the red square area of the mosaic in Fig. 12.1b, the crests of epidermis are prominent and elongated vessels from the dermis run upward through the crests of the epidermis (yellow arrows). The vessels are lined by endothelium with flat blue nuclei. Dermal inflammation is represented mainly by lymphocytes appearing as dark blue dots (red arrow). **c** Digitally zoomed H&E image (20X) of the red square area shown in Fig. 12.1c shows thickened epidermis with elongated rete ridges with intervening dermal papillae that contains dilated and tortuous capillaries (yellow arrows). Inflammatory cells, predominantly lymphocytes (red arrow), are observed in the superficial layer of the dermis

12.3 Lichen Planus

(LP) is an idiopathic inflammatory skin and mucous disease. Although its incidence varies depending on geographic location, cutaneous LP is reported to affect from 0,2 to 1% of the adult population [13]. Classic LP is characterized by pruritic, violaceous papules mostly located on the extremities. Clinically there are several variants: actinic, annular, atrophic, bullous, hypertrophic, inverse, pigmentosus, planopilaris, linear, oral, ulcerative, vulvovaginal. Histopathology is characterized by a dense, band-like lymphocytic infiltrate and keratinocyte apoptosis with destruction of the epidermal basal cell layer. Other diseases that show a lichenoid pattern on histopathology are lichen nitidus, lichen sclerosis, and lichenoid drug eruptions (Tables 12.3 and 12.4, Figs. 12.4, 12.5 and 12.6).

Table 12.3 Clinical (epidemiology, patient demographics, and genetics) and histopathological features of lichen planus

Basics of Lichen Planus
• **Epidemiology**: Incidence varies depending on geographic location. Incidence: 0,2-1% of adult population [13] • **Patient demographics**: Appears at all ages, but most common between 30-60 years of age. Sites of predilections: wrists, ankles, anterior skins, buttocks • **Genetics**: HLA-B27, -B51, -Bw57, -DR1, -DR9, -DR6 • **Histopathology**: Band-like lymphocytic infiltrate, irregular epidermal hyperplasia with a saw tooth appearance, orthokeratosis, and keratinocyte apoptosis with destruction of the epidermal basal cell layer

Table 12.4 Lichen planus: morphological features on FCM, DHE and corresponding H&E

Morphological features of lichen planus on FCM, DHE and corresponding conventional H&E		
Fluorescence Confocal Microscopy (FCM) mode (grayscale images) (Figs. 12.4a, 12.5a and 12.6a)	Digital H&E (combined FCM and RCM modes) (Figs. 12.4b, 12.5b and 12.6b)	Conventional H&E-stained images [9, 10] (Figs. 12.4c, 12.5c and 12.6c)
Fluorescence: Nuclei appears as hyperfluorescent bright dots with enhanced contrast compared to the surrounding structures	The bright white structures on FCM appear dark and purple-blue (high fluorescence signal; purple pseudo-color)	The fluorescent (on FCM) and purple pseudo-color (on DHE) structures correspond to hematoxylin-stained hyperchromatic nuclei of the cells
Dense band-like cellular infiltrate: The dermis immediately below the epidermis shows a dense band of lymphocytes with bright small and round nuclei	**Dense band-like cellular infiltrate**: inflammatory infiltrate forms a compact band immediately below the epidermis and is composed of blue small round dots that correspond to nuclei of the lymphocytes	**Dense band-like cellular infiltrate**: dense band of lymphocytes with dark blue small round nuclei in the dermis hugging the epidermis
Irregular epidermal hyperplasia with a saw tooth appearance: acanthosis with irregular and elongated rete ridges appear as a saw-tooth pattern. Nuclei in the epidermis are bright	**Irregular epidermal hyperplasia with a saw tooth appearance**: nuclei of epidermal cells are dark, blue-purple. Epidermis shows a saw-tooth appearance. Acanthosis appears as an increase of the stratification of the epidermal spinosum layer	**Irregular epidermal hyperplasia with a saw tooth appearance** is a characteristic feature of lichen planus. The intense fluorescent and purple pseudo colored structures correspond to hematoxylin-stained nuclei
Orthokeratosis: reflectant stratum corneum appears thickened and compact without nuclei (bright small spots)	**Orthokeratosis**: is represented by thickened light violet layers of keratin with no nuclei	**Orthokeratosis**: appears as a compact stratified layer of keratin without nuclei on traditionally stained H&E tissue section
Hypergranulosis: appears as thickening of the upper layer of epidermis composed of cells with mildly fluorescent nuclei containing sparse granules	**Hypergranulosis** appears as light violet small granules in the keratinocytes of stratum granulosum	**Hypergranulosis**: correspond to a thickened stratum granulosum with cells containing granules

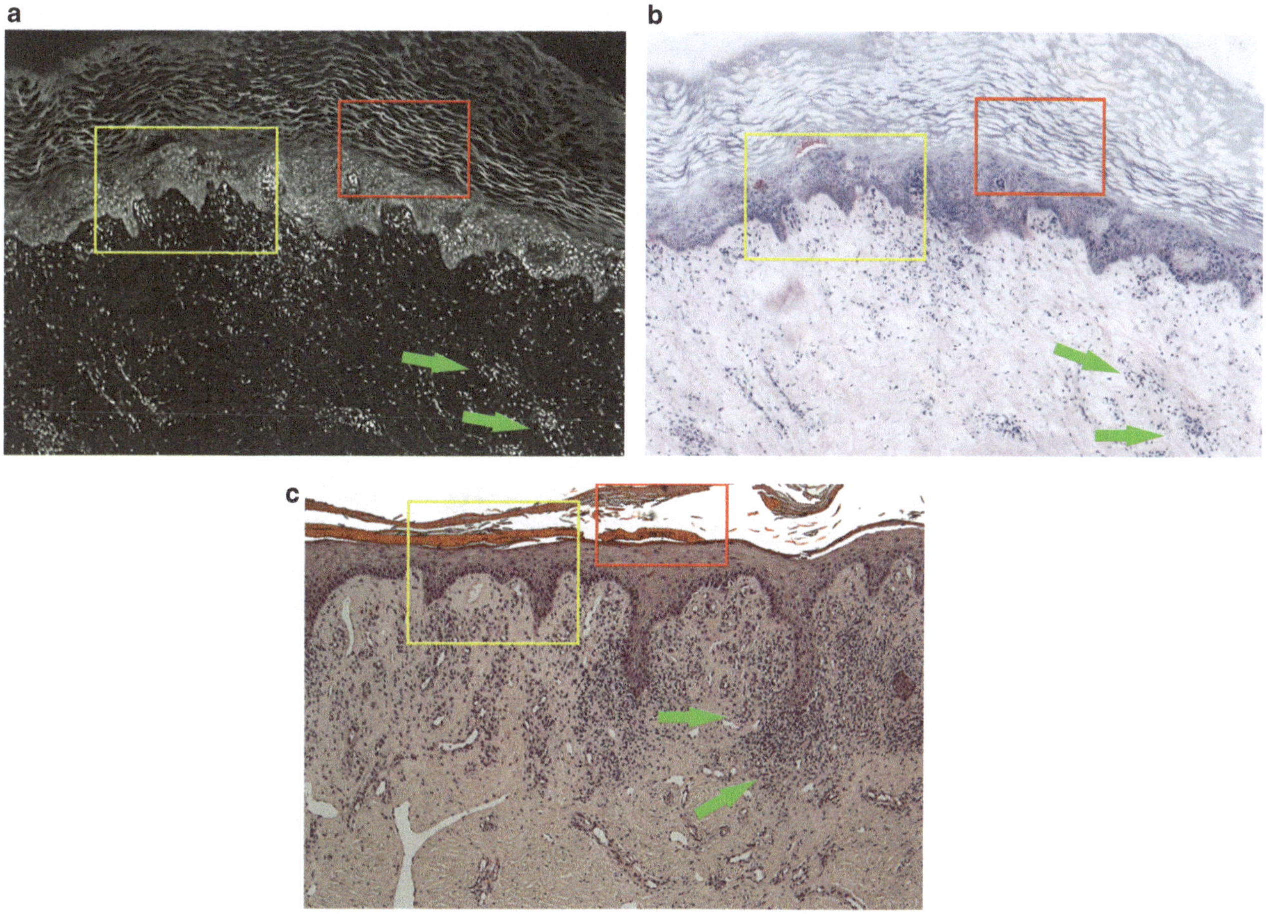

Fig. 12.4. Features on FCM compared with the corresponding DHE and conventional H&E-stained tissue section of a lichen planus lesion from right leg of a 70 years old man. **a** FCM mosaic of a lichen planus showing an irregular epidermis with acanthosis and hypergranulosis; the dermal papillae are elongated and show the typical saw-tooth appearance (yellow box area). The nuclei of the layered epidermal cells appear bright. Diffuse hyperkeratosis is evident (red box area). Inflammatory infiltrate forms a compact band immediately below the epidermis; it is composed of lymphocytes with bright small nuclei (green arrows). **b** Corresponding DHE image highlights the architecture of the epidermis and papillae that forms the classic saw-tooth pattern (yellow box area). Nuclei of epidermal cells are dark blue-purple. Hyperkeratosis and hypergranulosis are evident (red box area). Inflammatory infiltrate forms a compact band immediately beneath the epidermis and is composed of lymphocytes (green arrows). The lymphocytes have small round blue nuclei and inconspicuous cytoplasm. Dermis shows inflammation and edema. **c** Corresponding conventional H&E-stained tissue (10X) shows hyperkeratosis with the typical keratin slipping of the cut surface (red box area), acanthosis of epidermis, hypergranulosis, and the typical epidermic papillae distributed featuring saw-teeth (yellow box area). Epithelial nuclei are dark, blue-purple. Hyperkeratosis is evident. The dermis immediately below the epidermis shows a dense band of lymphocytes with dark blue small round nuclei and inconspicuous cytoplasm (green arrows). The infiltrate involves partially the deeper part of the dermis

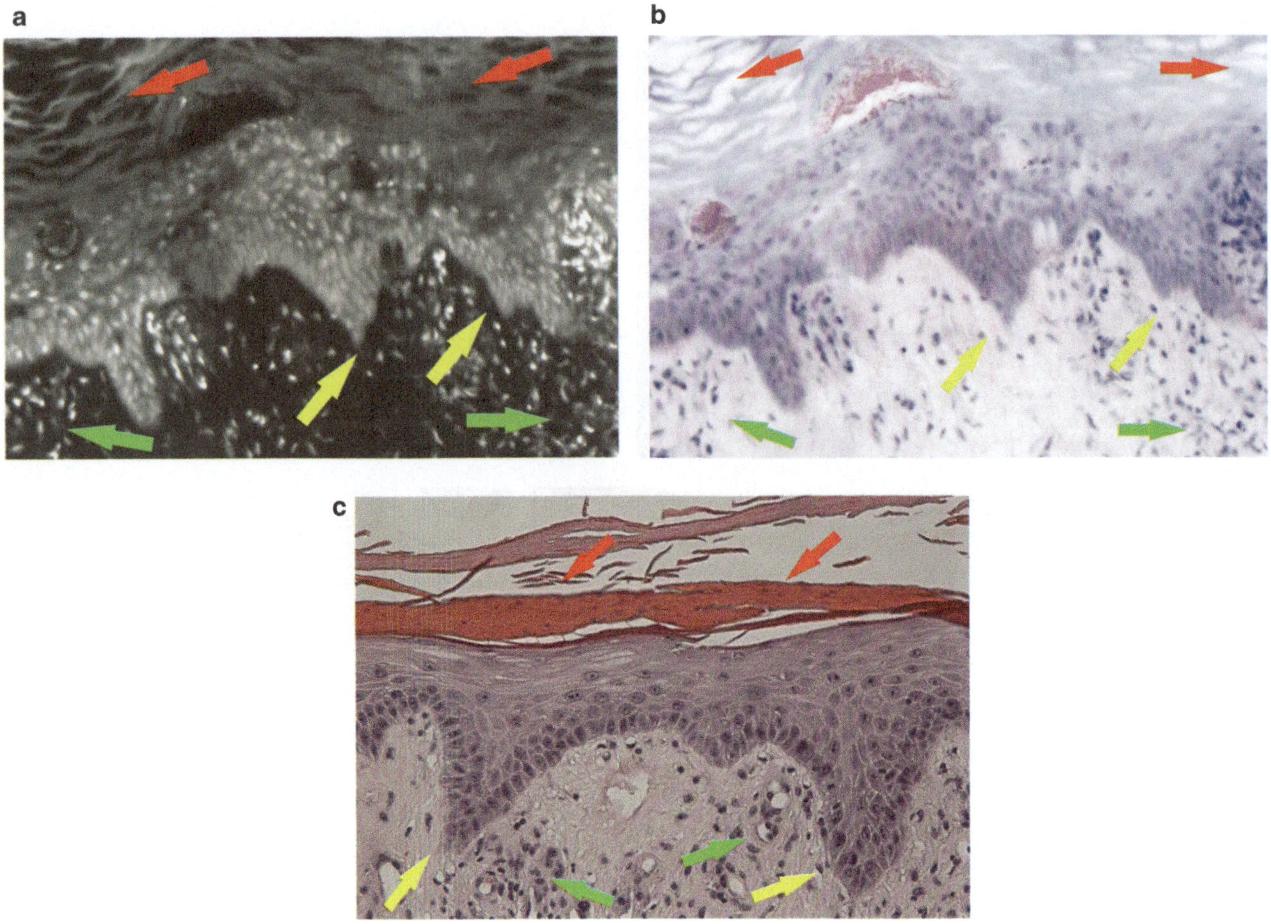

Fig. 12.5. Detailed features of lichen planus on FCM submosaic compared with the corresponding DHE and conventional H&E-stained tissue section obtained by digitally zooming in the yellow boxed areas in Fig. 12.4. **a** FCM submosaic image shows an acanthotic epidermis and dermal papillae forming the classic saw-tooth pattern (yellow arrows). Epithelial nuclei are bright. Hyperkeratosis is evident as a band of grayish keratin layers (red arrows). The dermis, immediately beneath the epidermis, shows a band of lymphocytes with bright small nuclei (green arrows). **b** In the digitally zoomed DHE submosaic, epidermis exhibits granular cell hyperplasia and acanthosis associated with orthokeratosis (red arrows). Saw-tooth epidermal pattern is evident (yellow arrows). Epithelial nuclei appear as dark blue-violet in color and are irregularly distributed in the hyperplastic epidermis. Dermis shows dense inflammation typically composed by lymphocytes (green arrows). Nuclei are small, round and dark blue-violet. **c** In the corresponding section of the H&E-stained tissue Sect. (20X), the irregular epidermal hyperplasia forms typical saw-tooth appearance (yellow arrows). The epidermis is acanthotic and hyper-ortho-keratosis (red arrows) is seen on the surface. The granular layer is focal hyperplastic. Lymphocytes are seen in the upper dermis (green arrows).

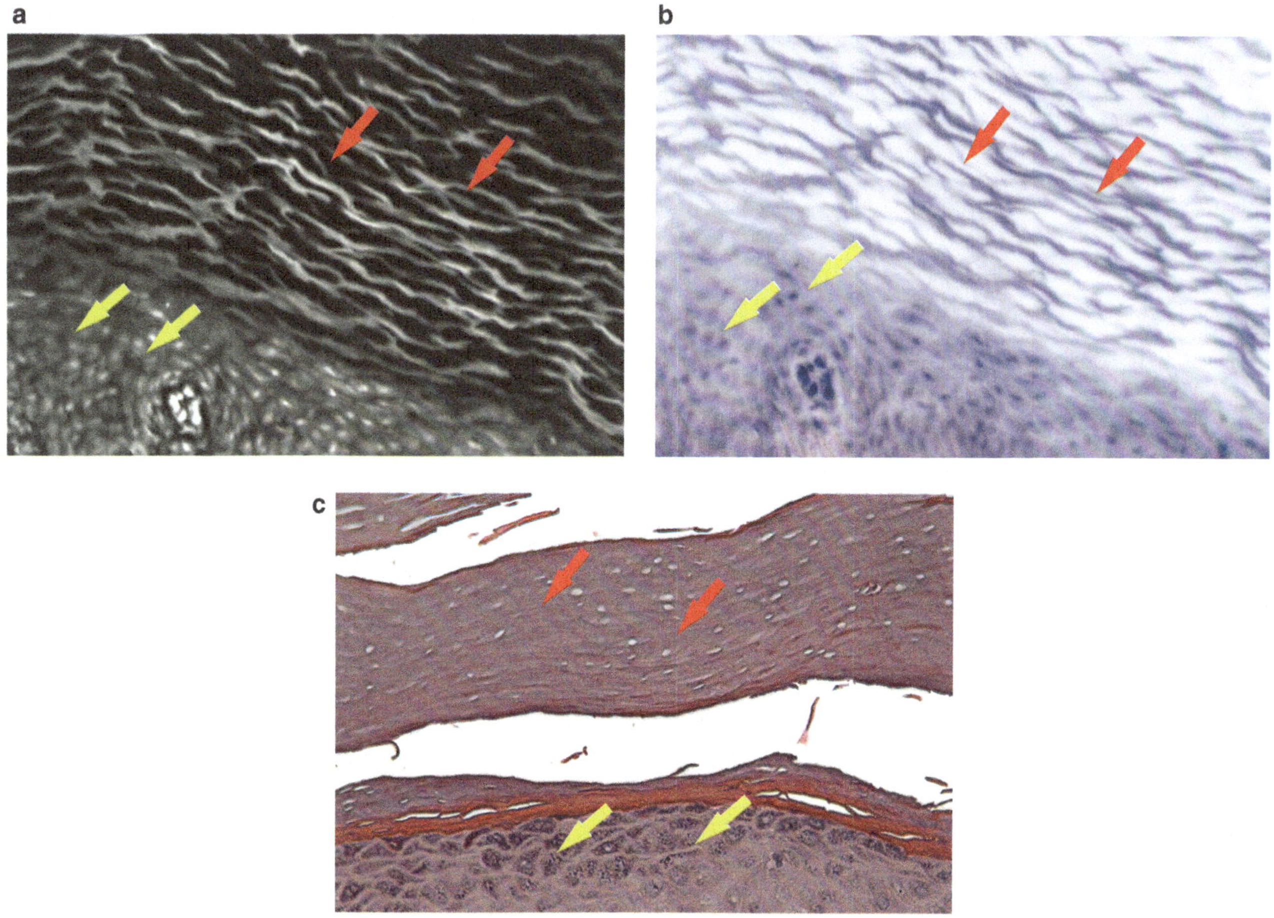

Fig. 12.6. Detailed features of lichen planus on FCM submosaic compared with the corresponding DHE and conventional H&E-stained tissue section obtained by digitally zooming in the red boxed area in Fig. 12.4. **a** In the FCM submosaic the stratum corneum is thickened and compact without nuclei, typically observed in compact orthokeratosis (red arrows). The cells of the stratum granulosum appear larger than the keratinocytes in the deeper layer. They have hypofluorescent big nuclei surrounded by small bright granules (yellow arrows). **b** DHE submosaic shows compact orthokeratosis that appears as light violet thickened layers of keratin with no nuclei (red arrows). The granulosum hyperplastic layer is composed of granulated large cells (yellow arrows). Deeper keratinocytes appear as dark blue nuclei. **c** In the corresponding H&E-stained tissue Sect. (20X) a compact stratified layer of keratin (orthokeratosis) is observed (red arrows). No nuclei are visible. Epidermis shows a hyperplastic granulosum layer composed by granulated large cells (yellow arrows).

References

1. Reggiani C, Pellacani G, Reggiani Bonetti L, et al. An intraoperative study with ex vivo fluorescence confocal microscopy: diagnostic accuracy of three visualization modalities. JEADV. 2021;35(1):e92–4. https://doi.org/10.1111/jdv.16831.
2. Longo C, Ragazzi M, Gardini S, et al. Ex vivo fluorescence confocal microscopy in conjunction with Mohs micrographic surgery for cutaneous squamous cell carcinoma. J Am Acad Dermatol. 2015;73(2):321–2.
3. Hartmann D, Krammer S, Vural S, et al. Immunofluorescence and confocal microscopy for ex-vivo diagnosis of melanocytic and non-melanocytic skin tumors: a pilot study. J Biophotonics. 2018;11(3). https://doi.org/10.1002/jbio.201700211.
4. Bertoni L, Azzoni P, Reggiani C, et al. Ex vivo fluorescence confocal microscopy for intraoperative, real-time diagnosis of cutaneous inflammatory diseases: a preliminary study. Exp Dermatol. 2018;27(10):1152–9. https://doi.org/10.1111/exd.13754.
5. Gordon M, Johnson WC. Histopathology and histochemistry of psoriasis: the active lesion and clinically normal skin. Arch Dermatol. 1967;95(4):402–7.
6. De Rosa G, Mignogna C. The histopathology of psoriasis. Reumatismo. 2007;59(Suppl 1):46–8.
7. Murphy M, Kerr P, Grant-Kels JM. The histopathologic spectrum of psoriasis. Clin Dermatol. 2007;25(6):524–8. https://doi.org/10.1111/cup.13334.
8. Gorouhi F, Davari P, Fazel N. Cutaneous and mucosal lichen planus: a comprehensive review of clinical subtypes, risk factors, diagnosis, and prognosis. Sci World J. 2014;2014:742826. https://doi.org/10.1155/2014/742826.
9. Elder DE, Elenitsas R, Johnson BL, Murphy GF, Xu X. Lever's Histopathology of the Skin. 10th ed. Philadelphia, Pa, USA: Lippincott Williams & Wilkins; 2008.

10. Joshi R. Interface dermatitis. Ind J Dermatol Venereol Leprol. 2013;79(3):349–59. https://doi.org/10.4103/0378-6323.110780.
11. Christophers E. Psoriasis-epidemiology and clinical spectrum. Clin Exp Dermatol. 2001;26:314–20.
12. Van de Kerkhof PCM, O Nestlé. Psoriasis. In Bolognia JL, Jorizzo JL, Schaffer JV. Dermatology. 3rd ed. Elsevier, 2012; vol.1, p. 135–55
13. Boyd AS, Neldner KH. Lichen planus. J Am Acad Dermatol. 1991;25:593–619.

Eczema: Features on FCM, Digital H&E, and Corresponding Conventional H&E

13

Laura Bertoni, Camilla Reggiani, Paola Azzoni, Luca Reggiani Bonetti, and Giovanni Pellacani

Abstract

Eczema is an inflammatory skin reaction characterized by itching, redness, scaling, and clustered papulo-vesicles. Histopathological features reflect the dynamism of the inflammatory sequencies occurring in the epidermis and dermal portion and include epidermal hyperplasia, ortho-parakeratosis, spongiosis, and inflammatory cells accumulation in the dermis. Ex vivo fluorescent confocal images show crowded bright nuclei of hyperplastic epidermis and bright nuclei of lymphocytes in the epidermal-dermal junction. Dark spaces in the basal layer of the epidermis are consistent with spongiosis. On digital H&E mode the bright white structures on FCM appear dark purple-blue, corresponding to hyperchromatic hematoxylin stained nuclei on histopathology.

13.1 Introduction

Eczema is an inflammatory skin reaction characterized by clinical features including itching, redness, scaling, and clustered papulo-vesicles. Eczema accounts for a large proportion of all skin diseases; the classification of many clinical forms is difficult, and multiple factors may be implicated. Worldwide, about 20% of children and up to 3% of the adult population have some form of eczema. Those who live in developed countries or colder climates seem to be more predisposed to developing eczema. Hand dermatitis accounts for 20–35% of all eczema cases. The occurrence of hand dermatitis was 1.5–2 times higher in females than males [1].

Histopathologic features reflect dynamic sequences of inflammation occurring in the epidermis and the underlying

L. Bertoni (✉) · C. Reggiani · P. Azzoni
Department of Surgical, Medical, Dental and Morphological Sciences with Interest in Transplant, Oncological and Regenerative Medicine, University of Modena and Reggio Emilia, Modena, Italy
e-mail: laura.bertoni@unimore.it

L. Reggiani Bonetti
Department of Medical and Surgical Sciences for Children and Adults, University of Modena and Reggio Emilia-AOU Policlinico of Modena, Modena, Italy

G. Pellacani
Dermatology Clinic, Department of Clinical Internal, Anesthesiological and Cardiovascular Sciences, Sapienza University of Rome, Modena, Italy

M. Jain et al. (eds.), *Cutaneous Atlas of Ex Vivo Confocal Microscopy*,
https://doi.org/10.1007/978-3-030-89316-3_13

Table 13.1 Clinical (epidemiology, patient demographics, and pathogenesis) and histopathological features of eczema

Basics of contact eczema
• **Epidemiology**: 2–5% of the population is affected; much higher in some occupational groups [5] • **Patient demographics**: Certain occupations have a very high prevalence of allergic contact dermatitis [5]. For hand dermatitis, women are affected twice as often as men, due to the effect of domestic work • **Pathogenesis**: In the sensitization phase of allergic contact dermatitis, allergens are taken up by Langerhans cells, processed and presented to the T-cells in the regional lymph node following migration. On re-exposure, the sensitized T-cells are activated and trigger skin inflammation at the site of exposure. In irritant contact dermatitis, cytokines are directly released by stimulated or damaged keratinocytes [6] • **Histopathology**: Thickening of the keratinocytic layer (epidermis hyperplasia) with spongiosis, ortho-parakeratosis, inflammatory infiltrate around blood vessels and skin appendages

Table 13.2 Eczema: morphological features on FCM, DHE and corresponding H&E

Eczema features on FCM, DHE, and corresponding conventional H&E		
Confocal microscopy fluorescence mode (FCM) [3] (Figs. 13.1A, 13.2A, 13.3A)	Digital H&E (combined FCM and RCM modes) (Figs. 13.1B, 13.2B, 13.3B)	Conventional H&E [3, 7, 8] (Figs. 13.1C, 13.2C, 13.3C)
Fluorescence: Nuclei are visualized as bright structures with enhanced contrast compared to the surrounding structures	**DHE pseudocolor**: The bright white structures on FCM appear dark purple-blue (high fluorescence signal; purple pseudo-color)	**Conventional H&E stained section**: The intense fluorescence (on FCM) and purple pseudo-color (on DHE) structures correspond to hematoxylin-stained hyperchromatic nuclei of the cells
Hyperplasia of the epidermis is recognizable as thickening of the keratinocytic layer. Presence of diffuse proliferation of the epidermis that determines increased stratified layers of crowded cells. Nuclei of keratinocytes appear hyper-reflective	**Hyperplasia of the epidermis** appears as thickened layer of keratinocytes forming slightly irregular epidermis with increased stratification of the cell nuclei which appear dark purple-blue	**Hyperplasia of the epidermis**: The intense fluorescent (on FCM) and purple pseudo-color (on DHE) nuclei corresponds to epidermis characterized by increased stratified layers of cells
Ortho-parakeratosis is observed as partially reflecting area; no nuclei in orthokeratosis, while cellular para-keratosis reflects light and nuclei are visible	**Ortho-parakeratosis**: orthokeratosis is characterized by light violet keratin plaques without nuclei, while parakeratosis is composed by purple-blue nucleated keratinized cells	**Ortho-parakeratosis**: the reflecting keratin with or without intense fluorescent nuclei (on FCM) and purple pseudo-color (on DHE) structures corresponds to hematoxylin-stained keratin plaques
Inflammatory infiltrate: lymphocytes are visible as cells of small size with round and well-defined bright nuclei around the vessels and appendages. Ectatic capillaries appear as dark empty spaces	**Inflammatory infiltrate**: lymphocytes appear as small cells with round purple-blue nuclei surrounded by a thin pink cytoplasmic rim. Lymphocytes aggregate around the blood vessels and skin appendages. Ectatic capillaries are visible as white spaces	**Inflammatory infiltrate**: mostly composed of lymphocytes, which are distributed in the dermis, around the blood vessels, and skin appendages. The lymphocytes have a dark round purple nucleus. Vessels in the dermis can be dilated
Spongiosis: appears as empty black spaces between the bright keratinocytes of the basal layer due to intercellular edema and detached cells	**Spongiosis**: The presence of empty white spaces between the purple-blue nuclei of keratinocytes of the basal layer is typical of spongiosis	**Spongiosis**: In the basal layer of the epidermis spongiosis appears as focal white spaces

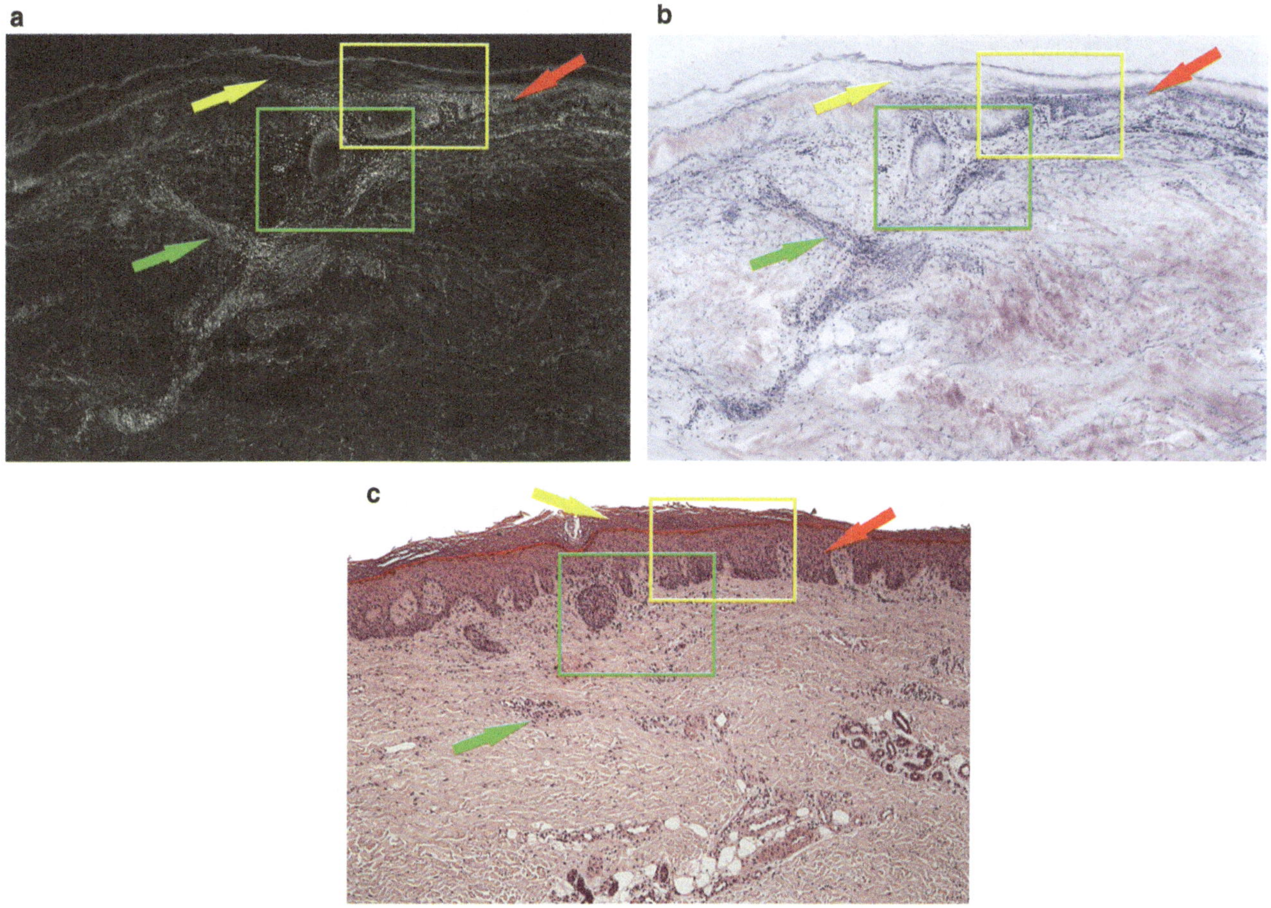

Fig. 13.1. Images of an eczema lesion from the leg of a 60 years old female. **a** FCM mosaic image allows typical features of the eczema. In particular, epidermis hyperplasia is evident and is recognizable as thickening of the keratinocytic layer. Note the presence of diffuse proliferation of the epidermis, which determines a stratification of the cell. Nuclei of keratinocytes appear hyper-reflective (red arrow). Thickened ortho-parakeratosis is observed as a partially reflecting layer: no nuclei in the entire orthokeratosis, while nuclei have light reflectance in the areas of parakeratosis (yellow arrow). The rete-ridges of the epidermis are elongated and irregular. In the dermis, there is the chronic inflammatory infiltrate consisting mainly of lymphocytes that are visible as cells of smaller size than the keratinocytes with small, round and well-defined bright nuclei (green arrow). Lymphocytes are aggregated around vascular spaces and surround cutaneous appendages. Empty spaces are ectatic capillaries visible in an edematous background (blue arrows). **b** Corresponding DHE image, closely resembling H&E-stained tissue histology (this figure), highlights epidermal hyperplasia that appears as a thickened layer of keratinocytes and slightly irregular epidermis. To note, the stratification of the cell nuclei, which appear dark purple-blue, is similar to H&E-stained section (red arrow). Orthokeratosis is characterized by keratin plaques without nuclei (yellow arrow) differently from parakeratosis which is composed of nucleated keratinized cells. The ridges of the epidermis are elongated and irregular. In the dermis it is possible to detect chronic inflammatory cells, consisting mainly of lymphocytes. They are visible as small cells with round nuclei surrounded by a thin pink cytoplasm rim (green arrow). Lymphocytes aggregate around blood vessels and skin appendages. Ectatic capillaries are visible in an edematous background (blue arrows) and adipose tissue is identified as optically empty spaces in the deep dermis. **c** Conventional H&E-stained section of eczema (×10 HPF) showing the hyperkeratosis layer (yellow arrow) with orthokeratosis (with no nuclei) and parakeratosis (with nuclei) and a thickened epidermis. The rete-ridges of the epidermis are different from each other and irregularly enlarged (red arrow). Epidermis is characterized by increased stratified lines of cells. Inflammatory cells (green arrow), mainly lymphocytes, are seen in the dermis around the blood vessels, and skin appendages with appear as dark round nuclei. Skin appendages and adipocytes are easily recognizable

dermal structures. These vary with the intensity and the stage of the eczematous process. This includes spongiosis, epidermal hyperplasia, associated with ortho-parakeratosis and an inflammatory infiltration in the dermis. Vascular dilatation in the dermis can be detected in all stages of diseases (Table 13.1) [2].

Fluorescence confocal microscopy (FCM), a novel optical technology, enables immediate digital image acquisition at an almost H&E-like resolution without requiring conventional tissue processing. Recently, ex vivo FCM has been applied to cutaneous inflammatory diseases, demonstrating its ability to diagnose main inflammatory diseases with a

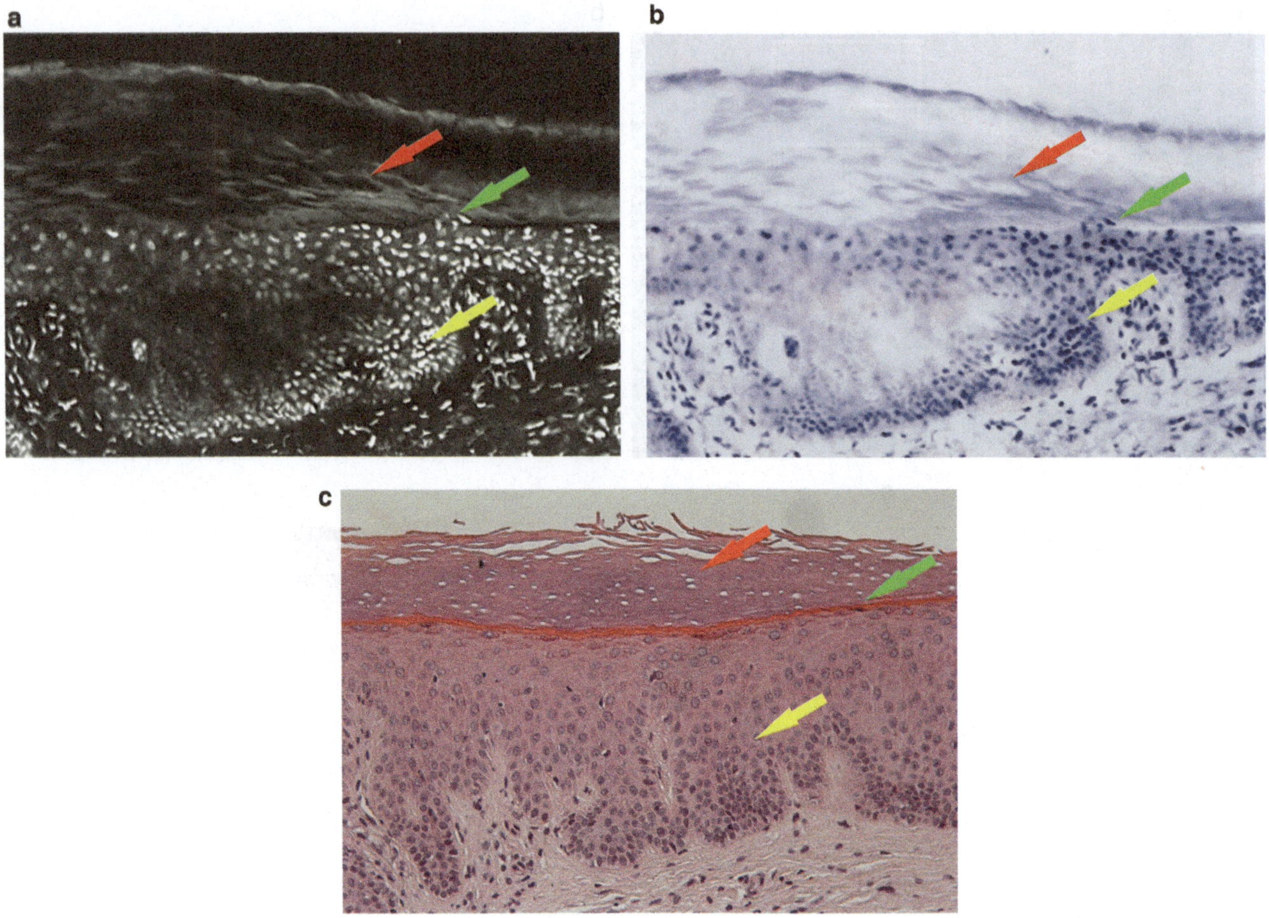

Fig. 13.2. Detailed features of eczema on FCM submosaic compared with the corresponding DHE and conventional H&E-stained images obtained from the yellow-boxed area in Fig. 13.1. **a** FCM image epidermal cells appear crowded with bright nuclei (yellow arrow) in the black background. Keratinocytic cell layers appear distinct. At this high magnification hyperkeratosis is clearly observed and can be distinguished into a more superficial orthokeratosis (red arrows) and in a deeper parakeratosis (green arrow). In orthokeratosis the nuclei are not seen, while in parakeratosis the nuclei are visible and bright. Bright nuclei of lymphocytes are visible at the dermal epidermal junction. **b** In the DHE image (5 × magnification) epidermal cells are clearly visible in the stratified epidermal layer; the nuclei are dark blue, round to oval and increased in number compared to a normal (yellow arrow). Individual cell layers appear distinct. Hyperkeratosis is clearly observed and can be distinguished into superficial orthokeratosis and deeper parakeratosis. In orthokeratosis (red arrows) nuclei are not visible, while in parakeratosis they are darkish blue and flat (green arrow). **c** Corresponding conventional H&E image (× 20 HPF) shows hyperkeratosis and the epidermal layer. Here, thickening of the keratin with a more superficial part of orthokeratosis without nuclei (red arrow) and a deeper part of parakeratosis with dark flat nuclei (green arrow) is visible. The epidermis is clearly visible and is thickened with multilayers of keratinocytes and a basal layer with empty spaces and distance between cells (yellow arrow)

substantial agreement to histopathological diagnoses [3]. FCM (Vivascope 2500 4th Gen®, MAVIG GmbH) acquires images in reflectance and fluorescence mode (black and white visualization). The visualization of images in pseudo-color purple and pink digital stain, similar to H&E (H&E-like mode) improves the diagnosis [4]. In Table 13.2, morphological features and terms used to describe eczema in FCM, digital H&E (DHE) mode and its histopathologic correlation are summarized.

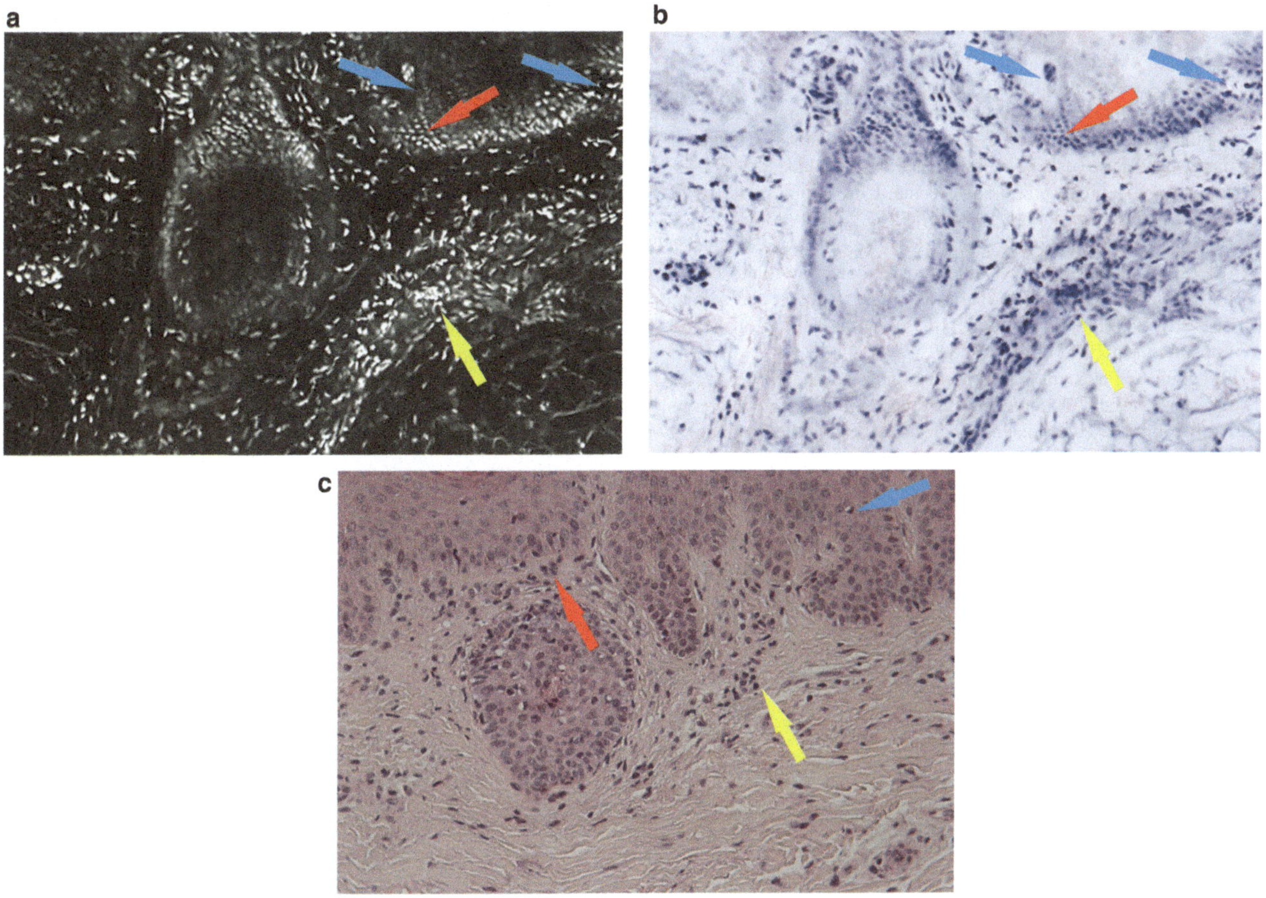

Fig. 13.3. Detailed features of eczema on FCM submosaic compared with the DHE and conventional H&E-stained image obtained from green-boxed area in Fig. 13.1. **a** FCM submosaic image shows mild/moderate chronic inflammatory infiltrate consisting mainly of lymphocytes (red arrow) with bright nuclei. These lymphocytes are seen hugging the dermal–epidermal junction, focally invading basal and suprabasal cells and in the edematous background of the dermis (yellow arrow). Dark spaces within the basal layer of the epidermis are consistent with spongiosis (blue arrow). **b** The DHE submosaic at the dermal–epidermal junction exhibits edematous dermis with mild/moderate chronic inflammatory mainly consisting of lymphocytes (red arrow). Nuclei of lymphocytes are dark blue and also seen invading basal and suprabasal layers of the epidermis. Inflammation also involves the lower dermis (yellow arrow) and is associated with an edematous background. Focal white spaces between the basal cells of the epidermis represent focal spongiosis (blue arrow). **c** Conventional H&E image (×20 HPF) shows that the basal layer of the epidermis has focal white spaces compatible with focal and mild spongiosis (blue arrow). Mild to moderate chronic inflammation consisting mainly of lymphocytes with small regular dark blue nuclei, is seen invading the basal and suprabasal layers of the epidermis (red arrow). Lymphocytes are also distributed deeper in the dermis (yellow arrow)

References

1. Quaade AS, Simonsen AB, Halling AS, Thyssen JP, Johansen JD. Prevalence, incidence, and severity of hand eczema in the general population: a systematic review and meta-analysis. Contact Dermatitis. 2021;84(6):361–74. https://doi.org/10.1111/cod.13804.
2. Jones RR. The histogenesis of eczema. Clin Exp dermatol. 1983;8 (3):213–25. https://doi.org/10.1111/j.1365-2230.1983.tb01774.x.
3. Bertoni L, Azzoni P, Reggiani C, Pisciotta A, Carnevale G, Chester J, et al. Ex vivo fluorescence confocal microscopy for intraoperative, real-time diagnosis of cutaneous inflammatory diseases: a preliminary study. Exp Dermatol. 2018;27(10):1152–9. https://doi.org/10.1111/exd.13754.
4. Reggiani C, Pellacani G, Reggiani Bonetti L, Zanelli G, Azzoni P, Chester J, et al. An intraoperative study with ex vivo fluorescence confocal microscopy: diagnostic accuracy of the three visualization modalities. J Eur Acad Dermatol Venereol. 2021;35(1):e92–4. https://doi.org/10.1111/jdv.16831.
5. McCall BP, Horwitz IB, Feldman SR, Balkrishnan R. Incidence rates, costs, severity, and work related factors of occupational dermatitis. Arch Dermatol. 2005;141(6):713–8. https://doi.org/10.1001/archderm.141.6.713.
6. Grabbe S, Schwarz T. Immunoregulatory mechanism involved in elicitation of allergic contact hypersensitivity. Immunol Today. 1998;19(1):37–44. https://doi.org/10.1016/s0167-5699(97)01186-9.
7. Buxton PK. ABC of dermatology eczema and dermatitis. Br Med J (Clin Res Ed). 1987;295(6605):1048–51. https://doi.org/10.1136/bmj.295.6605.1048.
8. Phelps RG, Miller MK, Singh F. The varieties of “eczema”: clinicopathologic correlation. Clin Dermatol. 2003;21(2):95–100. https://doi.org/10.1016/s0738-081x(02)00357-7.

14 Fluorescence Confocal Microscope for Identification of Discoid Lupus Erythematosus Histologic Features

Laura Bertoni, Camilla Reggiani, Paola Azzoni, Stefania Caramaschi, Luca Reggiani Bonetti, and Giovanni Pellacani

Abstract

Discoid lupus erythematosus is clinically represented by sharply demarcated erythematous scaly patches with follicular plugging occurring in the face, neck, scalp, eyelid, lips, oral mucosa and hands. Histopathology includes atrophic epidermis, hyperkeratosis, basal cell damage with vacuolar degeneration and apoptotic keratinocytes (Civatte bodies). Lymphocytes and histiocytes accumulation surround the periadnexal spaces. Ex vivo fluorescent confocal images show thick reflective stratum corneum overlying epidermis, bright doting apoptotic keratinocytes and hyperfluorescent dense nuclear accumulation around the periadnexal spaces. Basal spongiosis is represented by black spaces in the basal portion of the epidermis. On digital H&E mode, the fluorescent structures appear dark purple while reflecting ones are light pink. The fluorescent signals correspond in H&E images to hematoxylin-stained nuclei.

14.1 Introduction

Lupus erythematosus (LE) is a multisystem disorder that can affect the skin. It occurs most commonly in women during childbearing years [1]. The pathogenesis of LE is complex and involves genetic and environmental factors. LE can be classified in the following forms: discoid lupus erythematosus, subacute cutaneous lupus erythematosus, and acute cutaneous lupus [2]. Discoid lupus erythematosus (DLE) involves the epidermis, upper and lower dermis, and adnexal structures and it can evolve to scars [3]. The typical lesions of DLE are sharply demarcated, erythematous, scaly patches with follicular plugging. They usually occur in the skin of the face, the neck, scalp, eyelid, lips, oral mucosa, and hands [2].

Histopathologic findings in DLE are basal cell damage (also referred to as vacuolar degeneration) with scattered Civatte bodies (apoptotic keratinocytes), hyperkeratosis, orthokeratosis, and focal atrophic epidermis. In the dermis, lymphohistiocytic inflammatory infiltrate is evident with periadnexal inflammation, follicular plugging, and scarring (Table 14.1).

Ex vivo fluorescence confocal microscopy (FCM) is an innovative imaging tool that can be adopted in intraoperatively procedures to obtain real-time images of untreated excised tissue with almost histologic resolution. This technique has been validated on gray-scale (black and white) images using older generations' ex vivo FCM. Images of DLE were analyzed and their correlation with the corresponding histopathological specimens was evaluated. The results suggest that ex vivo FCM is a promising tool in assisting in DLE lesions diagnosis, with a level of accuracy quite close to that offered by histopathology [4]. Recently, digital (purple and pink) H&E images were obtained using newer generation ex vivo FCM (Vivascope 2500 4th Gen®, MAVIG GmbH) that combines FCM and reflectance mode [5]. The digital H&E images are more

L. Bertoni (✉) · C. Reggiani · P. Azzoni
Department of Surgical, Medical, Dental and Morphological Sciences with Interest in Transplant, Oncological and Regenerative Medicine, University of Modena and Reggio Emilia, Modena, Italy
e-mail: laura.bertoni@unimore.it

S. Caramaschi · L. R. Bonetti
Department of Medical and Surgical Sciences for Children and Adults, University of Modena and Reggio Emilia-AOU Policlinico of Modena, Modena, Italy

G. Pellacani
Dermatology Clinic, Department of Clinical Internal, Anesthesiological and Cardiovascular Sciences, Sapienza University of Rome, Modena, Italy

M. Jain et al. (eds.), *Cutaneous Atlas of Ex Vivo Confocal Microscopy*,
https://doi.org/10.1007/978-3-030-89316-3_14

Table 14.1 Clinical (epidemiology, patient demographics, genetics) and histopathological features of DLE

Basics of DLE
• **Epidemiology**: Discoid Lupus Erythematosus (DLE) occurs in approximately 12.4–15% of incident Systemic Lupus Erythematosus (SLE) cases and in 16.6–24.3% of prevalent cases. DLE can also occur in the absence of SLE [1] • **Patient demographics**: More common in women, usually appears in 15–60 years of age group. Sites of predilection include chronic sun-exposed areas (scalp, forehead, cheeks, ears, nose, upper lip, and chin) • **Genetics**: HLA-B7, -B8, -Cw7, -DR2, -DR3, -DQw1, -DRB1; various polymorphism-TNF-R gene, Fc-gamma receptor, CD19 gene [6] • **Histopathology**: Dense lymphohistiocytic infiltrate around the blood vessels of the superficial and deep plexuses and around pilosebaceous units, edema of the papillary dermis with widely dilated vascular spaces, vacuolar alteration at the dermo-epidermal interface, epidermal atrophy with apoptotic keratinocytes (Civatte bodies), and orthokeratosis

Table 14.2 DLE: morphological features on FCM, DHE and corresponding H&E

Morphological features of DLE on FCM, DHE, and corresponding conventional H&E		
Fluorescence Confocal Microscopy mode (FCM) (grayscale)	Digital H&E (combined FCM and RCM modes)	Conventional H&E stained images [2–8]
Fluorescence: the nuclei appear white bright spots (fluorescence) while other structures are gray (reflectance) (Figs. 14.1a, 14.2a and 14.3a)	**DHE pseudocolor**: The fluorescent structures appear dark purple while reflectant structures are light pink (Figs. 14.1b, 14.2b and 14.3b)	**Traditional H&E stained section**: These intense fluorescent (on FCM) and purple pseudo-color (on DHE) structures corresponds to hematoxylin-stained nuclei (Figs. 14.1c, 14.2c and 14.3c)
Deep periadnexal and perivascular infiltrate: appear as hyperfluorescent dense dots with characteristic localization near cutaneous adnexa and vessels	**Deep periadnexal and perivascular infiltrate**: deep dense small purple-blue nuclei are visible in perivascular and periadnexal location	**Deep periadnexal and perivascular infiltrate**: the intense fluorescent (in FCM) and purple pseudo colored (in DHE) nuclei correspond to the H&E stained inflammatory infiltrate
Thinned epidermis: is observed with bright small, regular, and round nuclei distributed in a thin layer of cells; apoptotic keratinocytes (Civatte's bodies) are visible as bright dots in the epidermis	**Thinned epidermis**: appears as a slim band of dark blue keratinocytes with a thick stratum corneum. Civatte's bodies can be identified as darkish nuclei within the basal layer	**Thinned epidermis**: is evident in H&E stained section with dark nuclei of the keratinocytes distributed in a reduced layer of cells. Some scattered eosinophilic apoptotic keratinocytes in the basal layer or in the epithelium are present
Orthokeratosis appears as a thick reflective stratum corneum overlying the epidermis	**Orthokeratosis** is characterized by a thick light violet stratum corneum without dark blue nuclei	**Orthokeratosis** with a dense stratum corneum is a characteristic microscopic feature in H&E stained images
Basal cell vacuolization is visible as black degeneration of the keratinocytes of the basal portion of the epidermis	**Basal cell vacuolization**: interface reaction appears as a white vacuolar change throughout the dermo-epidermal junction	**Basal cell vacuolization**: diffuse vacuolar degeneration of the keratinocytes of the basal portion of the epidermis is observed in the interface area of the dermo-epidermal junction

appealing and easier to recognize by dermatopathologists. Table 14.2 recapitulates morphological features and terms used to describe DLE in FCM and digital H&E modes and their histopathologic correlation (Figs. 14.1, 14.2 and 14.3).

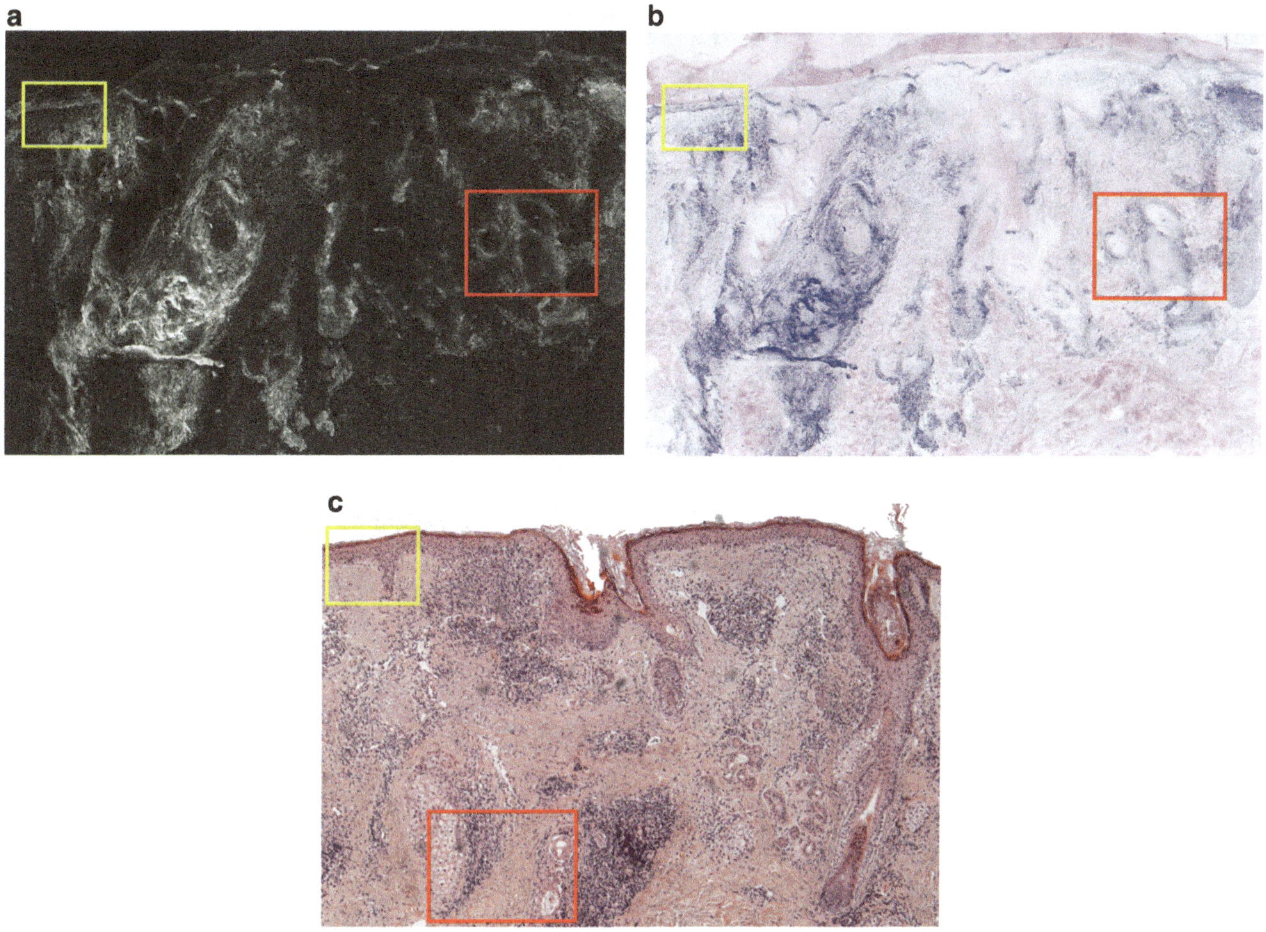

Fig. 14.1. Mosaic images of a DLE lesion from the forehead of a 55 years old female. **a** In FCM mosaic epidermis is thin (yellow box). Dermis shows diffuse chronic inflammatory infiltrate in the superficial and deep portions distributed in perivascular and periadnexal locations (red box). Bright nuclei of epidermal cells and lymphocytes are observed. The nuclei of epithelial cells and appendages are larger than the nuclei of the lymphocytes. **b** In the DHE image, thin epidermis (yellow box) and dermis are evident; it is possible to observe a diffuse chronic inflammatory infiltrate in the superficial and deep portion of the dermis, with specific distribution d in perivascular and periadnexal zones (red box). The nuclei of epidermal cells and lymphocytes are dark purple-blue. The dark dense blue portion of the image corresponds to dense inflammation around the adnexal structures. **c** Corresponding H&E stained image of the DLE shows prominent inflammation of the dermis distributed in superficial and deeper portions with periadnexal distribution (red box). The inflammation is also clustered around the pilosebaceous unit and sebaceous glands. Other features are thinned epidermis, orthokeratosis, and basal cell vacuolization (yellow box)

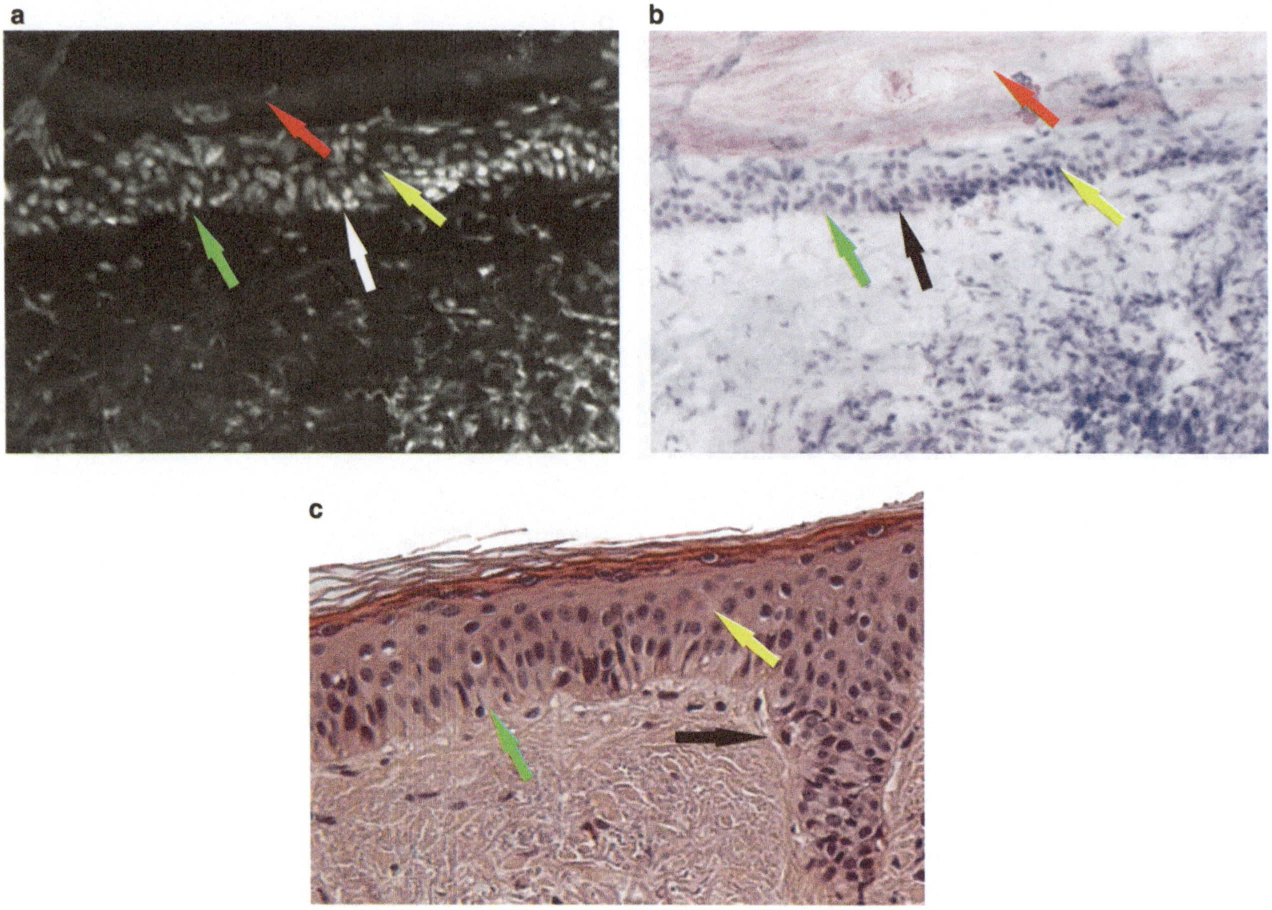

Fig. 14.2. Detailed features of DLE on FCM submosaic compared with the corresponding DHE and standard H&E stained tissue corresponding to the yellow-boxed area from Fig. 14.1. **a** In the submosaic FCM, a thinned epidermis is observed with bright nuclei distributed in a thin layer of cells (yellow arrow); it is associated with orthokeratosis, which appears as a thick stratum corneum overlying the epidermis (red arrow). The nuclei are small, regular, and round and show stratified distribution. Vacuolar degeneration of the keratinocytes of the basal portion of the epidermis is visible (green arrow). Moreover, Civatte's bodies can be identified as darkish nuclei within basal cells (white arrow). **b** DHE submosaic shows a thinned epidermal layer consisting of hyperplastic keratinocytes whose nuclei appear dark blue-purple (yellow arrow); orthokeratosis is observed which appears as a thick stratum corneum overlying the epidermis (red arrow). Diffuse vacuolar degeneration of the keratinocytes of the basal portion of the epidermis is observed, in the interface area of the dermo-epidermal junction (green arrow). Apoptotic keratinocytes (Civatte's bodies) can be identified as darkish nuclei within basal cells (black arrow). The dermis shows a chronic inflammatory infiltrate with a prevalence of lymphocytes, interstitial mucin deposition, and edema revisable as pale spaces within dermal fibers. **c** Traditional H&E stained section shows characteristic microscopic features of DLE like hyperkeratosis with thinning and flattening of the epithelium (yellow arrow), and hydropic degeneration of the basal layer (green arrow). In addition, there are scattered apoptotic keratinocytes (Civatte's bodies) in the basal layer or in the epithelium (black arrow). The basement membrane becomes thickened and more visible. There is interstitial mucin deposition and edema, and usually, no eosinophils and neutrophils are present

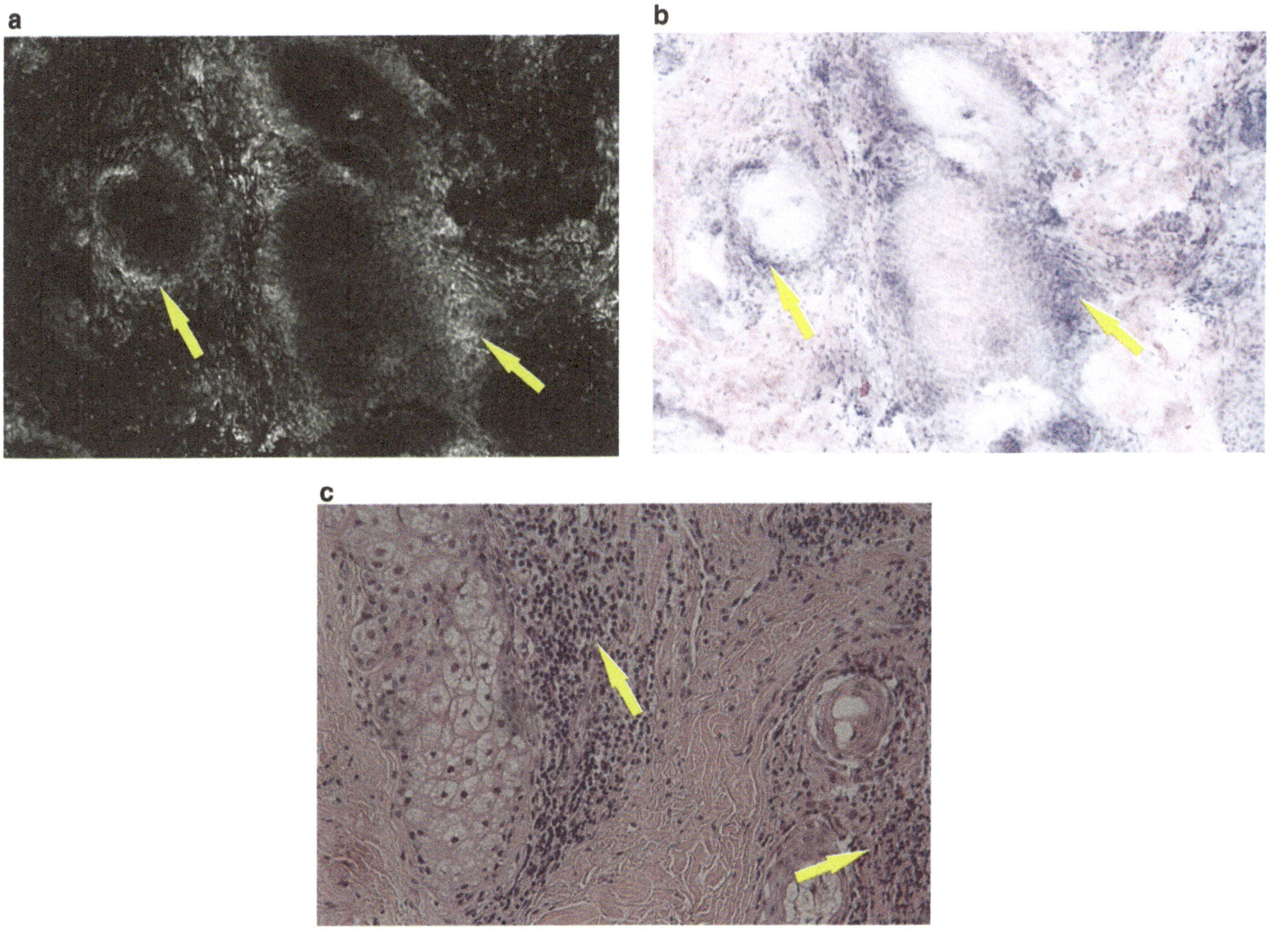

Fig. 14.3. Detailed features of DLE on FCM submosaic compared with DHE submosaic and standard H&E stained tissue corresponding to the red-boxed areas from Fig. 14.1. **a** In submosaic FCM annexes at the periphery with cells with large amounts of dark cytoplasm and bright nuclei show inflammation. In particular, it is possible to evidence an aggregate of lymphocytes with bright small nuclei (yellow arrows). **b** Digitally zoomed DHE submosaic (5× magnification) shows lymphocytes with dark small round nuclei and close to each other (yellow arrows) around the pilosebaceous follicles and sebaceous glands. **c** In H&E stained slide, pilosebaceous glands composed of large polygonal cells with clear cytoplasm and small nuclei are surrounded by aggregated inflammatory cells mainly represented by lymphocytes showing dark small round nuclei and inconsistent cytoplasm (yellow arrows)

References

1. Izmirly P, Buyon J, Belmont HM, et al. Population-based prevalence and incidence estimates of primary discoid lupus erythematosus from the Manhattan Lupus Surveillance Program. Lupus Sci Med. 2019;6(1):e000344. https://doi.org/10.1136/lupus-2019-000344.
2. Weedon D. Lupus erithematosus. In: Weedon's skin pathology, 3rd edn. Churchill Livingstone Elsevier; 2010. p. 57–59.
3. Bangert JL, Freeman RG, Sontheimer RD, et al. Subacute cutaneous lupus erythematosus and discoid lupus erythematosus comparative histopathologic findings. Arch Dermatol. 1984;120:332–7.
4. Bertoni L, Azzoni P, Reggiani C, Pisciotta A, Carnevale G, Chester J, et al. Ex vivo fluorescence confocal microscopy for intraoperative, real-time diagnosis of cutaneous inflammatory diseases: a preliminary study. Exp Dermatol. 2018;27(10):1152–9. https://doi.org/10.1111/exd.13754.
5. Bağcı IS, Aoki R, Vladimirova G, Ergün E, Ruzicka T, Sárdy M, et al. New-generation diagnostics in inflammatory skin diseases: immunofluorescence and histopathological assessment using ex vivo confocal laser scanning microscopy in cutaneous lupus erythematosus. Exp Dermatol. 2021;30(5):684–90. https://doi.org/10.1111/exd.14265.
6. Goodfield MJD, Jones SK, Veale DJ. Discoid lupus erithematosus. In: Rook's textbook of dermatology, vol. 3, 7thed. Blackwell Publishing; 2004. p. 56.5–56.8.
7. David-Bajar KM, Bennion SD, DeSpain JD, Golitz LE, Lee LA. Clinical, histologic, and immunofluorescent distinctions between subacute cutaneous lupus erythematosus and discoid lupus erythematosus. J Invest Dermatol. 1992;99(3):251–7. https://doi.org/10.1016/j.jaut.2014.01.021.
8. Elman SA, Joyce C, Nyberg F, Furukawa F, Goodfield M, Hasegawa M, et al. Development of classification criteria for discoid lupus erythematosus: results of a Delphi exercise. J Am Acad Dermatol. 2017;77(2):261–7. https://doi.org/10.1016/j.jaad.2017.02.030.

Immunofluorescence with Confocal Microscopy: Pemphigoid, Pemphigus, Cutaneous Vasculitis, Lichen Planus, and Cutaneous Lupus Erythematosus

15

Isin Sinem Bagci and Daniela Hartmann

15.1 Introduction

Ex vivo confocal laser scanning microscopy (ex vivo CLSM) has been recently shown to be useful in the examination of autoimmune blistering and inflammatory skin diseases. Besides providing conventional histology-like digital hematoxylin–eosin (H&E) images, ex vivo CLSM offers simultaneous direct immunofluorescence (DIF) examination of the same section. Using fluorescent labeled anti-human antibodies, ex vivo CLSM can detect autoantibody deposition in the epidermis, basement membrane, vessel wall, and cytoid bodies. This approach increases the diagnostic performance of ex vivo CLSM, on the other hand, decreases the number of biopsies as well as time and effort needed for these two different diagnostic methods. In addition, high-resolution images enable remote assessment via teledermatology. Herein, current knowledge on ex vivo CLSM characteristics in bullous pemphigoid (Tables 15.1 and 15.2), pemphigus vulgaris (Tables 15.3 and 15.4), cutaneous vasculitis (Tables 15.5 and 15.6), lichen planus (Tables 15.7 and 15.8), and cutaneous lupus erythematosus (Tables 15.9 and 15.10) are presented. Examples of typical ex vivo confocal features are displayed in Figs. 15.1, 15.2, 15.3, 15.4, 15.5, 15.6, 15.7, 15.8, 15.9, 15.10 and 15.11). VivaScope 2500 M-G4 (Lucid Inc., Rochester, New York) was used for imaging 300-μm-thick frozen sections. These sections were stained with fluorescein isothiocyanate (FITC)-labeled anti-human antibodies (IgG, IgM, IgA, C3, and fibrinogen). The staining was performed according to the staining protocol for DIF microscopy that is routinely used in our autoimmune laboratory

I. S. Bagci
Department of Dermatology, Stanford School of Medicine – Dermatology, Redwood City, CA, USA

D. Hartmann (✉)
Department of Dermatology and Allergy, University Hospital, LMU Munich, Frauenlobstrasse 9–11, 80337 Munich, Germany
e-mail: Daniela.Hartmann@med.uni-muenchen.de

M. Jain et al. (eds.), *Cutaneous Atlas of Ex Vivo Confocal Microscopy*,
https://doi.org/10.1007/978-3-030-89316-3_15

15.2 Bullous Pemphigoid

Table 15.1 Basics of bullous pemphigoid (BP = bullous pemphigoid, BM = basement membrane)

Basics of Bullous Pemphigoid (BP)
• **Epidemiology**: BP is the most common autoimmune blistering skin disease [1] • **Patient demographics**: Elderly patients, with an increasing frequency above the age of 70 [2] • **Pathomechanism**: Tissue-bound and circulating autoantibodies directed against two structural components of hemidesmosomes, namely BP180 and BP230, which attach basal keratinocytes to the underlying basement membrane (BM) • **Histopathology**: Subepidermal blister formation, eosinophilic spongiosis, upper dermal inflammatory infiltration of neutrophils, eosinophils, and lymphocytes [3] • **Direct immunofluorescence**: Linear deposition of IgG (80–90%) and/or complement C3 (almost all patients) along the BM of perilesional skin [4]

Table 15.2 Ex vivo CLSM features of bullous pemphigoid together with histological and DIF correlations (CLSM = confocal laser scanning microscopy, FCM = fluorescence confocal microscopy mode, RCM = reflectance confocal microscopy mode, DHE = digital hematoxylin–eosin-like staining, H&E = hematoxylin–eosin staining, FITC = fluorescein isothiocyanate, BM = basement membrane)

Bullous pemphigoid Features on FCM, RCM, DHE, and corresponding conventional H&E and DIF	
Ex vivo confocal microscopy reflectance (RCM), fluorescence (FCM), and digital H&E (DHE) modes [5]	Conventional H&E and DIF
Upper dermal infiltration: Accumulation of white or light gray colored hyperreflective (in RCM) (Fig. 15.1a)/ purple-colored (in DHE) cells (Fig. 15.1d) beneath the dermo-epidermal junction	These hyperreflective/purple-colored cells correspond to the subepidermal infiltration of inflammatory cells including neutrophils, eosinophils, and lymphocytes
Subepidermal cleavage: Separation of full thickness epidermis from the underlying dermis (in RCM and DHE) (Fig. 15.1)	This subepidermal cleavage corresponds to the subepidermal blister formation that can be detected in the samples taken from the edge of an intact blister (in H&E)
Fluorescence: Granular patchy fluorescence along the dermo-epidermal junction in sections stained with FITC-labeled anti-human IgG and/or C3 (in FCM) (Fig. 15.2)	This fluorescence corresponds to the linear deposition of IgG and/or C3 along the BM detected in the perilesional samples

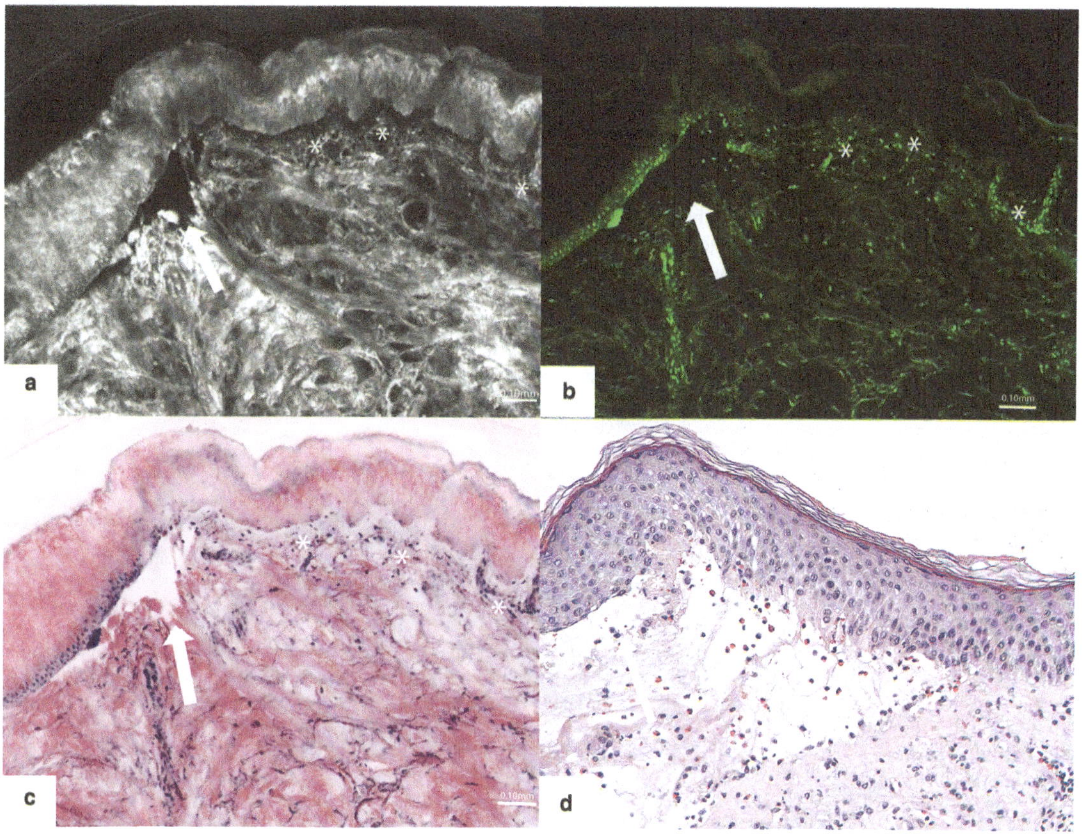

Fig. 15.1 Three imaging modes of the ex vivo CLSM examination (**a** reflectance, **b** fluorescence, **c** DHE) and corresponding histopathology findings (D, 20X) of BP showing inflammatory cells (asterisk) in the upper dermis and subepidermal blister formation (arrow), (BP = bullous pemphigoid, CLSM = confocal laser scanning microscopy, DHE = digital hematoxylin–eosin staining)

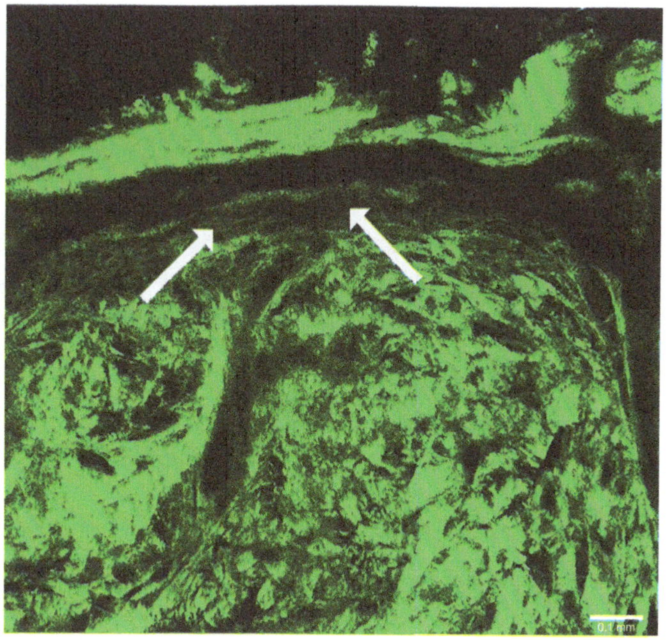

Fig 15.2 Fluorescence mode of the ex vivo CLSM examination of BP showing IgG deposition along BM (arrow), (CLSM = confocal laser scanning microscopy, BP = bullous pemphigoid, BM = basement membrane)

15.3 Pemphigus Vulgaris

Table 15.3 Basics of pemphigus vulgaris (PV = pemphigus vulgaris)

Basics of pemphigus vulgaris (PV)
• **Epidemiology**: PV is the most common subtype of the pemphigus group which is characterized by intraepidermal blister formation [6] • **Patient demographics**: Onset occurs mostly in 50–60 years of age group • **Pathomechanism**: Binding of IgG autoantibodies on epidermal desmoglein 3 and 1 results in the loss of cell–cell adhesion between keratinocytes [7] • **Histology**: Acantholysis, intraepidermal blister formation, sparse inflammatory infiltrate in the upper dermis • **Direct Immunofluorescence**: Intraepidermal intercellular IgG and/or C3 deposition in perilesional skin [8]

Table 15.4 Ex vivo CLSM features of pemphigus vulgaris together with histological and DIF correlations (CLSM = confocal laser scanning microscopy, FCM = fluorescence confocal microscopy mode, RCM = reflectance confocal microscopy mode, DHE = digital haematoxylin–eosin-like staining, H&E = haematoxylin–eosin staining, FITC = fluorescein isothiocyanate, BM = basement membrane)

Pemphigus vulgaris features on FCM, RCM, DHE, and corresponding conventional H&E AND DIF	
Ex vivo confocal microscopy reflectance (RCM), digital H&E (DHE), and fluorescence (FCM) modes [9,10]	Conventional H&E and DIF
Separation of epidermis above the basal layer of keratinocytes and scattered roundish cells within the separation (in RCM and DHE) (Fig. 15.3A and D)	Separation of the epidermis corresponds to the intraepidermal blister formation due to IgG autoantibodies. The remaining intact layer of basal keratinocytes is named as “tombstone appearance”. Scattered cells within the blister cavity correspond to the acantholytic keratinocytes
Fluorescence: Linear intraepidermal intercellular fluorescence in sections stained with FITC-labeled anti-human IgG and C3 (in FCM) (Fig. 15.3B and C)	This fluorescence corresponds to the intercellular IgG and C3 deposition in the perilesional sections

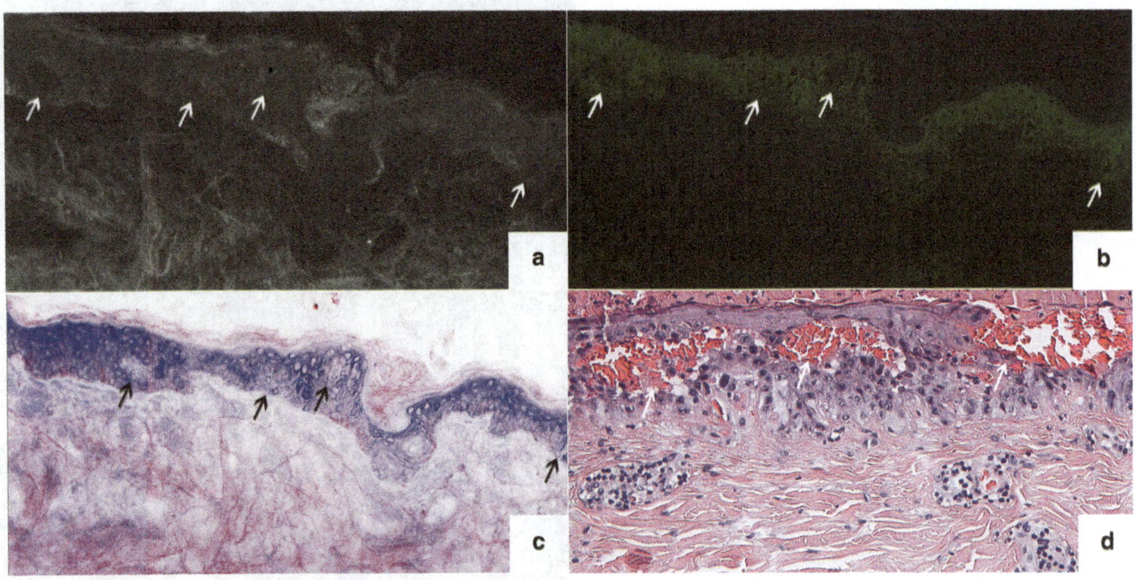

Fig. 15.3 Ex vivo CLSM examination of PV with intraepidermal intercellular deposition of IgG, as well as intraepidermal blister formation (arrow) in all three ex vivo CLSM imaging modes (**a** reflectance, **b** fluorescence, **c** digital hematoxylin–eosin-like staining) with corresponding histopathology findings (**d**, 20X), (CLSM = confocal laser scanning microscopy, PV = pemphigus vulgaris).

15.4 Cutaneous Vasculitis

Table 15.5 Basics of cutaneous vasculitis

Basics of Cutaneous Vasculitis
• **Epidemiology**: Cutaneous vasculitis represents a wide spectrum of diseases that are characterized by the inflammatory cell infiltration and destruction of blood vessels in the skin. Leukocytoclastic vasculitis represents the most common form of cutaneous vasculitis [11] • **Patient demographics**: Henoch-Schönlein purpura occurs more frequently in childhood, whereas leukocytoclastic vasculitis is seen mostly in adults • **Pathomechanism**: Deposition of antigen–antibody complex in blood vessels initiating the cascade of vascular injury secondary to infections, drugs, inflammatory or neoplastic disorders [12] • **Histopathology**: Neutrophil-rich infiltrates around the vessel wall, nuclear fragmentation, and erythrocyte extravasation [13] • **Direct immunofluorescence**: Deposition of IgA, IgM, IgG, C3, and fibrinogen along the vessel wall [14]

Table 15.6 Ex vivo CLSM features of cutaneous vasculitis together with histological and DIF correlations (CLSM = confocal laser scanning microscopy, FCM = fluorescence confocal microscopy mode, RCM = reflectance confocal microscopy mode, DHE = digital hematoxylin–eosin-like staining, H&E = hematoxylin–eosin staining, FITC = fluorescein isothiocyanate, BM = basement membrane)

Confocal Cutaneous Vasculitis features on FCM, RCM, and corresponding conventional H&E and DIF	
Ex vivo confocal microscopy reflectance (RCM), digital H&E (DHE), and fluorescence (FCM) modes [15]	Conventional H&E and DIF
Small, oval or round, bright, hyperreflective cells (white or light gray) in superficial and deep dermal zones arranged in an annular pattern, (in RCM and DHE) (Figs. 15.4 A and D)	These hyperreflective structures correspond to the inflammatory calls around the vessel walls of deep and superficial cutaneous vessels
Fluorescence: Granular, homogenous, or linear green fluorescence around the superficial or deep dermal vessels in sections stained with FITC-labeled anti-human IgG, IgM, IgA, C3, and fibrinogen (in FCM) (Fig. 15.5)	This fluorescence corresponds to the deposition of immune complexes of IgG, IgM, IgA, C3, and fibrinogen along the vessel wall

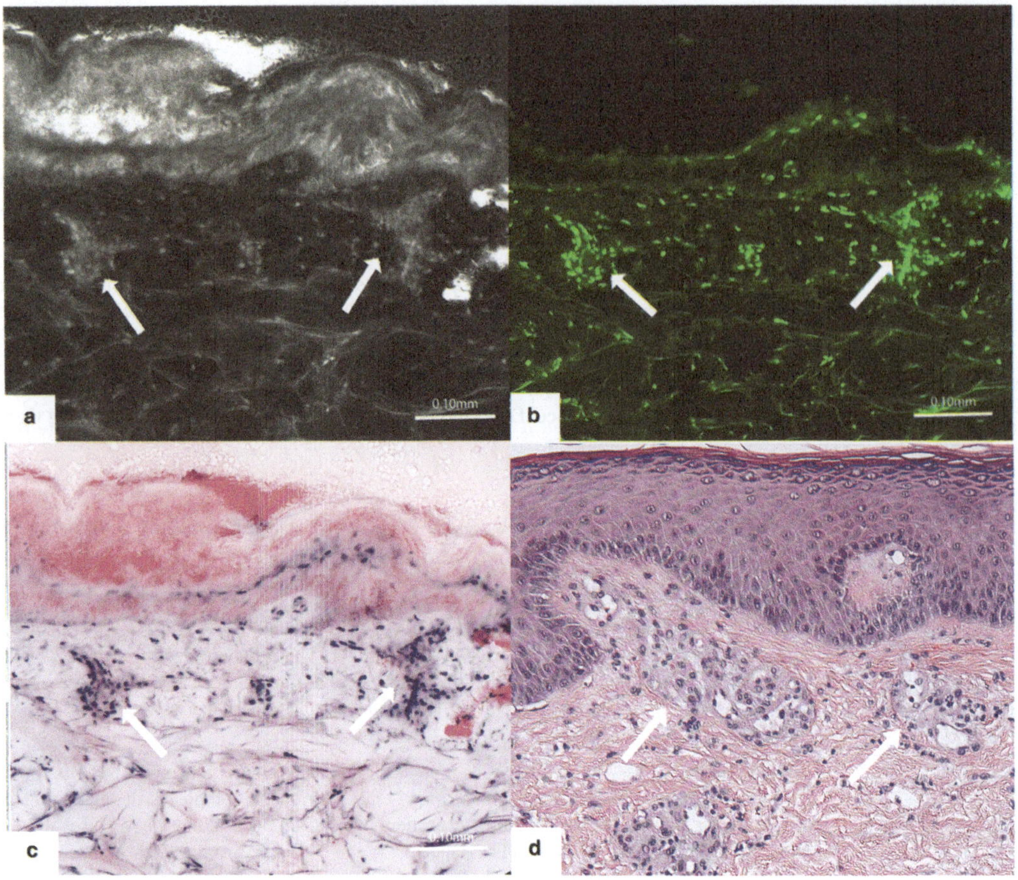

Fig. 15.4 Cutaneous vasculitis examined in the three modes of the ex vivo confocal laser scanning microscopy (**a** reflectance, **b** fluorescence, **c** DHE) and corresponding histopathology findings (**d**, 20X), presenting perivascular inflammatory infiltrates in the papillary dermis (arrows), (DHE = digital hematoxylin–eosin staining).

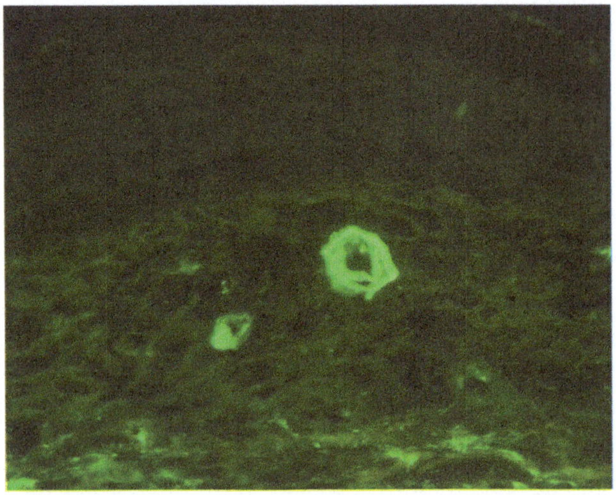

Fig. 15.5 Fluorescence mode of the ex vivo CLSM examination of cutaneous vasculitis section stained with anti-human fibrinogen antibodies showing fibrinogen deposition along the vessel wall of superficial dermal vessels (arrow), (CLSM = confocal laser scanning microscopy).

15.5 Lichen Planus

Table 15.7 Basics of lichen planus (LP = lichen planus)

Basics of Lichen Planus
• **Epidemiology**: LP is a chronic inflammatory disease of skin and mucosa with a reported incidence rate of 28–250/100.000 patients in different studies [16] • **Patient demographics**: LP mostly affects patients between 40–60 years of age [17] • **Pathomechanism**: T-cell mediated inflammation developed against an unknown antigen. Hepatitis-C infection, some drugs (captopril), and chemicals (thiomersal) can trigger the disease • **Histology**: Acanthosis, hypergranulosis, vacuolar degeneration, interface dermatitis, and a band-like infiltrate of inflammatory cells in the papillary dermis • **Direct immunofluorescence**: Shaggy deposition of fibrinogen along the basement membrane, cytoid bodies [18]

Table 15.8 Ex vivo CLSM features of lichen planus together with histological and DIF correlations (CLSM = confocal laser scanning microscopy, FCM = fluorescence confocal microscopy mode, RCM = reflectance confocal microscopy mode, DHE = digital hematoxylin–eosin-like staining, H&E = hematoxylin–eosin staining, FITC = fluorescein isothiocyanate, DIF = direct immunofluorescence)

Lichen planus features on FCM, RCM, DHE, and corresponding conventional H&E and DIF	
Ex vivo confocal microscopy reflectance (RCM), fluorescence (FCM), and digital H&E (DHE) modes [19]	Conventional H&E and DIF
Epidermal changes: Round/polygonal-shaped, larger appearing keratinocytes showing an increased intercellular distance (spongiosis), partly even resembling an initial blister formation in the basal layer of the epidermis and dermo-epidermal junction (in DHE, RCM) (Figs. 15.6a and d)	These larger appearing cells with increased intercellular space in the lower epidermis correspond to the spongiosis. A similar appearance of the basal layer of keratinocytes and dermo-epidermal junction corresponds to the vacuolar degeneration in conventional H&E
Dermal changes: Accumulation of hyperreflective (in RCM) / purple-colored (in DHE) round cells beneath the dermo-epidermal junction and around the vessels which partially disrupts the dermal papillae (in RCM and DHE) (Figs. 15.7 a and d)	The accumulation of the cells in the papillary dermis corresponds to the band-like infiltration and perivascular infiltration of inflammatory cells Destruction of papillary structures due to diffuse inflammatory cell infiltration corresponds to interface dermatitis in conventional H&E
Fluorescence: Shaggy fluorescence along the dermo-epidermal junction, fluorescence of roundish structures at and below the level of dermo-epidermal junction in sections stained with FITC-labeled fibrinogen (in FCM) (Fig. 15.8)	Shaggy fluorescence corresponds to the deposition of fibrinogen along the BM. Immunoreactive roundish structures correspond to the cytoid bodies which originate from apoptotic basal keratinocytes mainly in diseases characterized by interface dermatitis in conventional DIF

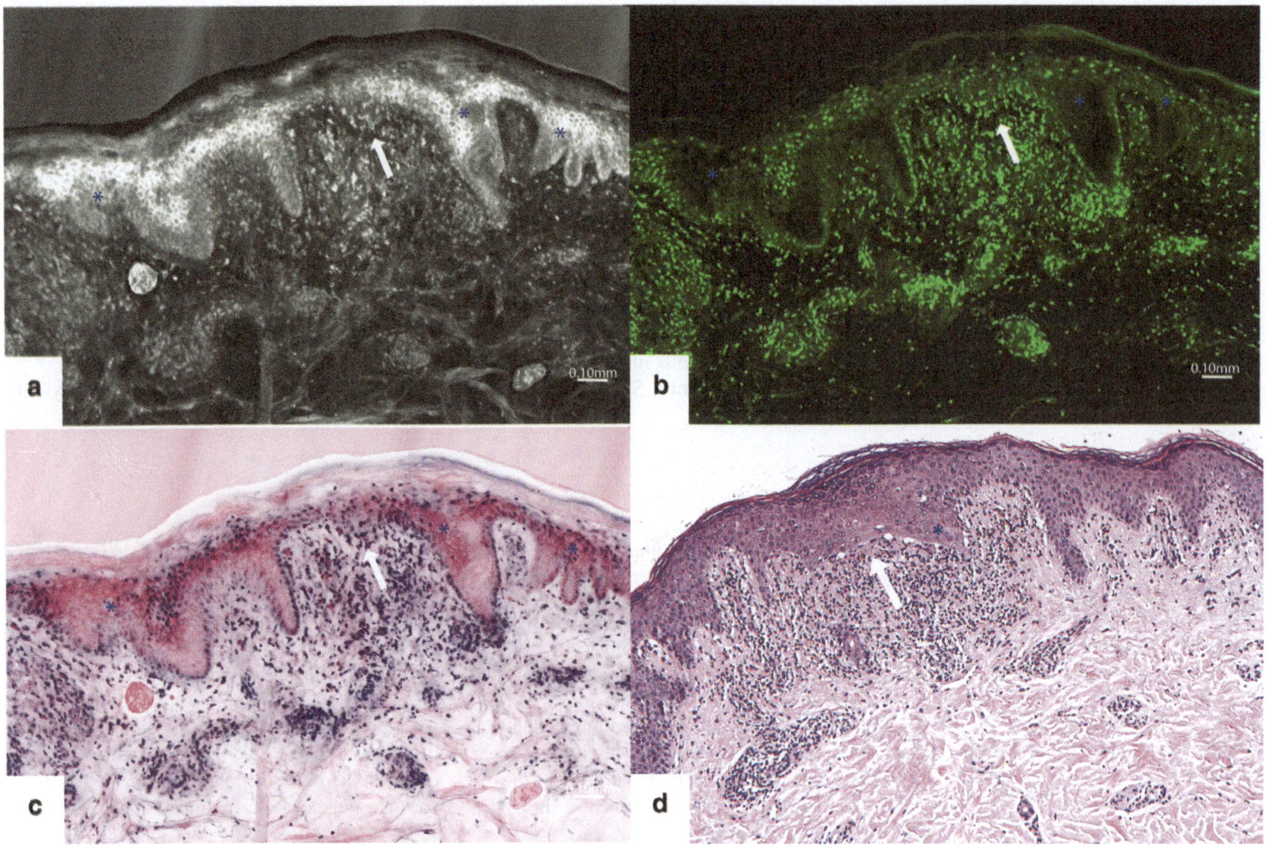

Fig. 15.6 Lichen planus examined in the three modes of the ex vivo confocal laser scanning microscopy (**a** reflectance, **b** fluorescence, **c** DHE) and corresponding histopathology findings (**d**, 10X), presenting band-like inflammatory infiltrate und vacuolar degeneration (arrow), as well as spongiosis (asterisk), (DHE = digital hematoxylin–eosin staining)

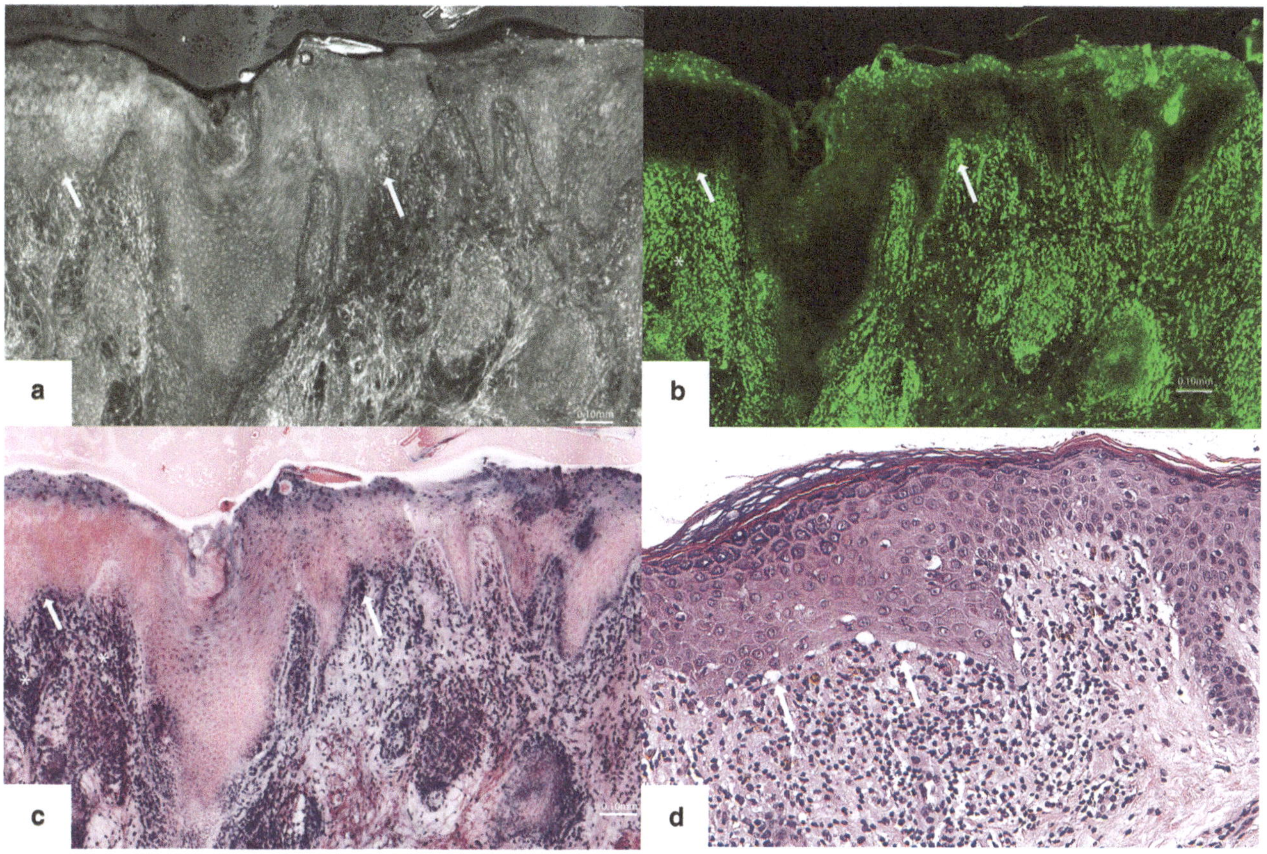

Fig. 15.7 Another example of lichen planus examined in the ex vivo confocal laser scanning microscopy (**a** reflectance, **b** fluorescence, **c** DHE) and corresponding histopathology findings (**d**, 20X), showing strong band-like infiltration of inflammatory cells (asterisk) in the dermis and interface dermatitis (arrow) in the dermo-epidermal junction, (DHE = digital hematoxylin–eosin staining)

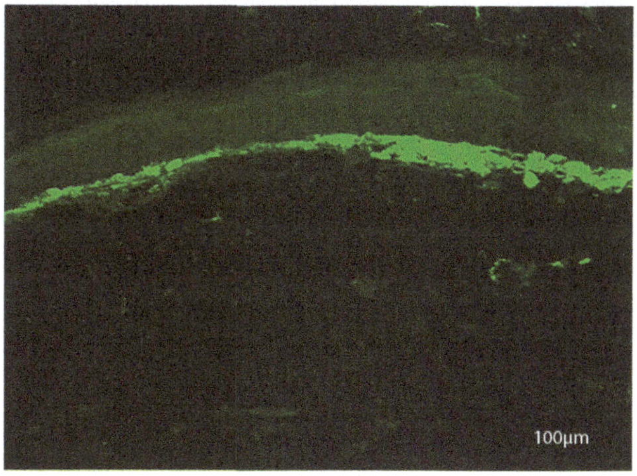

Fig. 15.8 Fluorescence mode of the ex vivo CLSM examination of lichen planus presenting homogeneous deposition of fibrinogen along the BM (arrow), (CLSM = confocal laser scanning microscopy, BM = basement membrane).

15.6 Cutneous Lupus Erythematosus

Table 15.9 Basics of cutaneous lupus erythematosus (CLE = cutaneous lupus erythematosus)

Basics of Cutaneous Lupus Erythematosus (CLE)
• **Epidemiology**: CLE is a rare autoimmune skin disease with an incidence of 4.3 cases per 100,000 persons [20]. CLE can manifest in acute, subacute, and chronic subforms • **Patient demographics**: CLE affects mostly females • **Pathomechanism**: Genetic, hormonal, and environmental (UV radiation) factors cause inflammation and autoantibody production [21] • **Histology**: Interface dermatitis, vacuolar degeneration, superficial and deep perivascular and periadnexal lymphocytic infiltrate [22] • **Direct immunofluorescence**: Homogeneous, granular or linear band of IgG, IgM, and IgA along the basement membrane [23]

Table 15.10 Ex vivo CLSM features of cutaneous lupus erythematosus together with histological and DIF correlations (CLSM = confocal laser scanning microscopy, FCM = fluorescence confocal microscopy mode, RCM = reflectance confocal microscopy mode, DHE = digital hematoxylin–eosin-like staining, H&E = hematoxylin–eosin staining, FITC = fluorescein isothiocyanate, DIF = direct immunofluorescence)

CLE features on FCM, RCM, and corresponding conventional H&E and DIF	
Ex vivo confocal microscopy reflectance (RCM), fluorescence (FCM), and digital H&E [24]	Conventional H&E and DIF
Epidermal changes: Round/polygonal-shaped and larger appearing keratinocytes showing an increased intercellular distance in the lower epidermis (in DHE, RCM) (*see* Fig. 15.9a and d)	These larger appearing cells in the lower epidermis correspond to the spongiosis. The same appearance limited to the basal layer keratinocytes corresponds to the vacuolar degeneration in conventional H&E
Dermal changes: Accumulation of hyperreflective (in RCM) / purple-colored (in DHE), round cells around the superficial and deep blood vessels and hair follicles (in RCM and DHE) (*see* Fig. 15.10a and d)	The accumulation of the cells around the blood vessels and hair follicles corresponds to the perivascular and periadnexal infiltration of inflammatory cells Destruction of the papillary structures due to diffuse inflammatory cell infiltration corresponds to interface dermatitis (in H&E)
Fluorescence: Homogeneous, granular, and linear fluorescence at the dermo-epidermal junction as well as in roundish structures in the papillary dermis in sections stained with FITC-labeled IgG, IgM, and IgA (in FCM) (Fig. 15.11)	This fluorescence at the dermo-epidermal junction corresponds to the lupus band which is defined as the deposition of IgG, IgM, and IgA along the basement membrane in conventional DIF. Roundish structures correspond to the cytoid bodies in the papillary dermis

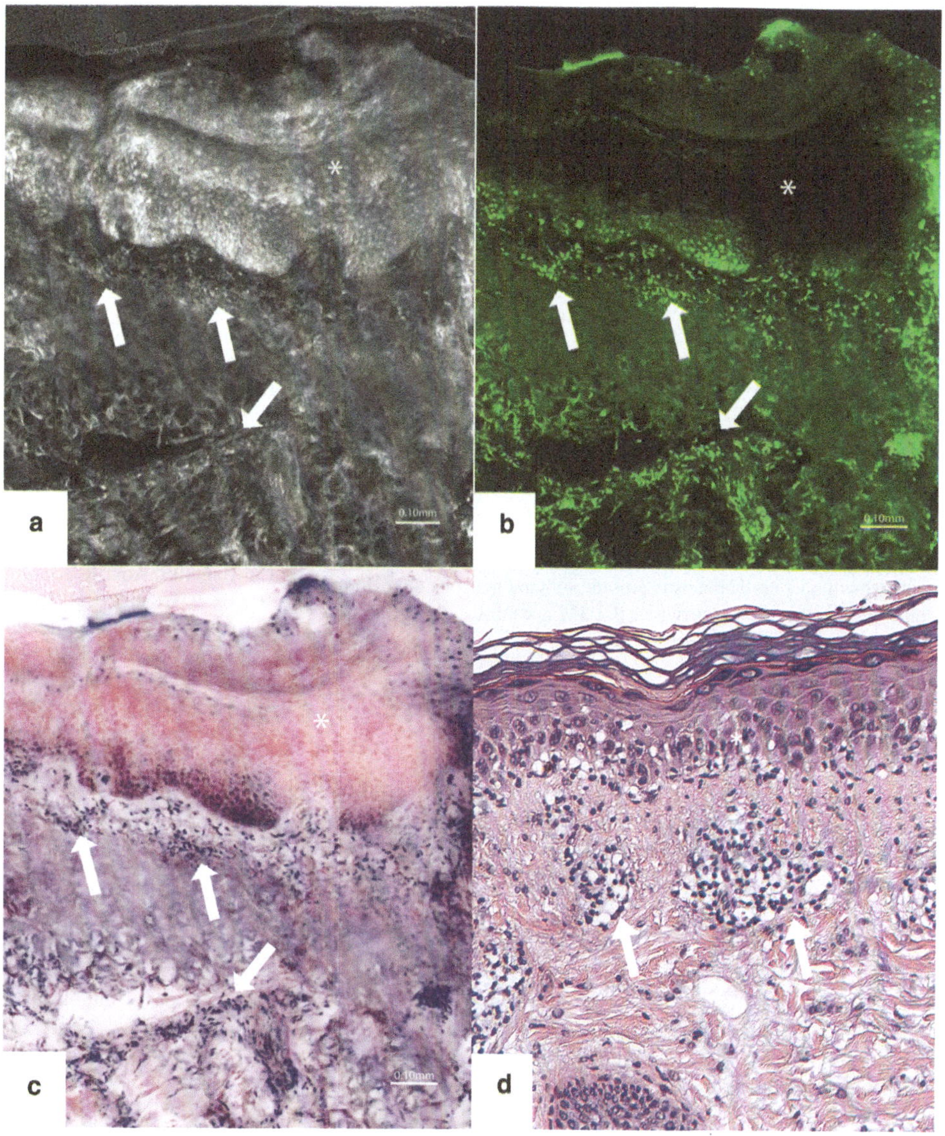

Fig. 15.9 Cutaneous lupus erythematosus examined in the three modes of the ex vivo confocal laser scanning microscopy (**a** reflectance, **b** fluorescence, **c** DHE) and corresponding histopathology findings (**d**, 20X), presenting perivascular infiltration of inflammatory cells in the papillary and reticular dermis (arrow), as well as acanthosis (asterisk), (DHE = digital hematoxylin–eosin staining).

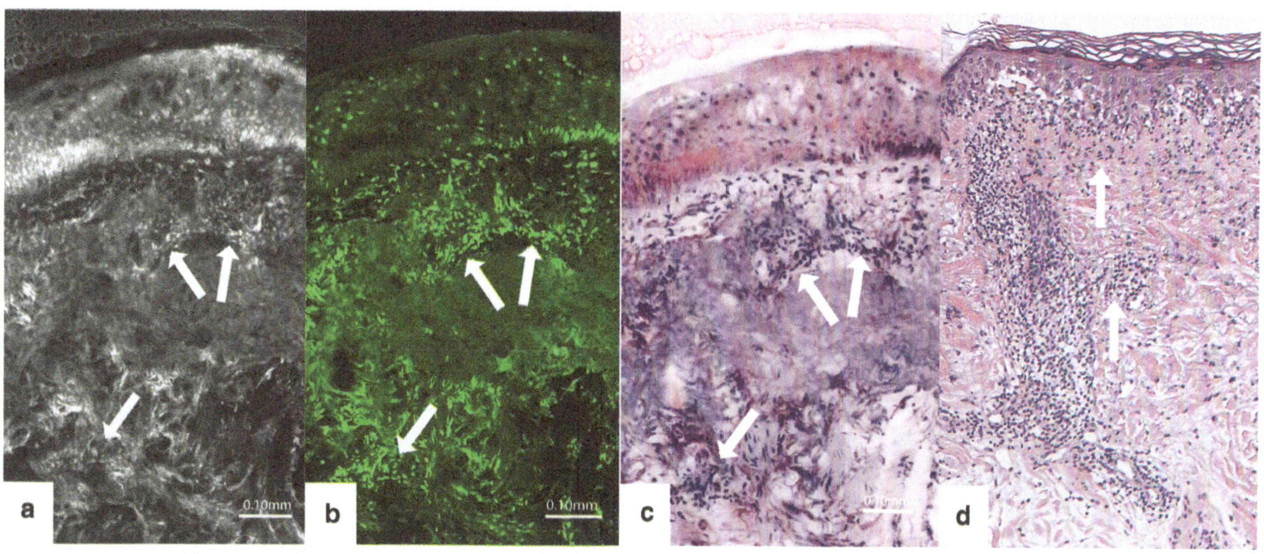

Fig. 15.10 Another example of cutaneous lupus erythematosus examined in the three modes of the ex vivo confocal laser scanning microscopy (**a** reflectance, **b** fluorescence, **c** DHE) and corresponding histopathology findings (**d**, 10X), presenting typical superficial and deep perivascular infiltration of inflammatory cells (arrow), (DHE = digital hematoxylin–eosin staining).

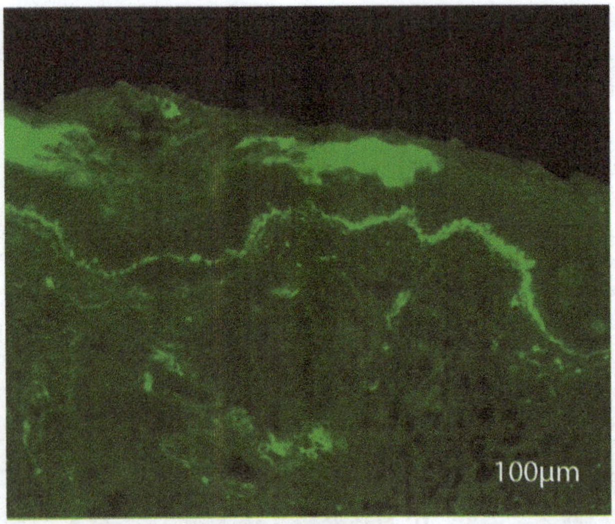

Fig. 15.11 Fluorescence mode of the ex vivo CLSM examination of cutaneous lupus erythematosus presenting homogeneous deposition of IgG along the BM (arrows), (CLSM = confocal laser scanning microscopy, BM = basement membrane)

References

1. Alpsoy E, Akman-Karakas A, Uzun S. Geographic variations in epidemiology of two autoimmune bullous diseases: pemphigus and bullous pemphigoid. Arch Dermatol Res. 2015;307:291–8.
2. Bağcı IS, Horváth ON, Ruzicka T, Sárdy M. Bullous pemphigoid. Autoimmun Rev. 2017;16:445–55.
3. Machado-Pinto J, McCalmont TH, Golitz LE. Eosinophilic and neutrophilic spongiosis: clues to the diagnosis of immunobullous diseases and other inflammatory disorders. Semin Cutan Med Surg. 1996;15:308–16.
4. Sadik C, Zillikens D, Hertl M. JA. Bullöse Autoimmunder-matosen. In Braun-Falco's Dermatologie, Venerologie und Aller-gologie (Ed. Plewig G, Ruzicka T, Kaufmann R, Hertl M). Springer, Berlin, 2018, p. 856.
5. Bağcı IS, Aoki R, Krammer S, Ruzicka T, Sárdy M, French LE, Hartmann D. Ex vivo confocal laser scanning microscopy for bullous pemphigoid diagnostics: new era in direct immunofluores-cence? J Eur Acad Dermatol Venereol. 2019;33:2123–30.
6. Kasperkiewicz M, Ellebrecht CT, Takahashi H, Yamagami J, Zillikens D, Payne AS, Amagai M. Pemphigus. Nat Rev Dis Primers. 2017;3:17026.
7. Amagai M, Klaus-Kovtun V, Stanley JR. Autoantibodies against a novel epithelial cadherin in pemphigus vulgaris, a disease of cell adhesion. Cell. 1991;67:869–77.
8. Saschenbrecker S, Karl I, Komorowski L, Probst C, Dähnrich C, Fechner K, Stöcker W, Schlumberger W. Serological Diagnosis of Autoimmune Bullous Skin Diseases. Front Immunol. 2019;20 (10):1974.
9. Bağcı IS, Aoki R, Vladimirova G, Sárdy M, Ruzicka T, French LE, Hartmann D. Simultaneous immunofluorescence and histology in pemphigus vulgaris using ex vivo confocal laser scanning microscopy. J Biophotonics. 2021 (accepted article).
10. Krammer S, Krammer C, Salzer S, Bağci IS, French LE, Hartmann D. Recurrence of Pemphigus Vulgaris Under Nivolu-mab Therapy. Front Med (Lausanne). 2019;12(6):262.
11. Carlson JA, Cavaliere LF, Grant-Kels JM. Cutaneous vasculitis: diagnosis and management. Clin Dermatol. 2006;24:414–29.
12. Lotti T, Ghersetich I, Comacchi C, Jorizzo JL. Cutaneous small-vessel vasculitis. J Am Acad Dermatol. 1998 ;39:667–87; quiz 688–90.
13. McLaren JS, McRorie ER, Luqmani RA. Diagnosis and assess-ment of systemic vasculitis. Clin Exp Rheumatol. 2002;20:854–62.
14. Barnadas MA, Pérez E, Gich I, Llobet JM, Ballarín J, Calero F, Facundo C, Alomar A. Diagnostic, prognostic and pathogenic value of the direct immunofluorescence test in cutaneous leuko-cytoclastic vasculitis. Int J Dermatol. 2004;43:19–26.
15. Bağcı IS, Aoki R, Krammer S, Ruzicka T, Sárdy M, Hartmann D. Ex vivo confocal laser scanning microscopy: An innovative method for direct immunofluorescence of cutaneous vasculitis. J Biophotonics. 2019;12:e201800425.
16. Halonen P, Jakobsson M, Heikinheimo O, Gissler M, Pukkala E. Incidence of Lichen Planus and Subsequent Mortality in Finnish Women. Acta Derm Venereol. 2020;100:adv00303.
17. Carbone M, Arduino PG, Carrozzo M, Gandolfo S, Argiolas MR, Bertolusso G, et al. Course of oral lichen planus: a retrospective study of 808 northern Italian patients. Oral Dis. 2009;15:235–43.
18. Buajeeb W, Okuma N, Thanakun S, Laothumthut T. Direct immunofluorescence in oral lichen planus. J Clin Diagn Res. 2015;9:ZC34–ZC37.
19. Bağcı IS, Aoki R, Krammer S, Vladimirova G, Ruzicka T, Sárdy M, French LE, Hartmann D. Immunofluorescence and histopatho-logical assessment using ex vivo confocal laser scanning microscopy in lichen planus. J Biophotonics. 2020;13: e202000328.
20. Durosaro O, Davis MD, Reed KB, Rohlinger AL. Incidence of cutaneous lupus erythematosus, 1965–2005: a population-based study. Arch Dermatol. 2009;145:249–53.
21. Achtman JC, Werth VP. Pathophysiology of cutaneous lupus erythematosus. Arthritis Res Ther. 2015;17:182.
22. Joshi R. Interface dermatitis. Indian J Dermatol Venereol Leprol. 2013;79:349–59.
23. Crowson AN, Magro CM. Cutaneous histopathology of lupus erythematosus. Diagn Histopathol. 2009;15:157–85.
24. Bağcı IS, Aoki R, Vladimirova G, Ergün E, Ruzicka T, Sárdy M, French LE, Hartmann D. New-generation diagnostics in inflam-matory skin diseases: Immunofluorescence and histopathological assessment using ex vivo confocal laser scanning microscopy in cutaneous lupus erythematosus. Exp Dermatol. 2020.

Part VII

Future of Ex Vivo Confocal Microscopy

16 Advances in Technology, Staining Protocol, and Flattening Devices in Ex Vivo Confocal Microscopy

Javiera Pérez-Anker, Susana Puig, and Josep Malvehy

16.1 Introduction

Since the development of confocal microscopy technology in 1957, a long journey has brought us to the analysis of digital images of hematoxylin and eosin, and artificial intelligence as applied to the field of histology. Progress in ex vivo confocal microscopy technology now allows immediate visualization of freshly excised tissue, in less than 3 min and at the bedside. This instant diagnosis has opened the door to significant progress in the field of dermatology and also in general pathology [1–4].

Not all the journey has been easy and represents more than 20 years of continuous progress. From a slow device with white and gray images divided into squares, to a fast device with real- time digital hematoxylin and eosin (H&E) images obtained by strip mosaicking, this road began with one modest first step (Fig. 16.1). Technology, as applied to ex vivo confocal microscopy, has steadily improved over the years. Furthermore, not only has the technology changed, but staining methods have also changed over time, along with methods of preparing the specimen to be scanned [4] (Table 16.1).

16.2 Reflectance Confocal Microscopy in Ex Vivo Confocal Microscopy

The first published results in ex vivo confocal microscopy (evCM) for clinical applications were in 2001, by Milind Rajadhyaksha and Salvador González [5]. A similar technique was described in 1995, for in vivo confocal microscopy (ivCM) by Rajadhyaksha et al. [6] and was applied for the analysis of evCM images.

These images were obtained by two reflectance confocal microscopes. The authors stained the nuclei with acetic acid at 5% in order to enhance the contrast by whitening the skin component and compacting the chromatin. A laser of 1064 nm wavelength allowed deep visualization of the tissue, in spite of a lower quality resolution of the structures.

In 2003, a new commercialized device only for ex vivo purposes (VivaScope 2000s, Lucid Inc., NY, USA), with an 830 nm diode laser (a reflectance laser) and a 30X objective lens, was utilized to evaluate discarded tissue from the first stage of Mohs surgery operations on basal cell carcinomas and squamous cell carcinomas. Acetic acid at 5% was utilized as a stain contrast with better results than reported in previous studies [7]. A real-time image visualization for Mohs surgery was achieved.

In 2007, the same pioneer group which created the first confocal created a mosaic of 15 × 15 mm (equivalent to a 2X view of macroscopic image) in 9 min [8]. They also realized that the flattening of the tissue was fundamental to visualizing all the structures during the scanning. A flattening device consisting of an Agarose 3% gel disk, 2 mm thick was proposed for the first time. The gel disk was placed on top of the specimen inside a piston housing of polycarbonate. A piston compressed the gel and the specimen on the slide, achieving a uniformly flattened specimen. Visualization of the tissue was greatly improved.

Staining methods also changed in 2009 when a German group proposed the use of Citric acid 10% for 2 min, scanning with the VS2500, observing the images in reflectance mode [9].

16.3 Fluorescence Confocal Microscopy for Ex Vivo Confocal Microscopy

Another radical change occurred in 2009 when the use of a fluorescent laser (445 nm) was proposed for Mohs surgery [10]. The staining method also varied, and the use of

J. Pérez-Anker (✉) · S. Puig · J. Malvehy
Dermatology Department, Hospital Clinic de Barcelona, Carrer de Villarroel, 170, Barcelona, 08036, España
e-mail: perez12@clinic.cat

M. Jain et al. (eds.), *Cutaneous Atlas of Ex Vivo Confocal Microscopy*,
https://doi.org/10.1007/978-3-030-89316-3_16

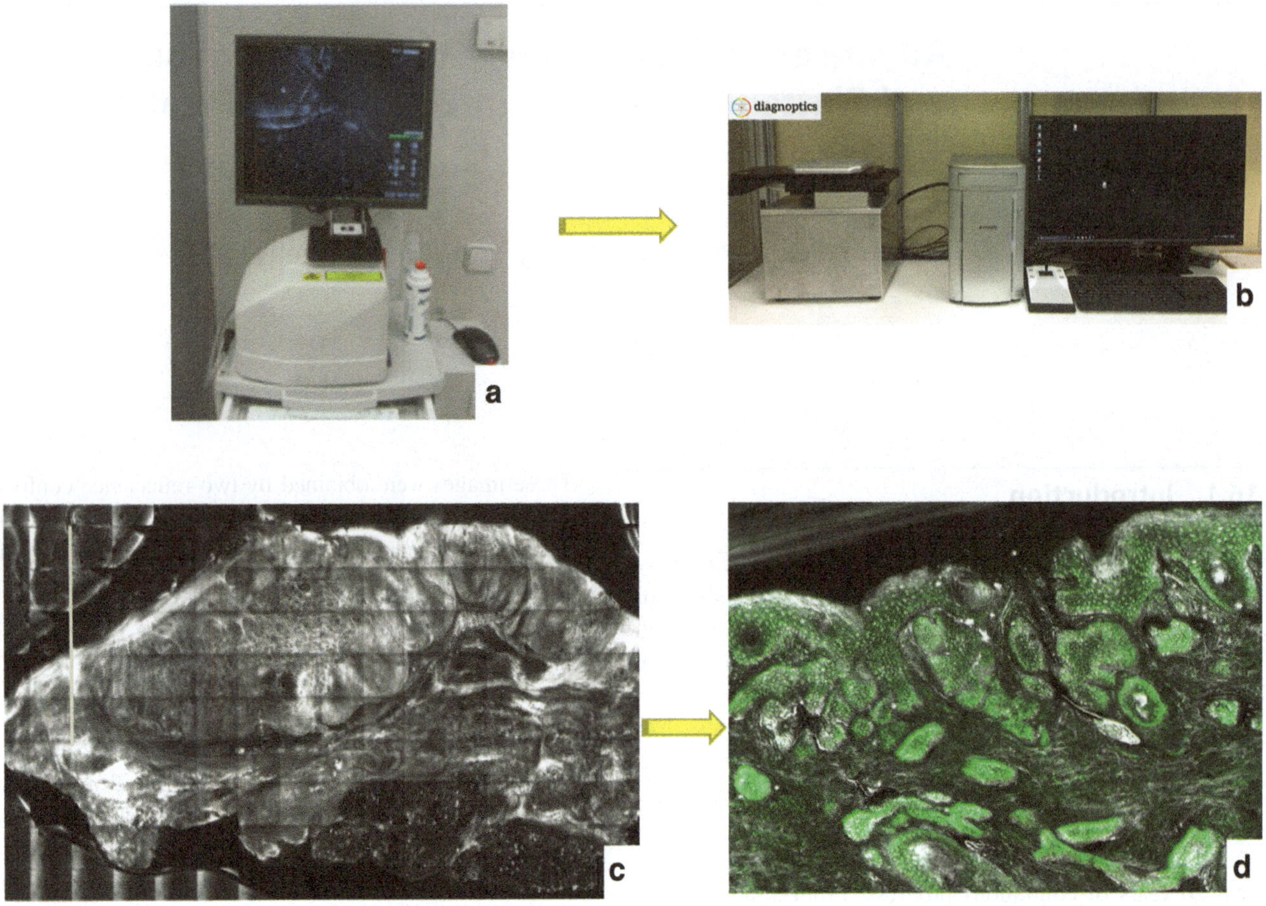

Fig. 16.1 Progression of evCM microscopes. **a** VS2500 Gen3. **b** Diagnoptics prototype. **c** Nodular basal cell carcinoma image obtained by VS2500 Gen3 with reflectance laser. **d** Infiltrating basal cell carcinoma image obtained by VS2500 Diagnoptics prototype with fusion mode and strip mosaicking.

Table 16.1 Evolution of Vivascope® devices

Device	Year	Laser	Mode	Scan area	Acquisition	Objective	FOV	Time
VS2000	2001	1064 nm	Reflectance		Mosaic	10, 20, 30, 60, 100X	150–2000 μm	5–6.5 min
VS2000	2003	830 nm	Reflectance	10 × 10 mm	Mosaic	30X	750 μm	
VS2500 Gen2	2009	830 nm 658 nm 445 nm	Reflectance Fluorescence Fluorescence	16 × 16 mm	Mosaic	25X	750 μm	9 min
VS2500 Gen2	2010	830 nm 658 nm 488 nm	Reflectance Fluorescence Fluorescence	16 × 16 mm	Mosaic	25X	750 μm	
VS2500 Gen3	2014	830 nm 658 nm 488 nm	Reflectance Fluorescence Fluorescence	20 × 20 mm	Mosaic	25X	635 μm	
VS2500 Gen4	2018	785 nm 488 nm	Reflectance Fluorescence	25 × 25 mm	Strip	38X	500 μm	
VS2500 Gen4	2020	638 nm 488 nm	Reflectance Fluorescence	25 × 25 mm	Strip	38X	500 μm	2–3 min

acridine orange was introduced in the preparation of the sample. Only a fluorescent laser was utilized.

Since 2010, a fluorescence 488 nm laser has been used in most published studies, and the technology has not changed significantly. Although the device could employ a fluorescent 658 nm laser or a reflectance 830 nm laser, investigators have, in general, preferred using acridine orange and a continuous wave 488 nm laser. A great deal of evidence has been accumulated in favor of safety in the use of this device and technique [3, 10–19].

Simultaneously, another device called HistoLog scanR FCM® was developed which allowed the visualization of larger images than the other devices 8 × 8 cm. The final image of the scanned tissue can be seen in digital purple, using only a 488 nm laser [4] (Table 16.2).

In 2007, efficacy in the visualization of tumoral islands of basal cell carcinoma was compared between the three different lasers available at that time: two fluorescent lasers (633 and 656 nm) and a reflectance laser (830 nm). In addition, new contrast agents were proposed: methylene blue 1% and toluidine blue 1%.

16.4 Fusion Mode: Simultaneous Fluorescence + Reflectance

In 2016, the authors of this chapter analyzed different benign and malignant lesions using fluorescence and reflectance lasers, with a new staining protocol of acetic acid and acridine orange, using the VS2500 Gen2 (Figs. 16.2 and 16.3) [20].

Since then, other European authors have also compared different visualization features in both lasers as applied to diverse skin lesions [21–23].

Regarding the flattening of the tissue, in 2017, another option appeared in an attempt to improve the representativeness of the sample. This system consisted of a plastic slide with grub screws that covered the specimen. Unfortunately, this method was time-consuming and rapid tissue visualization is essential [24].

The basis of the fusion mode 4th generation VivaScope device (the latest generation of confocal microscopy) consists in scanning with two different lasers (fluorescence and reflectance) at the same time, strip mosaicking and subsequently, the analysis of digital images of hematoxylin and eosin.

The first time that two different lasers, fluorescence (488 nm) and reflectance (830 nm), scanned at the same time was with a prototype developed in 2005 [25]. In this study, in vivo confocal images were correlated with ex vivo conventional microscopy images.

Another substantial change occurred in 2011 when the strip mosaicking in ratios of 10:1 instead of ~1:1 was obtained and stitched together using novel software [26]. With this technology, an area of 10 × 10 mm was scanned in about 3 min—3 times as fast asthe previous method.

16.5 Fusion Mode with Strip Mosaicking and Digital H&E

In 2009, the third revolutionary technological advance was initiated: for the first time, confocal ex vivo digital images of H&E could be observed, simulating real histopathology slides. The visualization was rendered by two lasers (according to the described prototype where both fluorescent and reflectance lasers were used simultaneously) and using software to change the initial colors of the scanned images: the reflectance mosaic was digitally changed to pink and the fluorescence mosaic to purple. This was the last significant step to the technology that is available today together with the improvement of the software compared to the first method of digital H&E staining in evCM [27, 28].

A further step was taken in 2016 when the authors of this chapter started to work with a prototype in the context of a European consortium called Diagnoptics (Fig. 16.4). The device was a prototype that scanned simultaneously in

Table 16.2 Comparison of evCM devices. Adapted from Malvehy et al. [4].

System	HistoLog scanR FCM®	Vivascope2500M-G4R FCM®
Modality of CM Image area (max)	FCM	FCM + +RCM
Image area (max)	8 × 8 cm	2.5 × 2.5 cm
Laser	488 nm	785 nm + 488 nm
Scanning characteristics	1600 × 1600 to 8000 × 8000 pixels	1024 × 1024 pixels Optical section thickness = 4 µm Depth of scanning <200 µm Magnification up to 540x
Display of images	Purple digital color	Green (FCM)/ Gray scale (RCM); Fusion green-gray; H&E digital staining Table

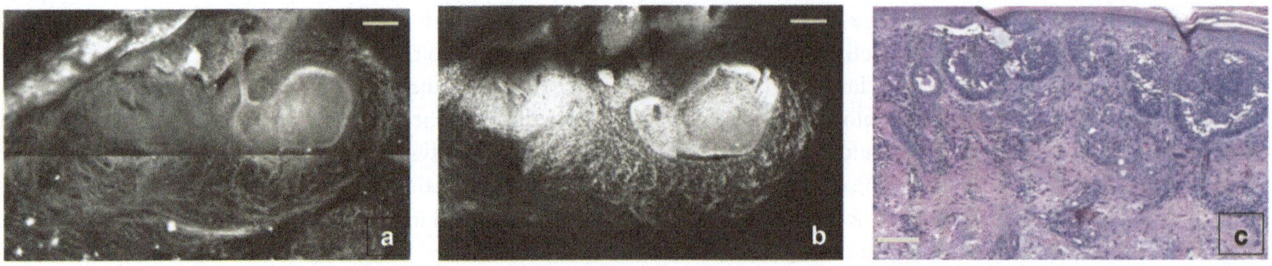

Fig. 16.2. VS2500 Gen2 scanned image of a superficial BCC stained with Acridine orange 0.1 mmol L^1 + acetic acid 50%. Different stromal reactions were observed in each laser. **a** Reflectance laser. **b** Fluorescence laser. **c** H&E [20]

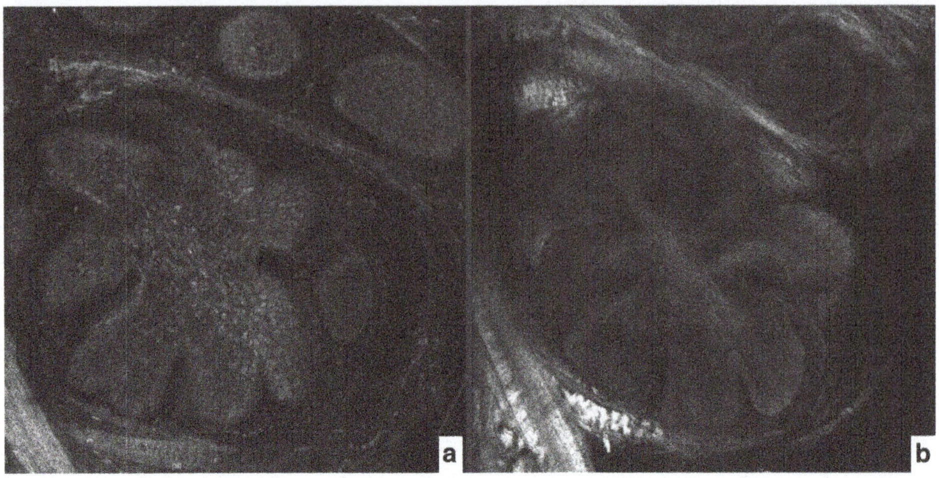

Fig. 16.3. Nuclear visualization with different stain protocols [30]. Nodular basal cell carcinoma using reflectance laser. Image obtained by VS2500 Diagnoptics prototype with fusion mode and strip mosaicking. **a** Acridine orange 0.1 mmol L^1 + acetic acid 50%. **b** Acridine orange 0.1 mmol L^1 [20]

reflectance and fluorescence modes and used the strip mosaicking software to integrate the captured images (Fig. 16.4) [29]. Differences between fluorescence, reflectance, and fusion laser mode and also a comparison between the different staining methods were described for basal cell carcinoma [20, 30].

With a few changes suggested by the study group, the algorithm put forward by Gareau et al. [27] was integrated into the new version of the commercial machine and the VS 4th Gen appeared on the market in 2018 (Figs. 16.4, 16.5, 16.6, 16.7, 16.8, 16.9, 16.10 and 16.11) [29–33].

16.6 New Staining Protocols

A new staining method was also suggested, combining the benefits of acetic acid and acridine orange staining (Table 16.3) (Figs. 16.3, 16.4, 16.5, 16.6, 16.7, 16.8, 16.9, 16.10 and 16.11).

The next significant change was suggested by the authors regarding flattening methods [34] and steps were suggested to obtain a good image with this technique (also described in Chap. 3) [35]. In addition to mechanical flattening methods, recently, Sendin-Martin et al. [36] demonstrated a digital approach to flatten excised Mohs margin tissue for complete visualization of epidermal margin.

The greatest advantage of applying digital H&E is that the diagnostic criteria normally applied in different histological diagnoses can also be visualized and applied in the diagnosis with ex vivo confocal microscopy [4, 37–39]. This has previously been demonstrated by the authors of this chapter, and the feature visualization of the structures, only in fluorescence or reflectance mode, is now utilized to differentiate doubtful areas of the image of the scanned tissue [4, 37–39].

The new generation of ex vivo confocal microscopy has led to a revolution in the field of diagnosis and histology-driven treatment. New applications in other fields of general pathology have also quickly begun to appear [1, 2, 40] and pathologists have been able to increase their confidence in diagnosis using ex vivo confocal microscopy with digital hematoxylin and eosin.

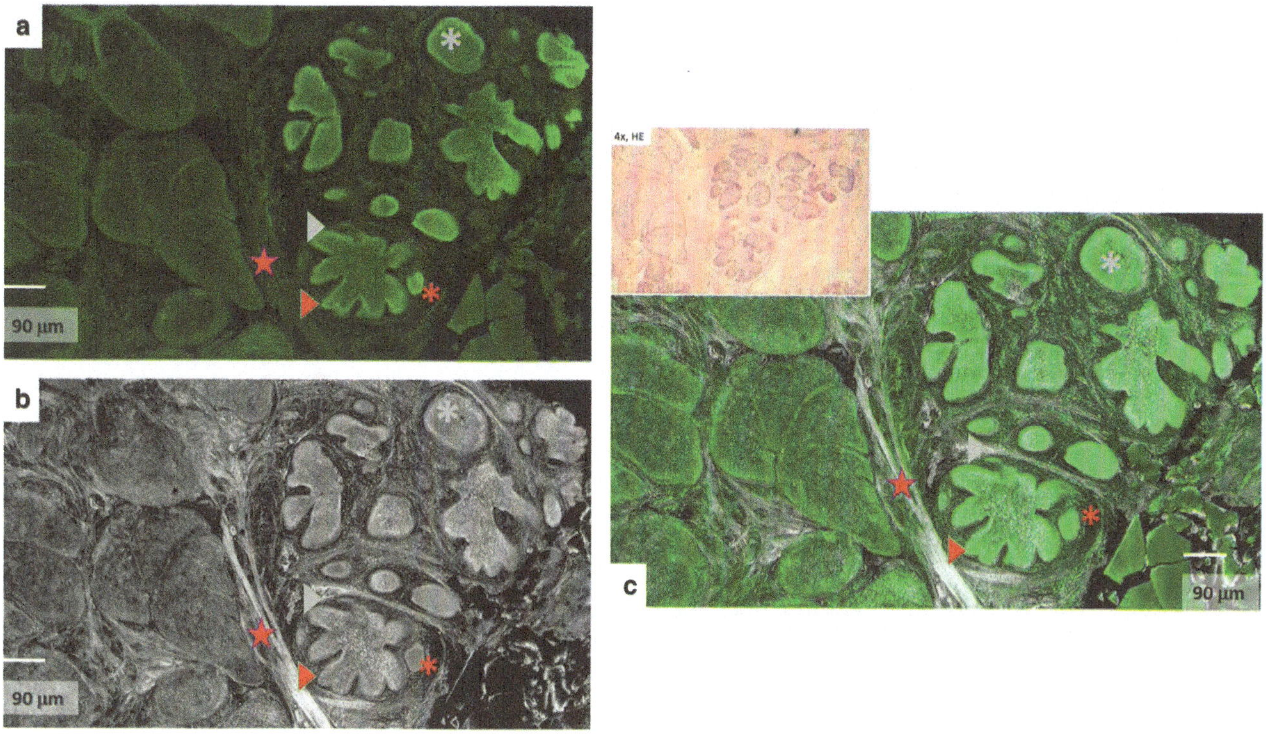

Fig. 16.4. Diagnoptics prototype scanned image of a nodular BCC stained with Acridine orange 0.1 mmol L[1] + acetic acid 50%. **a** Fluorescence laser. **b** Reflectance laser. **c** Fusion image (simultaneous fluorescence and reflectance lasers) and H&E. Histological features can be recognized: clefting (red asterisk), palisading (red triangle), nuclear pleomorphism and increased nuclear/cytoplasm ratio (gray asterisk), peri-tumoral stromal reaction (gray triangle), thickened collagen bundles (red star) [30]

Fig. 16.5. VivaScope® 2500 MM-Gen4 digital H&E scanned image of a nodular BCC, previously prepared with Acridine orange 0.1 mmol L[1] + acetic acid 50%. Histological features can be recognized: clefting (red asterisk), palisading (red triangle), thickened collagen bundles (red star) [4, 30, 36–38]

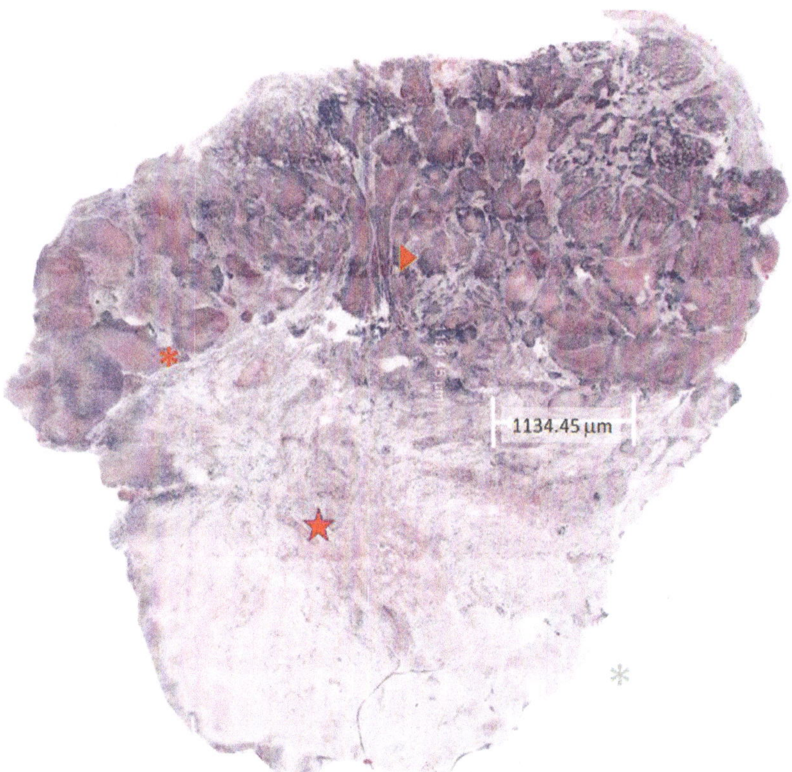

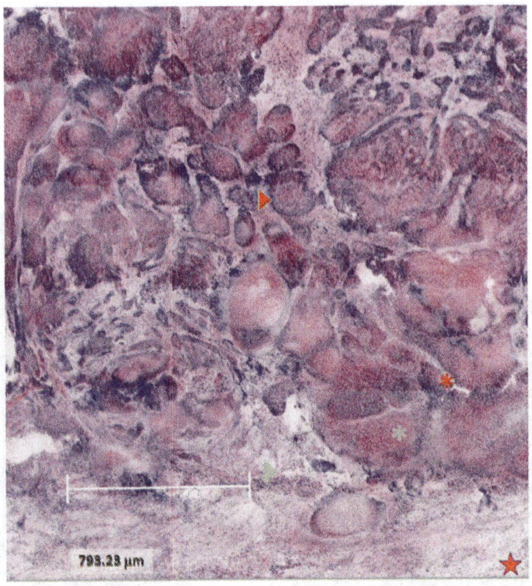

Fig. 16.6. VivaScope® 2500 MM-Gen4 digital H&E scanned detail image of a nodular BCC previously prepared with Acridine orange 0.1 mmol L1 + acetic acid 50%. Histological features can be recognized: clefting (red asterisk), palisading (red triangle), nuclear pleomorphism and increased nuclear/cytoplasm ratio (gray asterisk), peri-tumoral stromal reaction (gray triangle), thickened collagen bundles (red star) [4, 30, 36–38]

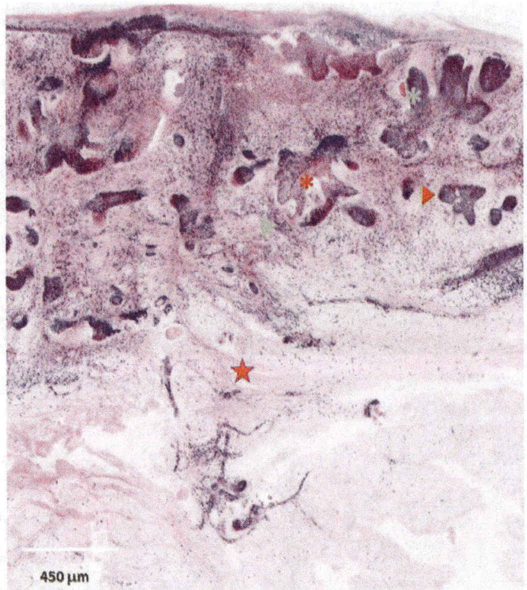

Fig. 16.7. VivaScope® 2500 MM-Gen4 digital H&E scanned detail image of an infiltrating BCC previously prepared with Acridine orange 0.1 mmol L1 + acetic acid 50%. Histological features can be recognized: clefting (red asterisk), palisading (red triangle), nuclear pleomorphism and increased nuclear/cytoplasm ratio (gray asterisk), peri-tumoral stromal reaction (gray triangle), thickened collagen bundles (red star) [4, 30, 36–38]

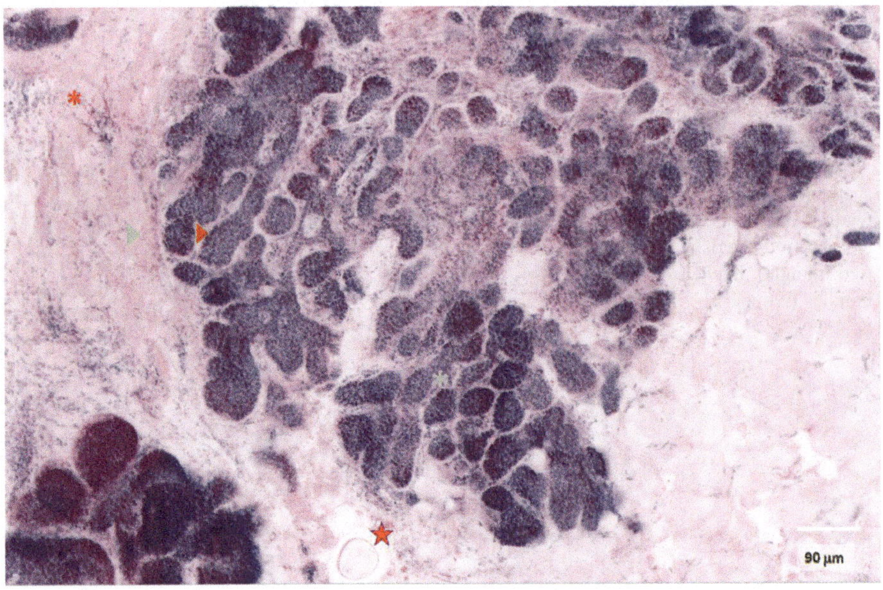

Fig. 16.8. VivaScope® 2500 MM-Gen4 digital H&E scanned detail image of a micronodular BCC previously prepared with Acridine orange 0.1 mmol L1 + acetic acid 50%. Histological features can be recognized: clefting (red asterisk), palisading (red triangle), nuclear pleomorphism and increased nuclear/cytoplasm ratio (gray asterisk), peri-tumoral stromal reaction (gray triangle), thickened collagen bundles (red star) [4, 30, 36–38]

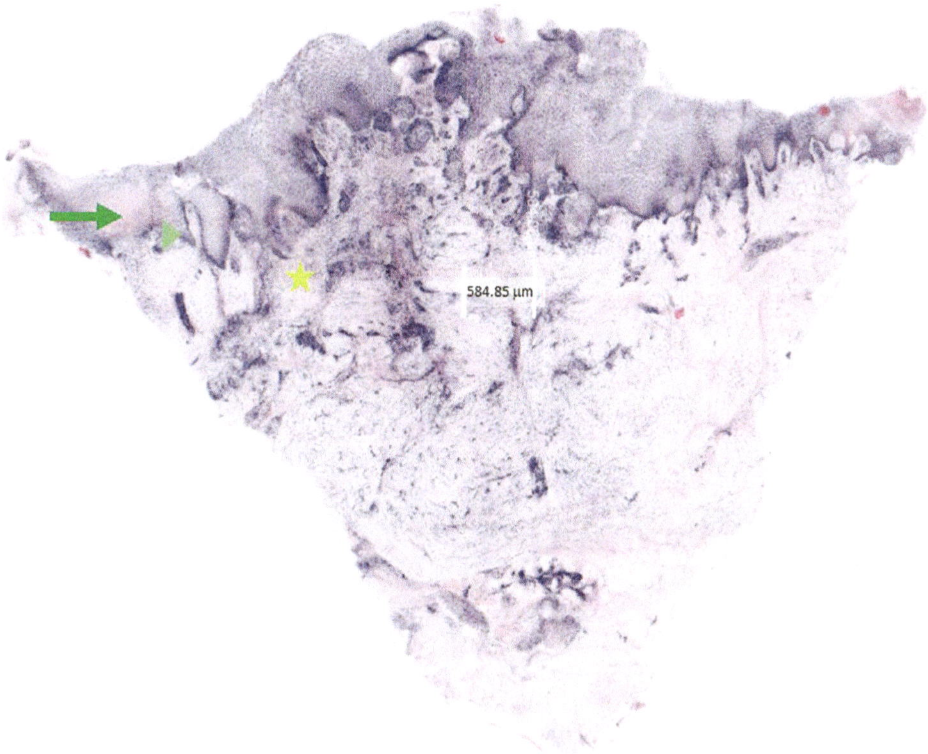

Fig. 16.9. VivaScope® 2500 MM-Gen4 digital H&E scanned image of an infiltrating SCC previously prepared with Acridine orange 0.1 mmol L^{1} + acetic acid 50%. Histological features can be recognized. Epidermis (dark green arrow), DEJ (light green triangle), dermis (yellow asterisk) [4, 30, 36–38]

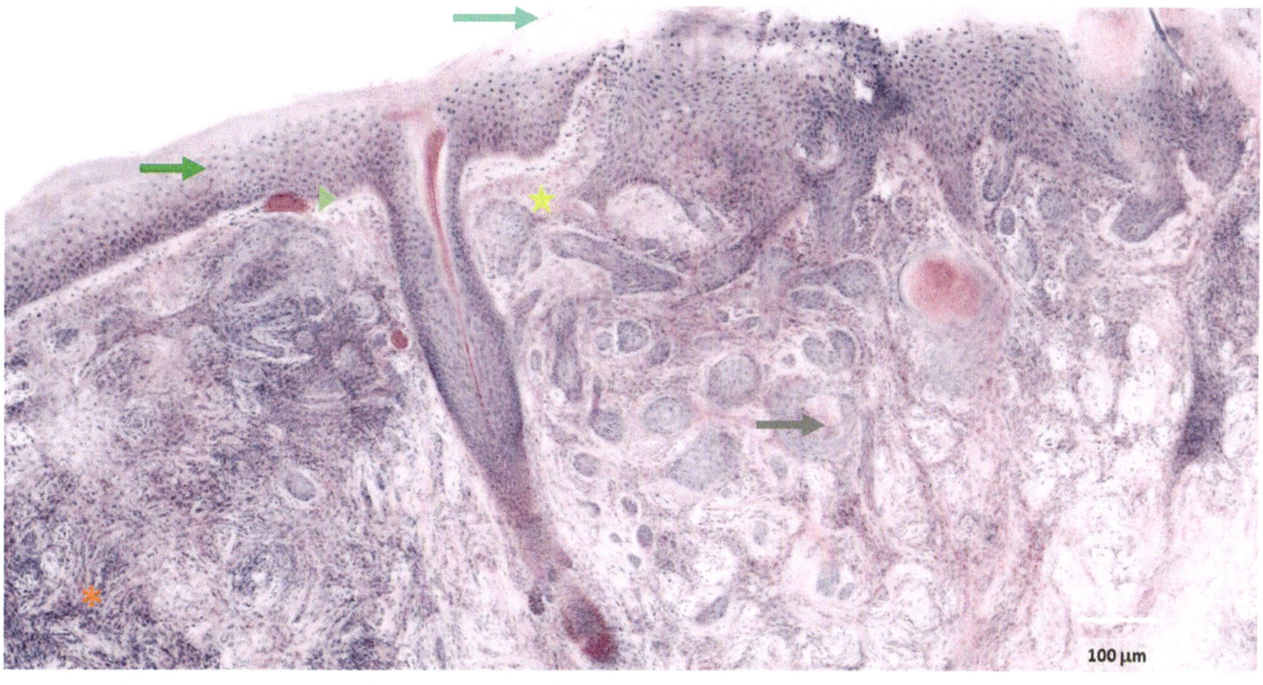

Fig. 16.10. VivaScope® 2500 MM-Gen4 digital H&E scanned detail image of an infiltrating SCC previously prepared with Acridine orange 0.1 mmol L^{1} + acetic acid 50%. Full macroscopic histological features can be seen. Epidermis (dark green arrow), DEJ (light green triangle), dermis (yellow asterisk), keratin pearls (light gray arrow), parakeratosis (turquoise arrow), inflammatory cells (dark orange asterisk) [4, 30, 36–38]

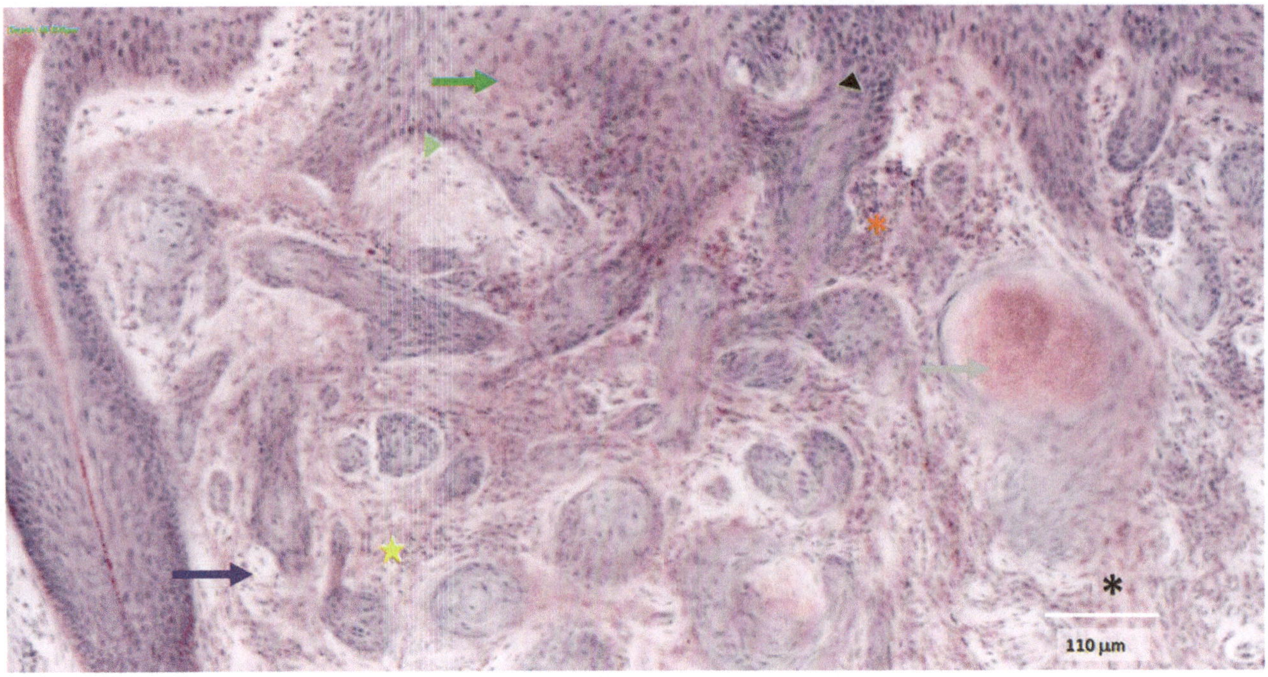

Fig. 16.11. Detail image of infiltrating SCC. Full microscopic histological features can be seen. Epidermis (dark green arrow), DEJ (light green triangle), dermis (yellow star), vessels (purple arrow), keratin pearls (light gray arrow), cellular atypia (black triangle), dyskeratotic cells (black asterisk), inflammatory cells (dark orange asterisk) [4, 36–38]

Table 16.3 Staining methods for ex vivo CM and application in dermatology. RCM: reflectance confocal microscopy; FCM: fluorescence confocal microscopy; M-CM: multimodal confocal microscopy. Adapted from Malvehy et al. [4].

Staining agent	Laser	Sample
Aluminum chloride, acetic acid, citric acid	RCM	Skin tumors
Acridine orange, methylene blue, and toluidine blue, fluorescein, Nile blue, or Patent Blue V	FCM	Skin tumors
Combinations	**Laser**	**Sample**
Methylene blue + toluidine blue	M-CM	Skin cancer
Fluorescence proflavine + acetic acid + oluidine blue	FCM	BCC
Acetic acid + Acridine orange	M-CM	BCC
Acridine orange + ethidium bromide	3-color FCM	BCC
Immunostaining	**Laser**	**Sample**
FITC-labeled S-100A10, Melan-A, and anti-Ber-EP4 antibodies, NPs10@D1_ICF_Alexa647_ DOTAGA Fe3 +	FCM	No-melanocytic and melanocytic tumors
Fluorescent-labeled IgG and C3 antibodies	M-CM	Bullous pemphigoid
IgG, IgM, IgA, C3, and Fibrinogen	M-CM	Cutaneous vasculitis

16.7 Artificial Intelligence Color and Image Interpretation

Progress toward this method started through experimenting with artificial intelligence. Marc Combalia et al. proposed digital staining through deep learning [41], with digital colors almost equal to regular H&E that may have clinical applicability in the future (Fig. 16.12). As with digital pathology, AI will have a role in diagnosis in evCM, assessment of tumor margins, and in other applications. However, validation of machine learning needs to be done in this field to prove its clinical usability.

To summarize, the technology for evCM has evolved steadily over recent years. The core technology has been miniaturized in different devices and new lasers have been developed, bringing fluorescence capability instead of only reflectance mode (830 nm). Fusion images of evRCM and

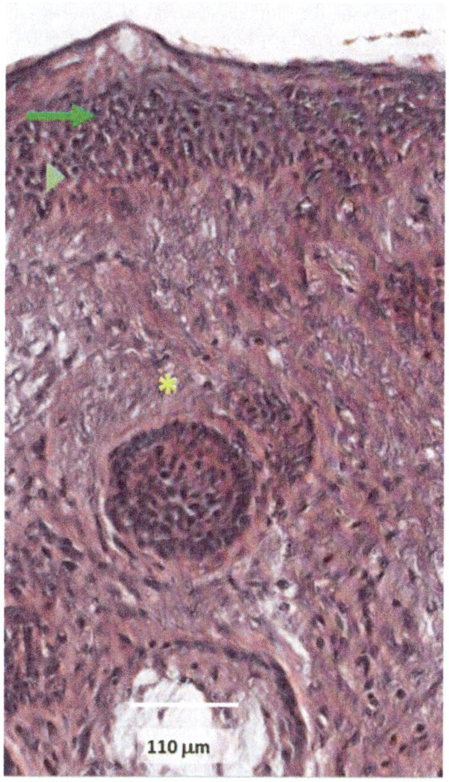

Fig. 16.12. VivaScope® 2500 M-Gen4 AI stained scanned detail image of normal skin previously prepared with Acridine orange 0.1 mmol L^1 + acetic acid 50%. Microscopic histological features can be recognized. Epidermis (dark green arrow), DEJ (light green triangle), dermis (yellow asterisk) [40]

evFCM and digital H&E staining render high-quality pathological images similar to standard digital pathology. Moreover, the most recent changes have been the incorporation of a macroscopic camera and also the implementation of strip mosaicking (V2500 Gen4). Furthermore, the first device (VS2000), built in 2003 from the prototype developed by Rajadhyaksha in 2001, had a 25X objective, and the latest generation uses 38X, increasing the resolution from previous models. Additionally, the field of view of the VS2000 was 750 µm and the maximum scan area was 10 × 10 mm compared to 30 × 3 0 mm.

The very first efforts made by Rajadhyaksha and González and their team in 2001 have, twenty years later, resulted in fast, efficient, revolutionary technology that is radically changing the traditional method of processing histological samples.

References

1. Panarello D, Compérat E, Seyde O, Colau A, Terrone C, Guillonneau B. Atlas of ex vivo prostate tissue and cancer images using confocal laser endomicroscopy: a project for intraoperative positive surgical margin detection during radical prostatectomy. Eur Urol Focus. 2019;1–18.
2. Villarreal JZ, Pérez-Anker J, Puig S, Pellacani G, Solé M, Malvehy J, et al. Ex vivo confocal microscopy performs real-time assessment of renal biopsy in non-neoplastic diseases. J Nephrol [Internet]. 2020;(0123456789). Available from: https://doi.org/10.1007/s40620-020-00844-8
3. Ragazzi M, Longo C, Piana S. Ex Vivo (fluorescence) confocal microscopy in surgical pathology: state of the art, vol. 23, Advances in Anatomic Pathology. 2016; pp. 159–69.
4. Malvehy J, Pérez-Anker J, Toll A, Pigem R, García A, Alós L, et al. Ex vivo Confocal Microscopy: revolution in fast pathology in dermatology. Br J Dermatol. 2020;183(6):1011–25.
5. Rajadhyaksha M, Menaker G, Flotte T, Dwyer PJ, González S. Confocal examination of nonmelanoma cancers in thick skin excisions to potentially guide mohs micrographic surgery without frozen histopathology. J Investig Dermatol. 2001;117(5):1137–43.
6. Rajadhyaksha M, Grossman M, Esterowitz D, Webb RH, Anderson RR. In vivo confocal scanning laser microscopy of human skin: Melanin provides strong contrast. J Invest Dermatol. 1995; 104:946–52.
7. Chung VQ, Dwyer PJ, Nehal KS, Rajadhyaksha M, Menaker GM, Charles C, et al. Use of ex vivo confocal scanning laser microscopy during Mohs Surgery for Nonmelanoma Skin Cancers. Dermatol Surg. 2004;30(12p1):1470–8.
8. Patel YG, Nehal KS, Aranda I, Li Y, Halpern AC, Rajadhyaksha M. Confocal reflectance mosaicing of basal cell carcinomas in Mohs surgical skin excisions. J Biomed Optics. 2007;12 (3):034027.
9. Schüle D, Breuninger H, Schippert W, Dietz K, Moehrle M. Confocal laser scanning microscopy in micrographic surgery (three-dimensional histology) of basal cell carcinomas. Br J Dermatol. 2009;161(3):698–700.
10. Karen JK, Gareau DS, Dusza SW, Tudisco M, Rajadhyaksha M, Nehal KS. Detection of basal cell carcinomas in Mohs excisions with fluorescence confocal mosaicing microscopy. Br J Dermatol. 2009;160(6):1242–50.
11. Longo C, Ragazzi M, Castagnetti F, Gardini S, Palmieri T, Lallas A, et al. Inserting ex vivo fluorescence confocal microscopy perioperatively in mohs micrographic surgery expedites bedside assessment of excision margins in recurrent basal cell carcinoma. Dermatology. 2013;227(1):89–92.
12. Dobbs JL, Ding H, Benveniste AP, Kuerer HM, Krishnamurthy S, Yang W, et al. Feasibility of confocal fluorescence microscopy for real-time evaluation of neoplasia in fresh human breast tissue. J Biomed Optics. 2013;18(10):106016.
13. Bennàssar A, Carrera C, Puig S, Vilalta A, Malvehy J. Fast evaluation of 69 basal cell carcinomas with ex vivo fluorescence confocal microscopy: criteria description, histopathological correlation, and interobserver agreement. JAMA dermatology (Chicago, Ill). 2013;149(7):1–8.

14. Bennàssar A, Vilata A, Puig S, Malvehy J. Ex vivo fluorescence confocal microscopy for fast evaluation of tumour margins during Mohs surgery. Br J Dermatol. 2014;170(2):360–5.
15. Longo C, Ragazzi M, Rajadhyaksha M, Nehal K, Bennassar A, Pellacani G, et al. In Vivo and Ex Vivo confocal microscopy for dermatologic and mohs surgeons. Dermatol Clin. 2016;34(4):497–504.
16. Longo C, Rajadhyaksha M, Ragazzi M, Nehal K, Gardini S, Moscarella E, et al. Evaluating ex vivo fluorescence confocal microscopy images of basal cell carcinomas in Mohs excised tissue. Br J Dermatol. 2014;171(3):561–70.
17. Cinotti E, Fouilloux B, Perrot JL, Labeille B, Douchet C, Cambazard F. Confocal microscopy for healthy and pathological nail. J Eur Acad Dermatol Venereol. 2014;28:853–8.
18. Kadouch DJ, Leeflang MM, Elshot YS, Longo C, Ulrich M, van der Wal AC, et al. Diagnostic accuracy of confocal microscopy imaging vs. punch biopsy for diagnosing and subtyping basal cell carcinoma. J Eur Acad Dermatol Venereol. 2017;31(10):1641–8.
19. Dika E, Patrizi A, Lambertini M, Scarfì F, Fanti PA. Diagnostic accuracy of ex vivo fluorescence confocal microscopy for Mohs surgery of basal cell carcinomas: a prospective study on 753 margins. Br J Dermatol. 2019; 180:1559.
20. Pérez-Anker J, Puig S, Malvehy J. Ex vivo confocal microscopy in vertical sections and histopathological correlation (Master of Science Degree Thesis). Barcelona, Spain: University of Barcelona; 2016.
21. Bertoni L, Azzoni P, Reggiani C, Pisciotta A, Carnevale G, Chester J, et al. Ex vivo fluorescence confocal microscopy for intraoperative, real-time diagnosis of cutaneous inflammatory diseases: a preliminary study. Exp Dermatol. 2018;27(10):1152–9.
22. Hartmann D, Ruini C, Mathemeier L, Bachmann MR, Dietrich A, Ruzicka T, et al. Identification of ex-vivo confocal laser scanning microscopic features of melanocytic lesions and their histological correlates. J Biophotonics. 2017;10(1):128–42.
23. Hartmann D, Krammer S, Bachmann MR, Mathemeier L, Ruzicka T, Bagci IS, et al. Ex-vivo confocal microscopy features of cutaneous squamous cell carcinoma. J Biophotonics. 2018;11:1–7.
24. Cinotti E, Grivet D, Labeille B, Solazzi M, Bernard A, Forest F, et al. The 'tissue press': a new device to flatten fresh tissue during ex vivo confocal microscopy examination. Skin Res Technol. 2017; 23:121–4.
25. Li Y, Gonzalez S, Terwey TH, Wolchok J, Li Y, Aranda I, et al. Dual mode reflectance and fluorescence confocal laser scanning microscopy for in vivo imaging melanoma progression in murine skin. J Investig Dermatol. 2005;125(4):798–804.
26. Abeytunge S, Li Y, Larson B, Toledo-Crow R, Rajadhyaksha M. Rapid confocal imaging of large areas of excised tissue with strip mosaicing. J Biomed Optics. 2011;16(5):050504.
27. Johnson S, Rabinovitch P. Ex vivo imaging of excised tissue using vital dyes and confocal microscopy. Curr Protocols Cytometry. 2012;(61).
28. Gareau DS. Feasibility of digitally stained multimodal confocal mosaics to simulate histopathology. J Biomed Optics. 2009;14 (3):034050.
29. Bini J, Spain J, Nehal K, Hazelwood V, DiMarzio C, Rajadhyaksha M. Confocal mosaicing microscopy of human skin ex vivo: spectral analysis for digital staining to simulate histology-like appearance. J Biomedical Optics. 2011;16(7):076008.
30. Anker JP, Ribero S, Yélamos O, García A, Alos L, Alejo B, et al. Basal cell carcinoma characterisation using fusion ex vivo confocal microscopy: a promising change in conventional skin histopathology. Br J Dermatol. 2020;182(2):468–76.
31. Yélamos O, Pérez-Anker J. Avances en el manejo del cáncer cutáneo : videomosaicos y microscopía confocal de fusión. Revista chilena de dermatología. 2018;34(1):6–8.
32. Pérez-Anker J, Malvehy J, Moreno-Ramírez D. Microscopia confocal ex vivo con método de fusión y tinción digital: cambiando paradigmas en el diagnóstico histológico. Actas Dermo-Sifiliográficas.
33. Reggiani C, Pellacani G, Reggiani Bonetti L, Zanelli G, Azzoni P, Chester J, et al. An intraoperative study with ex vivo fluorescence confocal microscopy: diagnostic accuracy of the three visualization modalities. J Eur Acad Dermatol Venereol. 2021;35(1):e92–4.
34. Pérez-Anker J, Puig S, Malvehy GJ. A fast and effective option for tissue flattening: optimizing time and efficacy in ex vivo confocal microscopy. J Am Acad Dermatol Am Dermatol . 2020;82(5): e157–8.
35. Pérez-Anker J, Toll A, Puig S, Malvehy J. Six steps to reach optimal scanning in ex vivo confocal microscopy. J Am Acad Dermatol. 2021 Jan 18:S0190–9622(21)00193–6. doi: https://doi.org/10.1016/j.jaad.2021.01.044. Epub ahead of print. PMID: 33476729.
36. Sendín-Martín M, Kose K, Harris U, Rossi A, Lee E, Nehal K, et al. Complete visualization of epidermal margin during ex vivo confocal microscopy of excised tissue with 3D-mosaicking and intensity projection. J Am Acad Dermatol. 2020 May 16:S0190–9622(20)30909–9. doi: https://doi.org/10.1016/j.jaad.2020.05.044. Epub ahead of print. PMID: 32428611; https://doi.org/10.1016/j.jaad.2020.05.044
37. Toloza Salech C, Pérez-Anker J, La Rotta Higuera E, Toll A, García A, Alós LL, et al. Tinción digital en diagnóstico histológico de tumors cutáneos con microscopía confocal ex vivo de fusión. Poster presented at: 47th Congreso nacional de Dermatología y Venerología (AEDV); June 5–8, 2019; Barcelona, Spain.
38. Pérez-Anker J, Toll A, Puig S, García A, Alós Ll, Malvehy J. Real time visualization of mitosis and nuclear pleomorphism in fresh tissue using ex vivo confocal microscopy. In: Poster presented at 28th Congress of the European Academy of Dermatology and Venerology (EADV); 9–13 Oct 2019; Madrid, Spain.
39. Toloza Salech C, Malvehy J, Puig S, García A, Alós Ll, Toll A, et al. A proof of concept: digital H & E in the histological diagnosis of cutaneous tumors with ex vivo fusion confocal microscopy. In: Poster presented at: 28th congress of the european academy of dermatology and venerology (EADV); 9–13 Oct 2019; Madrid, Spain.
40. Puliatti S, Bertoni L, Pirola GM, Azzoni P, Bevilacqua L, Eissa A, et al. Ex vivo fluorescence confocal microscopy: the first application for real-time pathological examination of prostatic tissue. BJU Int. 2019;0–3.
41. Combalia M, Pérez-Anker J, García-Herrera A, Alos L, Vilaplana V, Marqués F, et al. Digitally stained confocal microscopy through deep learning. In: Proceedings of machine learning research. 2019;1–9.

17 Epidermal Reconstruction During Ex Vivo Confocal Microscopy for Detection of Superficial Basal Cell Carcinoma with 3D-Mosaicking and Intensity Projection

Matthew Moronta, Ucalene Harris, Mercedes Sendín-Martín, Alex Bang, Anthony Rossi, Erica Lee, Kishwer Nehal, Chih-Shan Jason Chen, Milind Rajadhyaksha, Kivanc Kose, and Manu Jain

17.1 Epidermal Flattening

Ex vivo confocal microscopy (EVCM) allows rapid evaluation of freshly excised tissue margins with high sensitivity (73–100%) and specificity (89–100%) for residual basal cell carcinoma (BCC) during staged-excision in Mohs surgery [1–8]. A major limitation encountered in these studies has been the inability to flatten peripheral edges of excised tissue, affecting the evaluation of superficial BCCs (sBCC) [9, 10]. As seen in Fig. 17.1, sBCC is found along the basal layer of the epidermis which is why complete margin visualization is necessary for effective detection. A few mechanical devices [11, 12] have been used to physically flatten the tissue specimens, such as magnets, tissue presses, and tissue cassettes, but these have not been sufficient.

We have developed a digital approach to epidermal flattening incorporating real-time software-based reconstruction of overlaid mosaics, each with incomplete peripheral edges, to yield a final merged mosaic [13]. Around 8–10 mosaics, each with partial epidermis, are spaced $\sim$5 um in depth taking approximately 10–20 min in image acquisition. These mosaics are then merged by applying minimum intensity projection in MATLAB to yield a single digitally reconstructed image conferring visualization of the entire epidermal margin.

As seen in Fig. 17.2, the individual frames of a tissue negative for tumor, obtained from Mohs surgical margin, only show partial epidermis, but through digital reconstruction, the final image in Fig. 17.3a shows the full epidermal margin compared to its corresponding H&E-stained tissue section in Fig. 17.3b. As a result of this digital overlay, there is no loss of details of the dermal structures, such as eccrine and pilosebaceous glands even at high magnification (Fig. 17.4c–d). This approach is especially useful for the detection of sBCC (Figs. 17.2, 17.3, 17.5 and 17.6).

M. Moronta · U. Harris · A. Bang · A. Rossi · E. Lee · K. Nehal · C.-S.J. Chen · M. Rajadhyaksha · K. Kose · M. Jain (✉)
Department of Dermatology, Memorial Sloan Kettering Cancer Center (MSKCC), 530 E. 74th Street, New York, NY 10021, USA
e-mail: jainm@mskcc.org

M. Sendín-Martín
Dermatology Department, Hospital Universitario Virgen del Rocío, Sevilla, Spain

M. Jain et al. (eds.), *Cutaneous Atlas of Ex Vivo Confocal Microscopy*,
https://doi.org/10.1007/978-3-030-89316-3_17

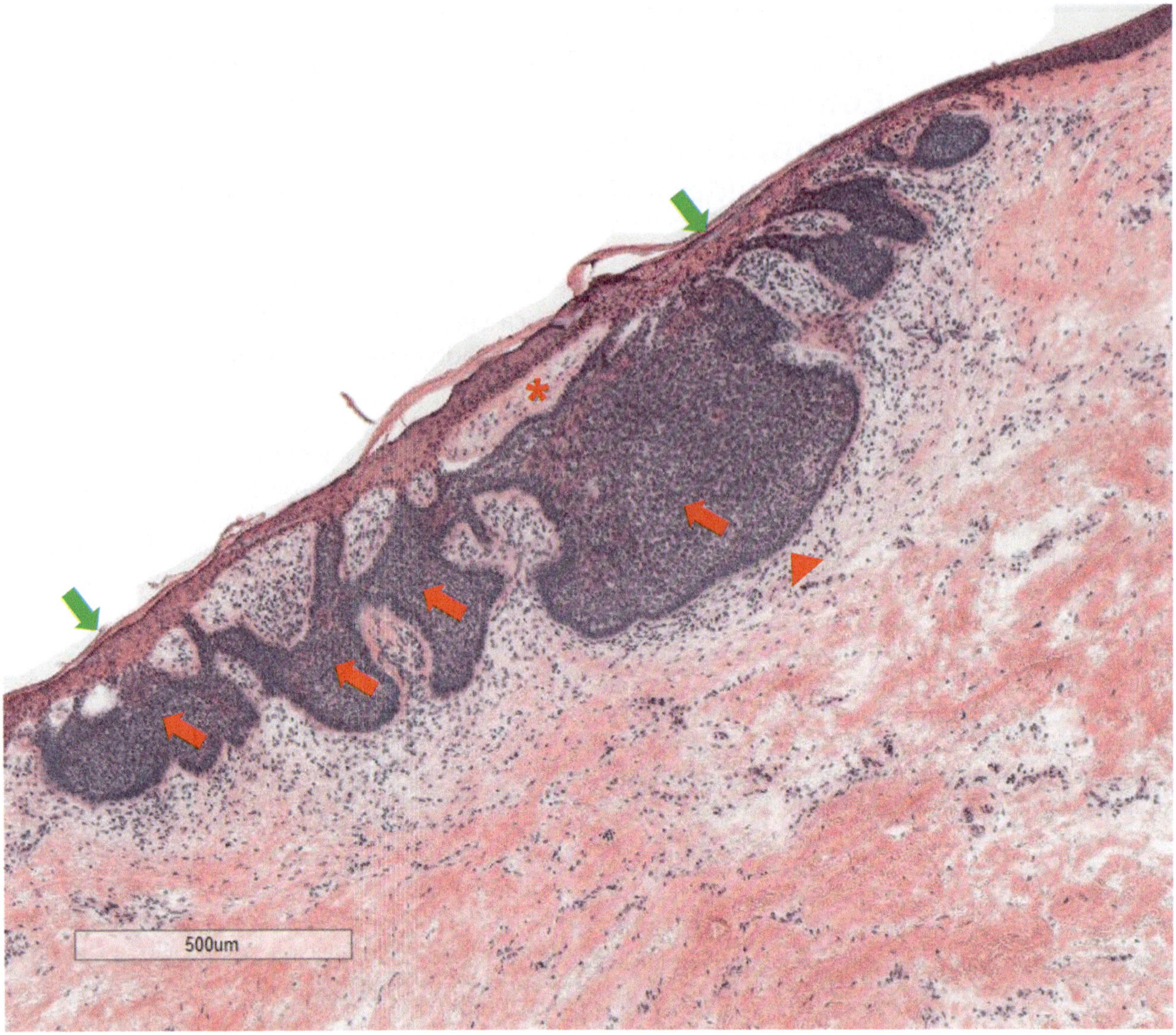

Fig. 17.1 Conventional H&E-stained tissue section of a superficial BCC. H&E image shows a sBCC (red arrow) with clefting (red asterisk) and palisading (red arrowhead) hanging off the epidermal margin (green arrow). Magnification = 10x

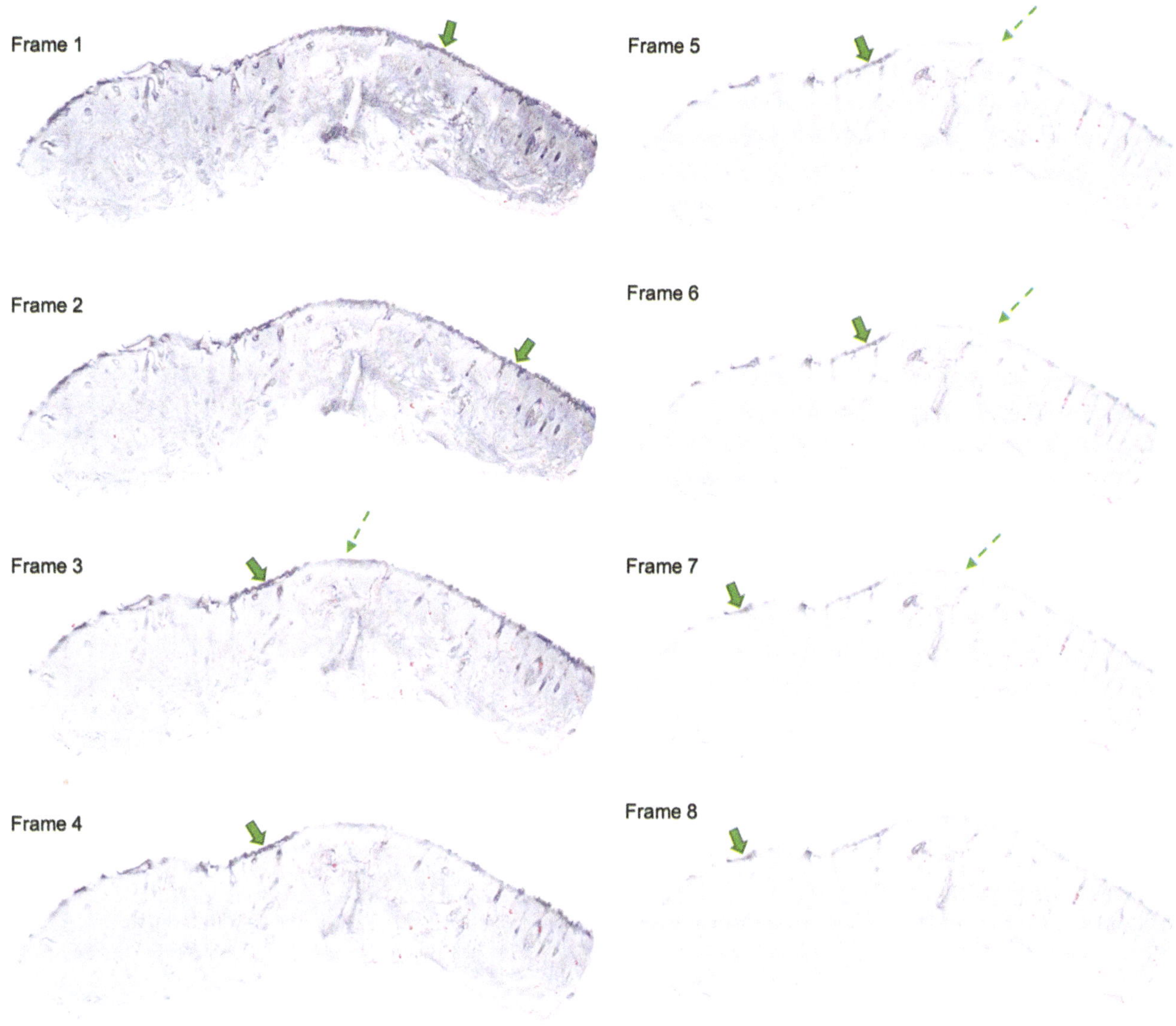

Fig. 17.2 Series of 10 individual digital H&E (DHE) mosaics at varying depths showing partial epidermis from a Mohs negative surgical margin. Digital H&E (DHE) individual mosaics (1.3 × 1.5 mm) spaced at 5 μm in depth show only partial epidermal margin (dashed green arrow) in each frame with areas of full epidermis (solid green arrow)

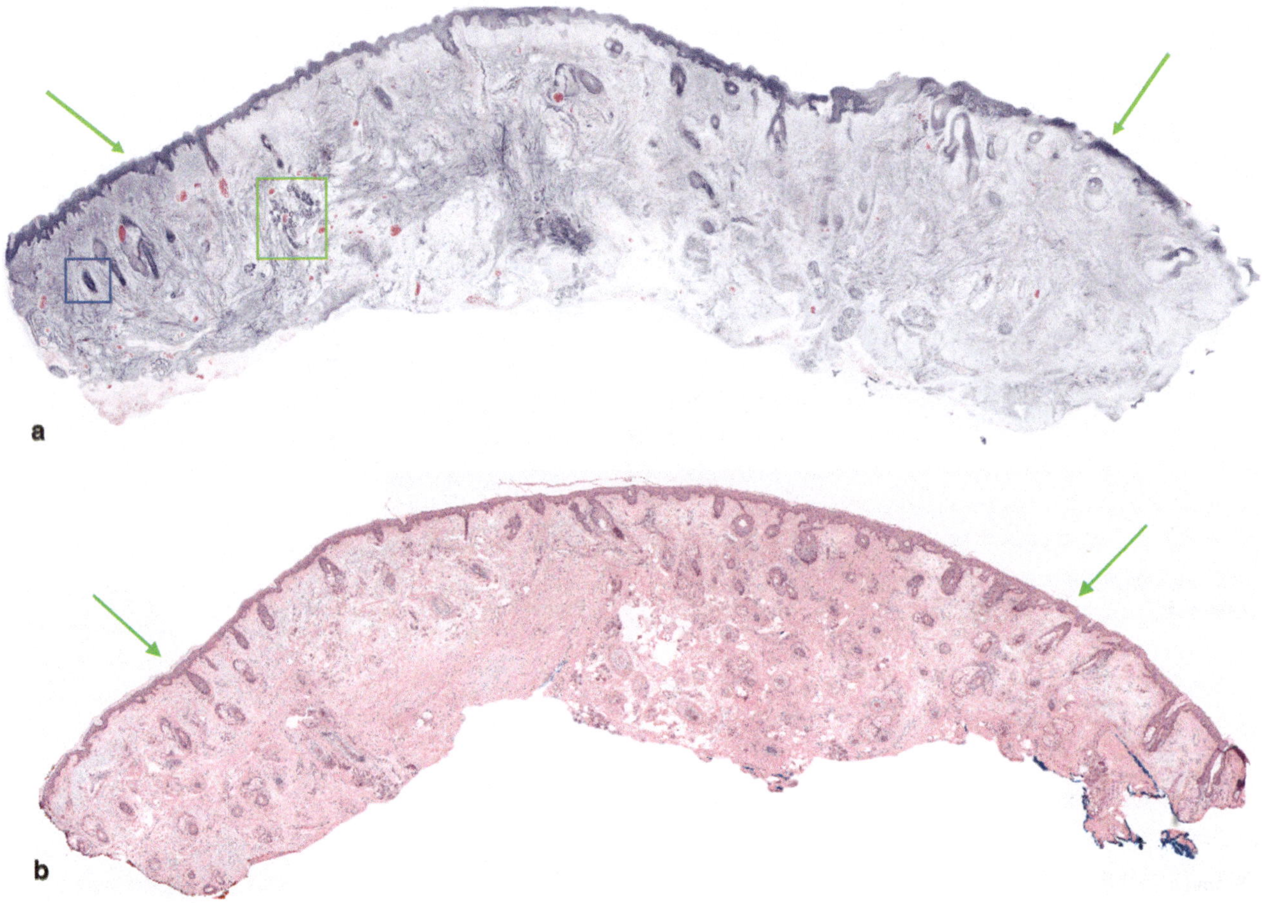

Fig. 17.3 Digitally reconstructed digital H&E (DHE) image of the above 10 individual mosaics (Fig. 17.3) compared to its corresponding H&E image. **a** A reconstructed DHE image was obtained by merging a series of mosaics (1.3 × 1.5 mm) through applying minimum intensity projection to show full epidermal margins (green arrow). Dermal structures such as hair follicle (blue square) and eccrine unit (fluorescent green square) are clearly visible even at this low magnification. **b** Corresponding H&E-stained tissue section (magnification 2x) shows a good correlation with the DHE image for the epidermis (green arrow), hair follicle (black square), and eccrine unit (blue square)

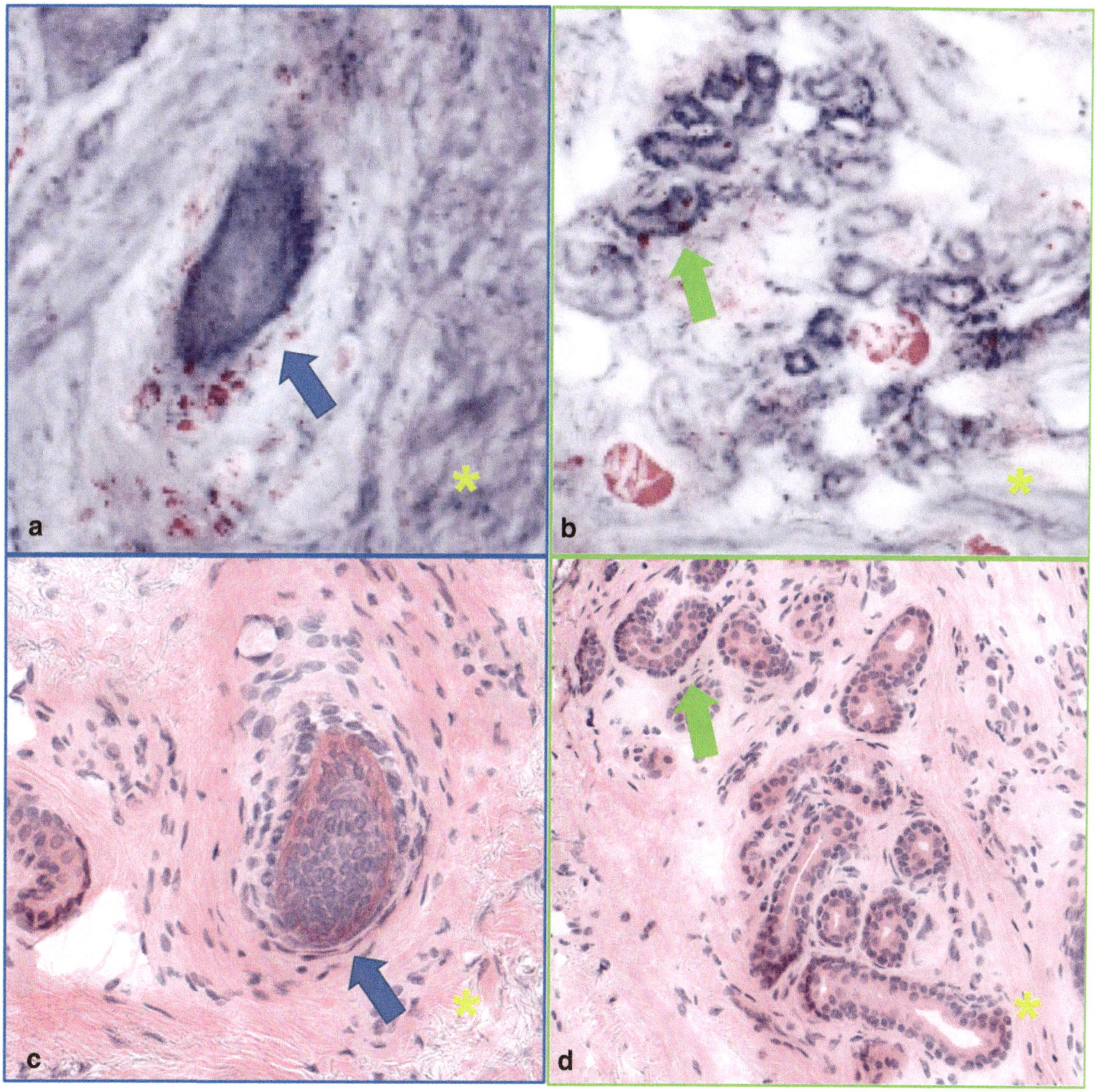

Fig. 17.4 Digitally reconstructed digital H&E (DHE) images of the dermal structures at higher magnification from Fig. 17.3 compared to its corresponding H&E image. **a**, **b** DHE high magnification images of a hair follicle (blue arrow) and an eccrine unit (fluorescent green arrow) within the dermal (yellow asterisk) shows cellular details. **c**, **d** Corresponding H&E images of a hair follicle (blue arrow) and an eccrine unit (fluorescent green arrow) within the dermal (yellow asterisk)

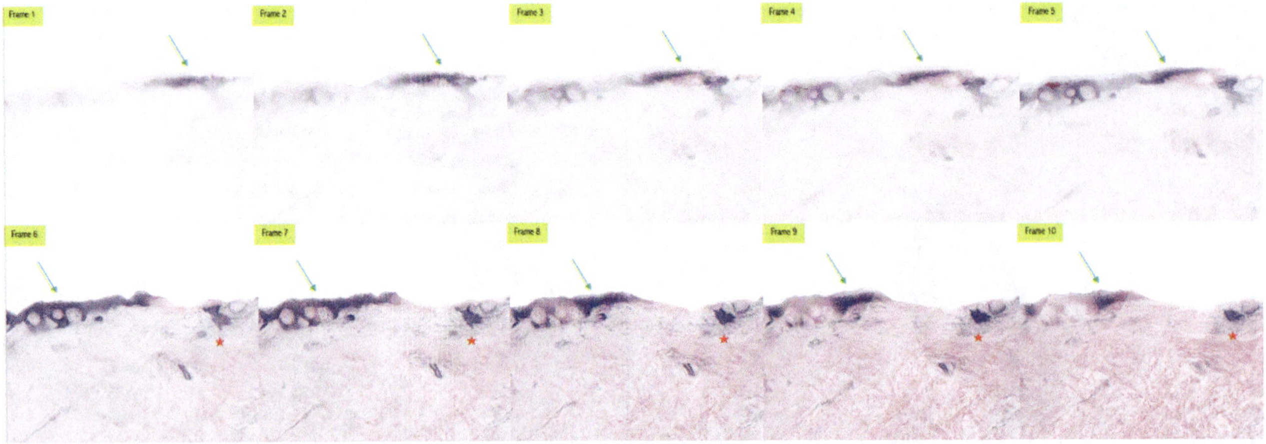

Fig. 17.5 Series of 10 individual sub-mosaics at varying depths showing partial epidermis of a superficial BCC. Digital H&E (DHE) individual mosaics (1.3 × 1.5 mm) spaced at 5 μm in depth show only partial epidermal details (green arrows) in each frame. Notice the presence of sBCC (red star) which is not fully visualized in individual frames. (Reprinted from Sendín-Martín et al. [13]; with permission from Elsevier.)

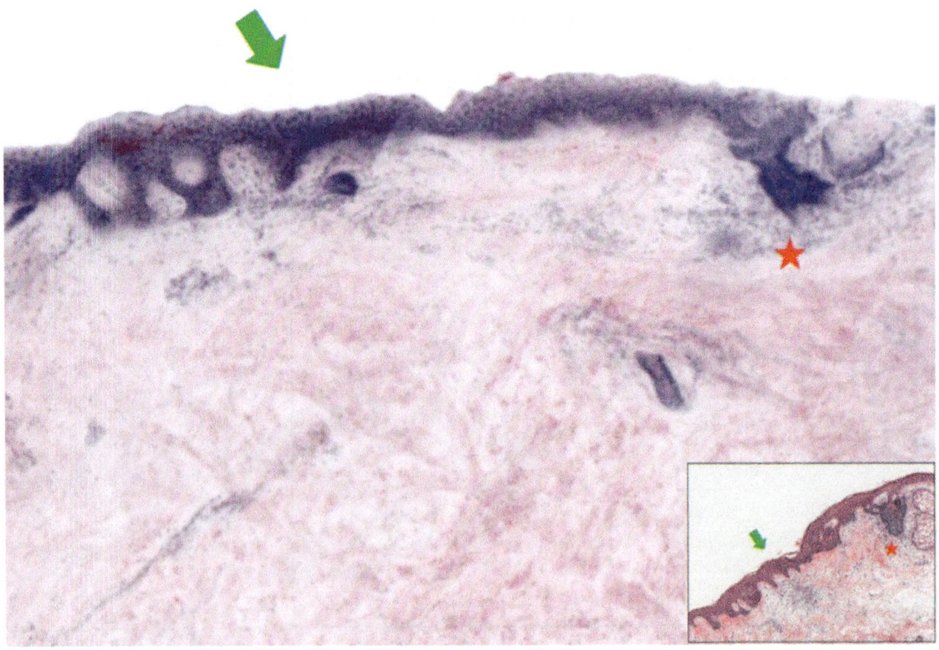

Fig. 17.6 Digitally reconstructed image of a superficial BCC. DHE image reconstructed by merging the series of individual mosaics (1.3 × .1.5 mm) from Fig. 17.5 by applying minimum intensity projection shows the entire epidermis (green arrow) with full visualization of sBCC (red star). The corresponding conventional H&E image can be viewed in the bottom right-hand corner for comparison. (Reprinted from Sendín-Martín et al. [13]; with permission from Elsevier.)

References

1. Karen, et al. Detection of basal cell carcinomas in Mohs excisions with fluorescence confocal mosaicking microscopy. Br J Dermatol. 2009;160(6):1242–50.
2. Larson et al. Detection of skin cancer margins in Mohs excisions with high-speed strip mosaicking confocal microscopy: a feasibility study. Br J Dermatol. 2013;169(4)
3. Bennàssar, et al. Ex vivo fluorescence confocal microscopy for fast evaluation of tumour margins during Mohs surgery. Br J Dermatol. 2014;170(2):360–5.
4. Espinasse, et al. 'En face' ex vivo reflectance confocal microscopy to help the surgery of basal cell carcinoma of the eyelid. Clin Exp Opthalmol. 2017;45:442–7.
5. Longo, et al. Diagnostic accuracy of ex vivo fluorescence confocal microscopy in Mohs surgey of basal cell carcinomas: a prospective study on 753 margins. Br J Dermatol. 2019;180:1473–80.
6. Mu, et al. Use of digitally stained multimodal confocal mosaic images to screen for nonmelanoma skin cancer. JAMA Dermatol. 2016;152(12):1335–41.
7. Peters, et al. Diagnostic accuracy of a new ex vivo confocal laser scanning microscope compared to H&E-stained paraffin slides for micrographic surgery of basal cell carcinoma. JEADV. 2019;33:298–304.
8. Jain M, Rajadhyaksha M, Nehal K. Implementation of fluorescence confocal mosaicking microscopy by "early adopter" Mohs surgeons and dermatologists: recent progress. J Biomed Opt. 2017;22(2):24002. https://doi.org/10.1117/1.JBO.22.2.024002. PMID:28199474;PMCID:PMC5310648.

9. Asgari MM, Moffet HH, Ray GT, Quesenberry CP. Trends in basal cell carcinoma incidence and identification of high-risk subgroups, 1998–2012. JAMA Dermatol. 2015;151:976–81. https://doi.org/10.1001/jamadermatol.2015.1188.
10. Malvehy, et al. Ex vivo confocal microscopy: revolution in fast pathology in dermatology. Br J Dermatol. 2020;2020. https://doi.org/10.1111/bjd.19017.
11. Pérez-Anker J, Puig S, Malvehy J. A fast and effective option for tissue flattening: optimizing time and efficacy in ex vivo confocal microscopy. J Am Acad Dermatol. 2019; 80(27):pii: S0190–9622(19)31040–0 https://doi.org/10.1016/j.jaad.2019.06.041.
12. Cinotti E, Grivet D, Labeille B, et al. The "tissue press": a new device to flatten fresh tissue during ex vivo confocal microscopy examination. Skin Res Technol. 2017;23(1):121–4. https://doi.org/10.1111/srt.12293.
13. Sendín M, Kose K, Harris U, Rossi A, Lee E, Nehal K, Rajadhyaksha M, Jain M. Complete visualization of epidermal margin during ex vivo confocal microscopy of excised tissue with 3D-mosaicking and intensity projection. J Am Acad Dermatol. 2020. https://doi.org/10.1016/j.jaad.2020.05.044.

Index

M. Jain et al. (eds.), *Cutaneous Atlas of Ex Vivo Confocal Microscopy*,
https://doi.org/10.1007/978-3-030-89316-3

The manufacturer's authorised representative in the EU is Springer Nature Customer Service Centre GmbH, Europaplatz 3, 69115 Heidelberg, Germany. If you have any concerns regarding our products, please contact ProductSafety@springernature.com

Printed and bound by CPI Group (UK) Ltd, Croydon, CR0 4YY
30/09/2026
02225516-0004